Multi-way Analysis with Applications in the Chemical Sciences

Multi-way Analysis with Applications in the Chemical Sciences

Age Smilde
University of Amsterdam, Amsterdam, and
TNO Nutrition and Food Research, Zeist, The Netherlands

and

Rasmus Bro
KVL, Frederiksberg, Denmark

and

Paul Geladi
SLU, Umeå, Sweden

John Wiley & Sons, Ltd

Other Wiley Editorial Offices

John Wiley & Sons Inc., 111 River Street, Hoboken, NJ 07030, USA

Jossey-Bass, 989 Market Street, San Francisco, CA 94103-1741, USA

Wiley-VCH Verlag GmbH, Boschstr. 12, D-69469 Weinheim, Germany

John Wiley & Sons Australia Ltd, 33 Park Road, Milton, Queensland 4064, Australia

John Wiley & Sons (Asia) Pte Ltd, 2 Clementi Loop #02-01, Jin Xing Distripark, Singapore 129809

John Wiley & Sons Canada Ltd, 22 Worcester Road, Etobicoke, Ontario, Canada M9W 1L1

Wiley also publishes its books in a variety of electronic formats. Some content that appears in print may not be
available in electronic books.

Library of Congress Cataloging-in-Publication Data

Smilde, Age K.
    Multi-way analysis with applications in the chemical sciences / Age Smilde and
Rasmus Bro and Paul Geladi.
       p.   cm.
    Includes bibliographical references and index.
    ISBN 0-471-98691-7 (acid-free paper)
    1. Chemistry – Statistical methods.   2. Multivariate analysis.   I. Bro, Rasmus.
II. Geladi, Paul.   III. Title.
    QD39.3.S7S65   2004
    540'.72 – dc22   2003027959

British Library Cataloguing in Publication Data

A catalogue record for this book is available from the British Library

ISBN 978-0-471-98691-1

CONTENTS

9 Preprocessing   221

10 Applications   257

References   351

Index   371

FOREWORD

The early days of chemometrics saw researchers from around the world developing and applying mathematical and statistical methods to a wide range of problems in chemistry. Chemometricians with analytical chemistry backgrounds were interested in such tasks as controlling or optimizing complicated analytical instruments, resolving spectra of complex mixtures into the spectra of pure components and, in general, getting more chemical information from quantitative measurements. Early chemometricians with physical or organic chemistry backgrounds were interested in relating chemical properties and molecular structural features of molecules to their chemical reactivity or biological activity. These scientists focused on the tools of classification and other areas of pattern recognition. The first few meetings of chemometricians tended to separate into problem areas until it was recognized that there was at least one common thread that tied us all together. That common thread can be summarized in a single word: multivariate. It gradually became clear that the power in our individual studies came from viewing the world as multivariate instead of univariate. At first this seemed to defy the scientific method we were taught that allows one variable at a time to be varied. Control engineers and some statisticians knew for some time that univariate experimental designs were doomed to fail in a multivariate world. While studying variance is useful, understanding covariance can lead one to move rapidly from data to information to knowledge.

As chemometrics matured, researchers discovered that psychometricians had moved beyond multivariate analysis to what they called multi-way data analysis. Psychometric studies involving multiple subjects (people) given several tests over periods of time lead to data structures that fit into three-dimensional computer arrays represented by blocks of data or what is called three-way arrays. Of course each subjects data (tests × time) could be analyzed separately by removing a matrix or slice from a block of data. Psychometricians recognized that this approach would lose the covariance among subjects so they began developing data models and analysis methodology to analyze entire blocks of data or three-way arrays at one time.

Chemometricians soon discovered that chemistry is rich with experimental designs and analytical tools capable of generating multi-way data. An environmental chemist, for example, may acquire rainwater samples from several selected locations (first way) and at multiple times (second way) over the course of a study and have each sample analyzed for several analytes of interest (third way). In analytical chemistry, an instrument capable of generating a three-way array of data from a single sample is called a third order instrument because a three-way array is a third order tensor. Such instruments (e.g. GC/GC/MS) have special powers not obtainable by second order (GC/MS) or first order (MS) instruments. It is not simply that so-called higher order instruments generate more data that makes them more powerful. But it is the structure or form of the data that is the key and to access the power the multi-way analysis tools described in this book must be employed.

This book is the first of its kind to introduce multi-way analysis to chemists. It begins in Chapter 1 by answering the question 'What is multi-way analysis?' and then carefully covers the possible data structures and the models starting with simple two-way data and extending on to four-way and higher data. Chapter 2 goes into definitions, properties and other mathematical details that must be appreciated before applying the tools of multi-way data analysis. The following chapters describe models and associated data analysis tools and algorithms for two- and three-way data.

Further chapters go into the details one must learn to implement multi-way analysis tools successfully. Several important questions are posed such as: How should data be preprocessed before analysis? How can models be validated and what diagnostics are available to detect problems? What visualization tools are available that attempt to show humans in our extremely limited three-dimensional world the powerful advantages of multi-way data?

The last chapter provides a look at selected applications of multi-way analysis that have appeared in the chemometrics literature. This section of the book is not meant to be a complete literature review of the subject. Rather, applications were selected to aid the reader in understanding what the tools can do and hopefully point the way for the reader to discover new applications in his/her own field of investigation.

Most chemists are certainly unaware of the power of multi-way analysis. Analytical chemists, with very few exceptions, are not developing new instruments that can take advantage of the second and third order advantages obtained from use of the tools described so well in this book. Paradigm shifts in a field can only take place when the field is looking for and can accept new ways to operate. This book opens the door to anyone looking for a paradigm shift in his or her field of study. Multi-way analysis provides a new way to look at experimental design and the scientific method itself. It can help us understand a world that is multivariate, often nonlinear and dynamic.

Bruce R. Kowalski
Somewhere in Hay Gulch, Hesperus, Colorado

PREFACE

Goal of the project

In chemistry and especially in chemometrics, there is a growing need for new data analysis technology. New data analysis technologies are usually first presented in papers in the literature, but after a while the volume grows and there is a need for completeness and a systematic overview. Review articles are written to give this needed overview. After a few years, a simple review article is not enough for containing the existing knowledge and a monograph is needed. The present monograph aims to give all this: an almost complete overview of all the aspects of multi-way analysis that is needed for users in chemistry. Applying multi-way analysis requires an in-depth understanding of techniques and methodologies, but also a wide overview of applications and the relation of the techniques to problem definition should not be forgotten. Many problems in chemical research and industry are such that three-way data result from the measurements, but simply putting the data in a three-way array and running an algorithm is not enough. It is important to understand the background of the problem in preprocessing, analyzing the data, presentation and interpretation of the model parameters and interpretation of the results. The user is provided with an overview of the available methods as well as hints on the pitfalls and details of applying these. Through numerous examples and worked-out applications, an understanding is provided on how to use the methods appropriately and how to avoid cumbersome problems.

History of the project

The idea for this book has existed for many years. The project was finally realized in 1995 when a contract was signed with a publisher. We knew that the task to be carried out was enormous, but we underestimated it. Years went by without much production, just discussion. In the meantime the volume of literature on three-way analysis grew and the need for this monograph became even more urgent. The result is that the structure of the book has undergone many improvements since the first synopsis and that the volume of examples has grown. In the process of writing, new ideas emerged and had to be tested and a deeper understanding of the literature was acquired. Some loose concepts had to be substantiated by rigorous proofs. A special development was that of notation and nomenclature. Much was borrowed from psychology and sociology, but much was also modified and adapted to chemical formulation.

Some topics were left out, mainly because they were too new and not fully developed. Other topics were considered too narrowly specialized to really fit in. The applications in this book do not give a complete overview of all the literature. They are selected examples of important developments that are helpful in realizing how three-way analysis can be useful also in other applications.

How to read this book

The book was written with chemistry or related fields (biology, biochemistry, pharmaceutical and clinical chemistry, process technology, quality control, etc.) in mind. The reader is expected to have a basic knowledge of statistics and linear algebra. Although two-way methods are introduced in Chapter 2, some background knowledge of latent variables and regression is useful. A course for newcomers would contain Chapters 1–4 and selected parts of Chapters 7–9, with Chapter 3 as a refresher for two-way methods. Such a course would probably start with selected parts of Chapter 10 and return to the same chapter in the end. The advanced reader can start with Chapter 4 and read through to Chapter 10.

Acknowledgements

We would like to thank numerous people who helped us in different ways. Three groups are mentioned separately: proofreaders of chapters, people who provided examples and those who were helpful as colleagues or management during the many years that we worked on the book project. They are given in alphabetical order.

Proofreaders were: Karl Booksh, Kim Esbensen, Neal Gallagher, Margriet Hendriks, René Henrion, Henk Kiers, Lars Nørgaard, Mary-Beth Seasholtz, Nikos Sidiropoulos, Riccardo Leardi and Peter Schoenmakers.

Those who provided examples are: Peter Åberg, Claus Andersson, Dorrit Baunsgaard, Helén Bergner, Jennie Forsström, Paul Gemperline, René Henrion, Phil Hopke, Josefina Nyström, Riccardo Leardi, Magni Martens, Torbjörn Lestander, Carsten Ridder, Lena Ringqvist and Barry Wise.

Helpful colleagues and management: Johanna Backman, Laila Brunes, Ulf Edlund, Bert Larsson, Calle Nilsson, Britta Sethson, Lars Munck, Ellen Abeling, Renate Hippert and Jan Ramaker.

NOMENCLATURE AND CONVENTIONS

x	a scalar
$\mathbf{x}$	a vector (always a column vector)
$\mathbf{X}$	a matrix / two-way array
$\mathbf{X}_{(I \times JK)}, \mathbf{X}_{(J \times IK)}, \mathbf{X}_{(K \times IJ)}$	matricized form of $\underline{\mathbf{X}}$ with mode A, B or C kept intact
$[\mathbf{X}_1 \, \mathbf{X}_2]$	concatenation of the matrices $\mathbf{X}_1$ and $\mathbf{X}_2$
$\underline{\mathbf{X}}$	a three-way array or multi-way array
$\mathbf{e}, \mathbf{E}, \underline{\mathbf{E}}$	vector, matrix or three-way array of residuals
$\mathbf{E}_x, \mathbf{E}_y$	residuals in X and Y (regression)
$\hat{\mathbf{x}}, \hat{\mathbf{X}}, \underline{\hat{\mathbf{X}}}$	vector, matrix or three-way array of fitted (model) values
$i, j, k, l, \ldots$	running indices
$I, J, K, L, \ldots$	maximum of index $i, j, k, l, ..$
I, J	preserved for size of $\mathbf{X}$
I, J, K	preserved for size of $\underline{\mathbf{X}}$
I, M	preserved for size of $\mathbf{Y}$
I, M, N	preserved for size of $\underline{\mathbf{Y}}$
p, q, r	running indices for number of latent variables in the different modes
P, Q, R	maximum values of p, q, r
$w, \mathbf{w}, \mathbf{W}$	weight, vector of weights, matrix of weights
$\mathrm{r}(\mathbf{A})$	rank of $\mathbf{A}$
$k_{\mathbf{A}}$	k-rank of $\mathbf{A}$
$\mathrm{vec}\,\mathbf{A}$	vectorization of $\mathbf{A}$
$\mathrm{tr}(\mathbf{A})$	trace of $\mathbf{A}$
$\mathbf{A}, \mathbf{B}, \mathbf{C}$	loading matrices
$\mathbf{D}$	diagonal matrix
$\underline{\mathbf{D}}$	three-way superdiagonal array
$\mathrm{diag}(\mathbf{a})$	matrix with the vector $\mathbf{a}$ on its diagonal and all off-diagonal elements zero
$\mathbf{I}$	identity matrix
$\underline{\mathbf{I}}$	superdiagonal three-way array with ones on the diagonal

$\mathbf{G}$	matricized core-array
$\underline{\mathbf{G}}$	three-way core-array
$\Re(\mathbf{X})$	range or column-space of $\mathbf{X}$
$\aleph(\mathbf{X})$	null space of $\mathbf{X}$
SS_{tot}	total sum of squares in a data set
SS_{mod}	sum of squares explained by the model
SS_{res}	residual sum of squares
2D plot	two-dimensional plot
3D plot	three-dimensional plot
$*$	Hadamard or Direct product
$\odot$	Khatri–Rao product
$\otimes$	Kronecker product
$\triangle$	Tensor product
$\|\cdot\|$	Frobenius or Euclidian norm of vector or matrix

1

INTRODUCTION

1.1 What is Multi-way Analysis?

Multi-way analysis is the analysis of multi-way data. In this book a distinction will be made between one-way, two-way, three-way and higher-way data. If a single measurement on a chemical system generates a single number and this measurement is repeated, then a sequence of numbers is the result. For example, a titration is repeated four times, resulting in the volumes x_1, x_2, x_3 and x_4. This is a sequence of numbers and this sequence has to be analyzed with one-way tools, such as calculating the mean and the standard deviation of these four numbers.

Going up one level to instruments that generate sequences of numbers, e.g. a UV–Vis spectrum consisting of absorptions at J wavelengths. Upon taking I of those measurements, a matrix $\mathbf{X}$ of dimensions $I \times J$ results. This is a two-way array and such a matrix can be analyzed appropriately with two-way analysis tools, such as principal component analysis.

If a single instrument generates a table of numbers for each sample, e.g., a fluorescence emission/excitation landscape, this results in a matrix $\mathbf{X}$ ($J \times K$), where J excitation wavelengths are used for measuring emission at K wavelengths. Taking I of these measurements, for example at I different occasions, generates a three-way array of size $I \times J \times K$. Such arrays can be analyzed with three-way analysis methods, which is the issue of this book. It is even possible to generate four-way, five-way, in general multi-way data. Methods that deal with four- and higher-way data will also be discussed, although in less detail than the more commonly used three-way analysis tools.

1.2 Conceptual Aspects of Multi-way Data Analysis

The goal of this section is to introduce data, data bases and arrays. Data are measured; such as a cardiologist measuring electrocardiogram data and a radiologist measuring computer aided tomography scans. This is data collection and is practiced a lot in academia and industry. Putting the data together patient by patient will lead to construction of a data base or data file. This data base may still be analog or it may be in digital form as computer

Multi-way Analysis With Applications in the Chemical Sciences.   A. Smilde, R. Bro and P. Geladi
© 2004 John Wiley & Sons, Ltd   ISBN: 0-471-98691-7

files. Some simple statistical and bookkeeping operations may be carried out on such a data base. An array is only made when a statistician or data analyst constructs it from data in the base, usually with a specific purpose. In analytical chemistry all these steps are often done automatically, especially with homogeneous variables in spectroscopy, chromatography, electrophoresis etc. The instrument collects the data in digital form, keeps the files in order and simply constructs the desired array when needed, but this is still an exception.

Multi-way analysis is not only about algorithms for fitting multi-way models. There are many philosophical and technical matters to be considered initially or during analysis. The data have to be collected. Problems such as detection limits, missing data and outliers must be handled. Preprocessing might be needed to bring the data in a suitable form. This gets even more complicated when qualitative and quantitative data are mixed. Some of the points to be considered are:

- the nature of a way/mode and how different combinations of these coexist;
- mean-centering and linear scaling of ways;
- nonlinear scalings;
- the role of outliers;
- missing data;
- type of data: real, complex, integer, binary, ordinal, ratio, interval;
- qualitative versus quantitative data;
- rank and uniqueness of multi-way models.

Many of these aspects will be encountered in subsequent chapters and others are treated in standard statistical literature.

Problem definition

Apart from these data analytical issues, the problem definition is important. Defining the problem is the core issue in all data analysis. It is not uncommon that data are analyzed by people not directly related to the problem at hand. If a clear understanding and consensus of the problem to be solved is not present, then the analysis may not even provide a solution to the real problem. Another issue is what kind of data are available or should be available? Typical questions to be asked are: is there a choice in instrumental measurements to be made and are some preferred over others; are some variables irrelevant in the context, for example because they will not be available in the future; can measurement precision be improved, etc. A third issue concerns the characteristics of the data to be used. Are they qualitative, quantitative, is the error distribution known within reasonable certainty, etc. The interpretation stage after data analysis usually refers back to the problem definition and should be done with the initial problem in mind.

Object and variable 'ways'

In order to explain the basics of three-way analysis, it is easiest to start with two-way data and then extend the concepts to three-way. Two-way data are represented in a two-way data matrix typically with columns as variables and rows as objects (Figure 1.1). Three-way

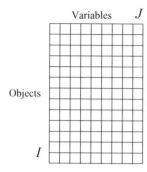

Figure 1.1. A two-way array has, e.g., the objects as rows and the variables as columns. There are I objects and J variables.

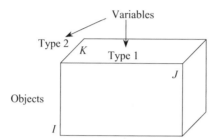

Figure 1.2. A typical three-way array has I objects, J type 1 variables and K type 2 variables.

data are represented as a three-way array (box) (Figure 1.2). The columns and rows of the two-way array are now 'replaced' by slices, like slicing a rectangular bread or cake in different ways. This is shown in Figure 1.3. Each horizontal slice (also sometimes called horizontal slab) holds the data for one object in a three-way array. Each vertical slice holds data of one specific variable of one type (say absorbances at different wavelengths or retention times) and the back-to-front slices the variables of the other type, say judges or time slots. Other types of arrangements can exist as well. Some three-way arrays have two types of object modes and only one type of variable mode, or even three objects modes and no variable mode. An example is in multivariate image analysis where the data are three-way arrays with x-coordinate, y-coordinate of the pixels as object ways and wavelength as the variable way. See Figure 1.4. The whole discussion of what the nature of an object or variable is may not be clear on all occasions. Basically an object is a physical or mental construct for which information is sought. A variable provides means for obtaining that information. For example, the physical health of persons might be of interest. Different persons are therefore objects in this respect. The blood pressure is a variable, and measuring the blood pressure of the different objects provides information of the health status of the objects.

For a three-way array of liquid-chromatography–ultraviolet spectroscopy data, the samples are the objects whereas the absorbance at different retention times (of the chromatography mode) and the wavelengths (of the ultraviolet spectroscopy mode) are the two variable modes. Hence, absorbance is measured as a function of two properties: retention time and wavelength. For a three-way array of food products × judges × sensory properties, the judge mode can be an object mode or a variable mode. In a preliminary analysis the judges

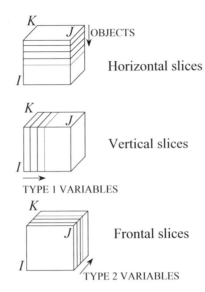

Figure 1.3. Slicing of a three-way array can be done objectwise or along type 1 or type 2 variables. The slices are two-way arrays. This results in horizontal, vertical and frontal slices.

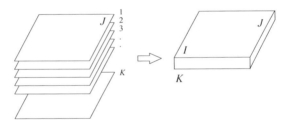

Figure 1.4. An image of $I \times J$ pixels for K different wavelengths gives a multivariate image which is an $I \times J \times K$ array. In this array there are two object ways.

can be seen as objects, where the purpose of the analysis is to assess if the judges agree. In a subsequent analysis with the judges as variables, the properties of the products are investigated, hence the products are the objects.

As can be understood from the above sensory problem, data alone do not determine what is an object and what is a variable. This goes back to the problem definition. Only when a stated purpose of the data is provided can the nature of the ways be deduced.

Some three-way models do not take the nature of a way into account, but the concepts of objects and variables are important, e.g., for defining how an array is preprocessed and for interpretation of the results. The same naturally holds for two-way analysis methods.

Types of objects and variables

A variable mode in a two-way or three-way array can consist of homogeneous variables, heterogeneous variables or a mixture of the two. Homogeneous variables have the same physical units and are often measured on the same instrument. The variables of a two-way array may be absorbance measured in a spectrometer at different wavelengths. This means that all variables have the same scale (absorbance).

There is sometimes an intrinsic natural order in the variables. Spectral variables can for example be meaningfully ordered according to wavelength/energy/wavenumber; elution profiles have an intrinsic order in time; geographical measurements have a two-dimensional order in space, etc. For homogeneous variables the order is often easy to detect visually because of autocorrelation. For chemical reasons, spectra are often smooth, showing distinct peaks and valleys. This property can be used in the analysis, for verifying and validating the results and it makes it easy to study the data or loadings of the way, either by themselves or in scatter plots. Putting the wavelengths in random order will destroy much of the understanding of the data, even if the data analysis gives numerically identical results.

Also for objects there can be intercorrelations, e.g., related to distance or time, as in environmental sampling and sequential sampling in industrial processes. This property is important for the choice of model, validation, interpretation etc.

Heterogeneous variables are measured in different units, or measured with different instruments, often according to a large variety of principles. The variables in an investigation may be:

- temperature in $°C$;
- wind speed in m s^{-1};
- wind direction in degrees/radians;
- humidity in %;
- pressure in pascals;
- CO concentration in mg m^{-3};
- particulate matter in mg m^{-3}.

The different units used for measuring these variables require attention. For example, scaling is an important issue for such variables, because the possible widely different scales lead to variables of very different numerical values. Using unscaled data would lead to models fitting and reflecting mainly those specific variables that have high numerical values, rather than the whole data set as such.

Mixtures of homogeneous and heterogeneous variables also occur. Whole body impedance data for patients are used for measuring fat, water and bone content. The impedances are measured at different frequencies. The frequencies are homogeneous variables, usually tabulated as increasing frequencies, but to these data, the heterogeneous variables age, sex, drug dosage, length and weight are usually added. Note that sex is a categorical variable, age is usually tabulated as an integer and drug dosage is a real. The impedances may be expressed as complex numbers. Again, scaling of such data needs to be considered.

1.3 Hierarchy of Multivariate Data Structures in Chemistry

In order to set the stage for introducing different types of methods and models for analyzing multi-way data it is convenient to have an idea of what type of multivariate data analysis problems can be encountered in chemical practice. A categorization will be developed that uses the types of arrangements of the data set related to the chemical problem.

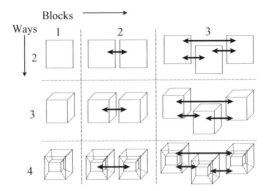

Figure 1.5. Relationship between data analysis methods for multi-block and multi-way arrangements of the data. Blocks refer to blocks of data, and ways to the number of ways within a block.

A data set can have different features. Two general features of a data set will be distinguished and used to categorize data sets. The first feature is whether the data set is two way, three-way, or higher-way. In Section 1.1 this distinction has already been explained. Another feature is whether the data set can be arranged as a single block of data, two blocks or multiple blocks of data. Crossing both features leads to a categorization as visualized in Figure 1.5. The arrows in this figure indicate the relationship that is sought. Some problems may not fit exactly into this categorization, but it provides a simple overview of the bulk of problems encountered in data analysis. For two-block data, both blocks must have at least one mode (or way) in common, e.g., the rows in the first block of data contain measurements on the same objects as the corresponding rows in the second block of data. For multi-block data, different relational schemes can be formulated; this is problem and context dependent. Note that a two-block problem does not necessarily need two blocks of the same order. For example, relationships between a three-way array and, e.g., a vector are positioned in the middle of Figure 1.5. This is seen by setting the dimension in the second and third mode to one in one of the three-way arrays in the figure.

In the following a brief description is given of some methods available to investigate the typical data analysis problems, as indicated in Figure 1.5.

Two-way one-block data

The simplest multivariate data set is a data set consisting of measurements (or calculated properties) of J variables on I objects. Such a data set can be arranged in an $I \times J$ matrix $\mathbf{X}$. This matrix $\mathbf{X}$ contains variation which is supposed to be relevant for the (chemical) problem at hand. Several types of methods are available to investigate this variation depending on the purpose of the research and the problem definition.

EXPLORATION

If the purpose of analyzing the data set is to find patterns, relations, differences and agreements between objects and/or variables, then decomposition methods can be used to summarize the data conveniently and explore the data set using plots and figures. Typically, subspace-based methods (projection methods) are used to handle collinearity problems. A well-known technique to do this is principal component analysis (PCA). PCA will be described in more detail in Chapter 3. It suffices to say here that PCA is a dimension reducing technique, that projects objects and variables to low dimensional spaces [Pearson 1901]. Plots of these projections can then be used for exploratory purposes.

While PCA is a linear projection method, there also exist nonlinear projection methods, e.g. multidimensional scaling [Mardia *et al.* 1979] and nonlinear PCA [Dong & McAvoy 1996]. A good overview of nonlinear multivariate analysis tools is given by [Gifi 1990].

ELABORATION **1.1**

Exploratory and Confirmatory Analysis

There are two basic modes of analysis: exploratory and confirmatory analysis. Confirmatory analysis is usually related to formal statistical hypothesis testing. A model is hypothesized, data generated, and the significance of the model tested. In exploratory analysis the main purpose of the analysis is to learn from the data about interrelationships between variables and objects. Exploratory analysis is typically a process where several models provide new insight in and knowledge of what to do. Initially, data are generated according to a hypothesis based on a very general idea. The results and information arising from the data give rise to a new set of ideas, which lead either to generation of new data or different analyses of the current data. By repeated analyses generating new hypotheses, new data and new models, the knowledge of the problem increases and the analysis may become more focused. Thus, in exploratory analysis the main hypotheses comes from the data and the models, as opposed to confirmatory and traditional deductive thinking where the hypothesis is posed before data are even acquired.

CLASSIFICATION PROBLEMS

Techniques that are suitable to detect grouping in the objects can be divided in supervised and unsupervised classification methods. The purpose of classification is typically to assign some sort of class membership to new samples. Mostly there are several classes, but, e.g., in raw material identification, the purpose is to establish whether or not a new batch of raw material is within the specified limits or not. Hence, there is only one well-defined class, being good raw material. In the case of supervised classification the data analyst knows beforehand which object in the data set belongs to which group. The purpose of the data analysis is then to verify whether this grouping can be found in the data set and to establish means of discriminating between the groups using a classification rule. Methods to serve this purpose are, e.g., discriminant analysis [Mardia *et al.* 1979].

In unsupervised classification, the grouping of the objects is not known beforehand and finding such groupings is the purpose of the analysis. Classical methods to perform unsupervised classification include variants of cluster analysis [Mardia *et al.* 1979].

ESTIMATING UNDERLYING PHENOMENA

An important class of problems in chemistry is the estimation of underlying phenomena or latent variables. Suppose, e.g., that the $I \times J$ matrix $\mathbf{X}$ contains spectroscopic measurements (absorbances at J wavelengths) made on I mixtures of R chemical constituents. Curve resolution techniques try to resolve the matrix $\mathbf{X}$ as a product $\mathbf{CS}'$ based on Beer's Law [Christian & O'Reilly 1986] where $\mathbf{S}$ $(J \times R)$ contains the pure spectra of the R chemical constituents and $\mathbf{C}$ $(I \times R)$ contains the relative concentrations of the R chemical constituents in the mixtures.

There is a wide variety of curve resolution techniques available [Lawton & Sylvestre 1971, Liang *et al.* 1993, Malinowski 1991, Tauler 1995, Vandeginste *et al.* 1987, Windig & Guilment 1991]. In these techniques, constraints can be included to direct the solution to meaningful spectra and concentration estimates.

Two-way two-block data

PREDICTION PROBLEMS

An enormous variety of problems in chemistry can be formulated in this type of data arrangement. Multivariate calibration, in which one data block contains spectral measurements and the other block the known concentrations of the absorbing constituents, is one example [Martens & Næs 1989]. The purpose is then to build a calibration model, with which it is possible to predict the concentrations of the constituents in a new sample using the spectral measurement of that sample and the calibration model (the relationship, as indicated with an arrow in Figure 1.5).

Another example is Quantitative Structure–Activity Relationships (QSAR) in which the first block contains measurements or calculated values of compounds and the second block contains measured biological activities. Here the purpose is to predict activities of new compounds using (simple) measurements or calculated values of that new compound and the model. Sometimes, the reverse is also sought: given a certain desired activity, what should a compound look like? This is known as the inverse problem and is considerably more complicated than predicting activity.

The aim of regression methods in general is to predict one block of measurements (predictands) using another block of measurements (predictors). Examples of methods are Multiple Linear Regression [Draper & Smith 1998], Principal Component Regression [Martens & Næs 1989], Partial Least Squares Regression (PLS) [Wold *et al.* 1984], Ridge Regression [Hoerl & Kennard 1970], Projection Pursuit Regression [Friedman & Stuetzle 1981], Multivariate Adaptive Regression Splines [Friedman 1991], Principal Covariates Regression [De Jong & Kiers 1992]. Clearly, there is an enormous number of regression methods.

EXPLORATION AND OPTIMIZATION PROBLEMS

For illustrating optimization problems, an example is given from the field of designed experiments. If the yield of an organic synthesis reaction must be optimized, then the variables that influence this yield, e.g. pH and temperature, can be varied according to an experimental design [Box *et al.* 1978]. The experiments are carried out according to this design and the measured yields are obtained in the second block of data. The first block of data contains the settings of the controlled variables; pH and temperature. The effect on the yield of varying the controlled variables can be estimated from the data, thus providing means for optimizing the yield [Box *et al.* 1978, Montgomery 1976].

The methods to use for these types of problems are again the regression methods, as mentioned in the previous section. Methods also exist that are especially developed to deal with designed data [Box *et al.* 1978].

In the above examples there is a natural way to order the complete data set in two blocks, where both blocks have one mode in common. In Chapter 3 the methods of multiple linear regression, principal component regression, and partial least squares regression will be discussed on an introductory level.

Two-way multiblock data

EXPLORING, PREDICTING AND MODELING

Examples of this type of data are, e.g., from the field of process modeling and multivariate statistical process control. Suppose that process measurements are taken from a chemical reactor during a certain period of time. In the same time period, process measurements are taken from the separation column following that reactor as a unit operation. The composition of the product leaving the column is also measured in the same time period and with the same measurement frequency as the process measurements. This results in three blocks of data, with one mode in common. Relationships between these blocks can be sought and, e.g., used to develop control charts [Kourti *et al.* 1995, MacGregor *et al.* 1994].

Methods to deal with this type of problems are Multiblock Partial Least Squares [Wangen & Kowalski 1988], Hierarchical PLS [Westerhuis *et al.* 1998, Wold *et al.* 1996] and Latent Path Modeling [Frank & Kowalski 1985, Jöreskog & Wold 1982]. The latter method is not very popular in chemistry, but is common in the social sciences.

COMPARING CONFIGURATIONS

In sensory analysis different food products are subjected to judgements by a sensory panel. Suppose that several food products are judged by such a panel and scored on different variables (the meaning and number of variables may even differ for different assessors). In order to obtain objective descriptions of the samples and not subjective reflections of the individual judges, it is necessary to use tools that correct for individual differences between judges and extract the consensus judgements. Techniques to tackle these types of problems are, e.g., Procrustes Analysis [Gower 1975, Ten Berge 1977] and Hierarchical PCA [Westerhuis *et al.* 1998, Wold *et al.* 1996].

Three-way one-block data

EXPLORATION

One of the typical purposes of using three-way analysis on a block of data is exploring the interrelations in those data. An example is a three-way environmental data set consisting of measured concentrations of different chemical compounds on several locations in a geographical area at several points in time. Three-way analysis of such a data set can help in distinguishing patterns, e.g., temporal and spatial behavior of the different chemical compounds. Another example is from the field of chromatography where retention values of different solutes are measured on different chromatographic columns at different mobile phase compositions. Again, three-way analysis can help in finding patterns [Smilde *et al.* 1990], such as systematic differences between the chromatographic columns. Another application area is image analysis. Two-dimensional images of an object taken at different wavelengths are stacked on top of each other and analyzed with three-way tools [Geladi & Grahn 1996].

CALIBRATION AND RESOLUTION

One of the earliest applications of three-way analysis in chemistry on a single block of data is in second-order calibration. Calibration is concerned with estimating the concentration of a certain analyte in a mixture. In the case of second-order calibration, an instrument is used that generates a matrix of measurements for a single chemical analysis. If the standard $\mathbf{X}_1$ contains the pure response of the measured analyte and $\mathbf{X}_2$ is the measurement of the mixture containing that analyte, then under certain conditions it is possible to quantify the analyte in the mixture, even if this mixture contains unknown interferents. This is done by stacking $\mathbf{X}_1$ and $\mathbf{X}_2$ on top of each other and building a three-way model of that stacked array. Not only the concentration estimate of the analyte is obtained but also estimates of the pure response profiles, e.g., pure spectra and chromatograms of the analyte [Sanchez & Kowalski 1988]. Second-order calibration will be explained in more detail in Chapter 10.

Three-way two-block data

Three-way two-block data can be encountered, e.g., in modeling and multivariate statistical process control of batch processes. The first block contains the measured process variables at certain points in time of different batch runs. The second block might contain the quality measurements of the end products of the batches. Creating a relationship between these blocks through regression analysis or similar, can shed light on the connection of the variation in quality and the variation in process measurements. This can be used to build control charts [Boqué & Smilde 1999, Kourti *et al.* 1995]. Another application is in multivariate calibration where, for example, fluorescence emission/excitation data of samples are used to predict a property of those samples [Bro 1999].

Three-way multiblock data

An example of a three-way multiblock problem was published in the area of multivariate statistical process control of batch processes [Kourti *et al.* 1995]. Suppose, e.g., that the feed of a batch process is characterized by a composition vector of length L. If I different batch runs have been completed this results in a matrix $\mathbf{X}$ ($I \times L$) of feed characteristics. The process measurements are collected in $\underline{\mathbf{Z}}$ ($I \times J \times K$) having I batches, J variables and measured at K time points each. The M different end product quality measurements are collected in $\mathbf{Y}$ ($I \times M$). Investigating if there is a connection between $\mathbf{X}$, $\underline{\mathbf{Z}}$ and $\mathbf{Y}$ is a common task in process analysis.

Four-way data and higher

Applications and examples of multi-way data and multi-way analysis with four- and higher-way data are scarce. One application is in 3D-QSAR based on Comparative Molecular Field Analysis (COMFA), where a five-way problem is tackled [Nilsson *et al.* 1997]. In COMFA a molecule is placed in a hypothetical box with a grid of points. The energy of the molecule in a grid point is calculated when a hypothetical probe is placed at a certain position in the box. The five ways are: different probes, three different spatial directions in the molecule, and the number of molecules. Five-way analysis has been used to study the variation in the resulting five-way array. Moreover, for each molecule an activity is measured. A predictive relationship was sought between the five-way array and the vector of activities.

In image analysis, it has become possible to produce three-way images with rows, columns and depth planes as voxel coordinates. Measuring the images at several wavelengths and following such a system over time will also give a five-way array. This array has image coordinates $H \times I \times J$, K wavelength variables and M time samples. Other examples on higher-way data occur, e.g., in experimental design, spectroscopy and chromatography [Durell *et al.* 1990, Bro 1998].

1.4 Principal Component Analysis and PARAFAC

A short introduction to principal component analysis for two-way arrays and PARAFAC for three-way arrays will be given here. These methods are used in the next chapter, and therefore, a short introduction is necessary. Definitions, notations and other details on these methods are explained in later chapters.

In principal component analysis (PCA), a matrix is decomposed as a sum of vector products, as shown in Figure 1.6. The 'vertical' vectors (following the object way) are called scores and the 'horizontal' vectors (following the variable way) are called loadings. A similar decomposition is given for three-way arrays. Here, the array is decomposed as a sum of triple products of vectors as in Figure 1.7. This is the PARAFAC model. The vectors, of which there are three different types, are called loadings.

A pair of a loading vector and a matching score vector is called a component for PCA and similarly, a triplet of loading vectors is called a PARAFAC component. The distinction between a loading and a score vector is relatively arbitrary. Usually, the term score vector is

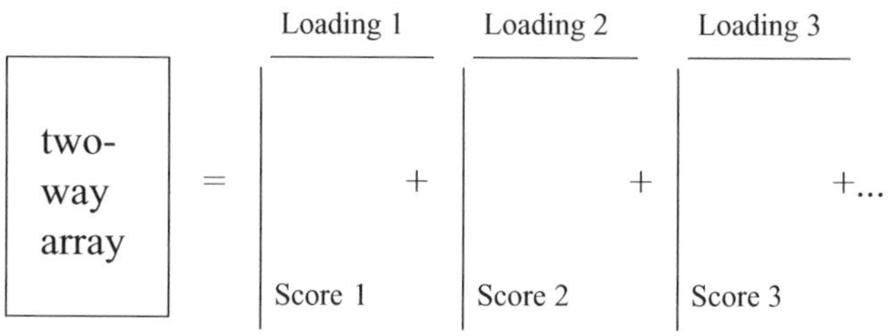

Figure 1.6. In principal component analysis, a two-way array is decomposed into a sum of vector products. The vectors in the object direction are scores and the ones in the variable direction are loadings.

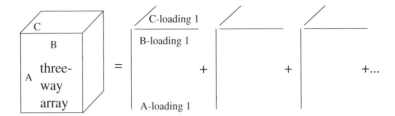

Figure 1.7. In PARAFAC, a three-way array is decomposed into a sum of triple products of vectors. The vectors are called loadings. There are three different types: called A,B and C for convenience.

only used if a mode is considered an object mode and loadings are used for variable modes. In some cases the distinction between objects and loadings, however, is not clear from the data alone and the more broad term loading vector is then generally used. Furthermore, since most decomposition methods to be discussed make no distinction in the decomposition of the array, it has become customary in three-way analysis to only use the term loading vector. As seen in the Figures 1.6 and 1.7, a number of components can be found for an array, numbered 1, 2, 3, 4 etc. If the data can be well-modeled by a small number of components the interpretation of the variations in the data can be analyzed through these components instead of the raw data.

1.5 Summary

This chapter introduces multi-way analysis and describes important concepts of data analysis in general. In data analysis, sometimes a distinction is made between objects and variables. Such distinctions are not always clear. Moreover, there is a large variety of types of variables, such as homogenous or heterogenous. It is important to keep these things in mind when doing multi-way analysis.

An overview is given of the different kinds of possible multi-way analyses and existing methods to perform such analyses. A categorization is presented in which different multi-way analyses are divided into classes using two criteria: (i) the number of blocks involved, and (ii) the order of the arrays involved. Practical examples are given for each class, showing the relevance of the multi-way analysis problem.

Finally, two important decomposition methods for two-way analysis (PCA) and three-way analysis (PARAFAC) are introduced briefly, because these methods are needed in the following chapter.

2

ARRAY DEFINITIONS AND PROPERTIES

2.1 Introduction

There are different ways to represent two- and multi-way models. One way is to use rigorous matrix algebra, another way is to use a pictorial description. When exactness is needed, matrix algebra is necessary. However, in order to understand the structure of the models, a pictorial description can be more informative. Hence, in this book both types of representations are used. In this section the pictorial representation will be introduced. The nomenclature used for vectors, matrices and multiway arrays is given in the Nomenclature Section. Moreover, notational conventions for multi-way arrays are taken from Kiers [2000] with a few exceptions.

The rules for the pictorial description are: numbers (scalars) are represented by small circles; vectors by line segments; matrices by rectangles; and three-way arrays by rectangular boxes. This is shown in Figure 2.1. These graphics also allow for a convenient representation of vector and matrix multiplication (see Figure 2.2). As is shown later in this chapter, more advanced concepts as dyads, triads and matricizing can also be visualized with these rules.

The matrix product is well known and can be found in any linear algebra textbook. It reduces to the vector product when a vector is considered as an $I \times 1$ matrix, and a transposed vector is a $1 \times I$ matrix. The product $\mathbf{a}'\mathbf{b}$ is the inner product of two vectors. The product $\mathbf{ab}'$ is called the outer product or a dyad. See Figure 2.2. These products have no special symbol. Just putting two vectors or matrices together means that the product is taken. The same also goes for products of vectors with scalars and matrices with scalars.

2.2 Rows, Columns and Tubes; Frontal, Lateral and Horizontal slices

For two-way arrays it is useful to distinguish between special parts of the array, such as rows and columns. This also holds for three-way arrays and one such a division

Multi-way Analysis With Applications in the Chemical Sciences.   A. Smilde, R. Bro and P. Geladi
© 2004 John Wiley & Sons, Ltd   ISBN: 0-471-98691-7

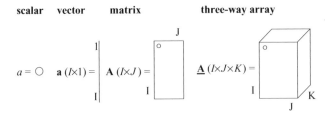

Figure 2.1. Pictorial representation of scalars, vectors, matrices and three-way arrays. The circle in the matrix indicates the (1,1) element and in the three-way array indicates the (1,1,1) element; I, J and K are the dimensions of the first, second and third mode, respectively.

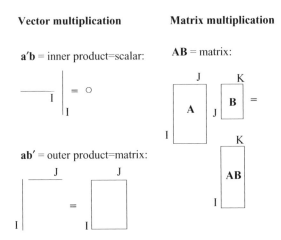

Figure 2.2. Pictorial representation of some vector and matrix manipulations. Note that for multiplication no special symbol is used.

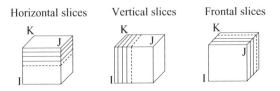

Figure 2.3. Partitioning of a three-way array in slices (two-way arrays).

in parts was already introduced in Chapter 1: frontal, horizontal and vertical slices or slabs. For convenience, the figure is repeated here (Figure 2.3). There are three different kinds of slices for the three-way array $\underline{\mathbf{X}}\,(I \times J \times K)$. The first ones are the horizontal slices: $\mathbf{X}_1, \ldots, \mathbf{X}_i, \ldots, \mathbf{X}_I$; all of size $(J \times K)$. The second ones are the vertical slices: $\mathbf{X}_1, \ldots, \mathbf{X}_j, \ldots, \mathbf{X}_J$; all of size $(I \times K)$. The last ones are the frontal slices: $\mathbf{X}_1, \ldots, \mathbf{X}_k, \ldots, \mathbf{X}_K$; all of size $(I \times J)$. This shorthand notation is convenient but not always unambiguous, e.g. $\mathbf{X}_2$ might mean three different things. Such ambiguity is removed in the text where necessary, e.g., using $\mathbf{X}_{i=2}$ for the first mode. It is also possible to define column, row and tube vectors in a three-way array. This is shown in Figure 2.4. In a three-way

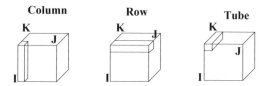

Figure 2.4. Definition of columns, rows and tubes in a three-way array.

array $\underline{\mathbf{X}}$ ($I \times J \times K$) there are $J \times K$ columns $\mathbf{x}_{jk}$ ($I \times 1$); $I \times K$ rows $\mathbf{x}_{ik}$ ($J \times 1$) and $I \times J$ tubes $\mathbf{x}_{ij}$ ($K \times 1$). Again, ambiguities are resolved in the text where necessary.

2.3 Elementary Operations

Vec-operator

If $\mathbf{A}$ is an $I \times J$ matrix with columns $\mathbf{a}_j$ ($j = 1, \ldots, J$), then vec $\mathbf{A}$ is the IJ vector

$$\text{vec } \mathbf{A} = \begin{bmatrix} \mathbf{a}_1 \\ \mathbf{a}_2 \\ \cdot \\ \cdot \\ \cdot \\ \mathbf{a}_J \end{bmatrix} \tag{2.1}$$

Hence, vec $\mathbf{A}$ is obtained by stacking all the column vectors of $\mathbf{A}$ underneath each other.

EXAMPLE 2.1

Vec-operation

An example of vec-operation on a matrix $\mathbf{A}$ is given here. The operation looks trivial and is very simple, but it is useful for writing out complicated equations in a simpler form.

$$\text{vec } \mathbf{A} = \text{vec} \begin{bmatrix} 1 & 2 \\ 3 & 4 \end{bmatrix} = \begin{bmatrix} 1 \\ 3 \\ 2 \\ 4 \end{bmatrix}$$

The vec-operator obeys some simple rules [Magnus & Neudecker 1988]:

(i) vec $(\mathbf{a}') = $ vec $\mathbf{a} = \mathbf{a}$

(ii) vec $(\mathbf{ab}') = \mathbf{b} \otimes \mathbf{a}$

(iii) vec $(\mathbf{A})'$ vec $\mathbf{B} = $ tr $\mathbf{A}'\mathbf{B}$; if $\mathbf{A}$ and $\mathbf{B}$ are of the same order

(iv) vec $(\mathbf{ABC}) = (\mathbf{C}' \otimes \mathbf{A})$vec $\mathbf{B}$; if $\mathbf{ABC}$ exists

$$\tag{2.2}$$

Property 2.2(ii) connects Kronecker products (see later) with the vec-operator and 2.2(iii) connects the vec-operator with the trace (tr).

Kronecker, Hadamard and Khatri–Rao products

In order to concisely describe multi-way models, the usual matrix product is not sufficient. Three other types of matrix products are introduced: the Kronecker ($\otimes$), Hadamard (*) and Khatri–Rao ($\odot$) product [McDonald 1980, Rao & Mitra 1971, Schott 1997]. The Kronecker product allows a very efficient way to write Tucker models (see Chapter 4). Likewise, the Khatri–Rao product provides means for an efficient way to write a PARAFAC model (see Chapter 4). The Hadamard product can, for instance, be used to formalize weighted regression (see Chapter 6).

The Kronecker product of two matrices $\mathbf{A}\,(I \times J)$ and $\mathbf{B}\,(K \times M)$ is defined as

$$
\mathbf{A} \otimes \mathbf{B} =
\begin{bmatrix}
a_{11}\mathbf{B} & \cdot & \cdot & \cdot & a_{1J}\mathbf{B} \\
\cdot & \cdot & & & \cdot \\
\cdot & & \cdot & & \cdot \\
\cdot & & & \cdot & \cdot \\
a_{I1}\mathbf{B} & \cdot & \cdot & \cdot & a_{IJ}\mathbf{B}
\end{bmatrix}
\tag{2.3}
$$

from which it follows that the size of $\mathbf{A} \otimes \mathbf{B}$ is $(IK \times JM)$ and the Kronecker product is also defined for two matrices for which the usual matrix product does not exist (if $J \neq K$).

EXAMPLE 2.2

Kronecker product

A simple example of the Kronecker product is the following:

$$
\mathbf{A} =
\begin{bmatrix} 1 & 2 \\ 3 & 4 \end{bmatrix}; \quad
\mathbf{B} =
\begin{bmatrix} 0 & 7 \\ -1 & 8 \end{bmatrix}
$$

$$
\mathbf{A} \otimes \mathbf{B} =
\begin{bmatrix} 1\mathbf{B} & 2\mathbf{B} \\ 3\mathbf{B} & 4\mathbf{B} \end{bmatrix} =
\begin{bmatrix}
0 & 7 & 0 & 14 \\
-1 & 8 & -2 & 16 \\
0 & 21 & 0 & 28 \\
-3 & 24 & -4 & 32
\end{bmatrix}
$$

The Kronecker product obeys some simple multiplication laws:

$$
\mathbf{A} \otimes \mathbf{B} \otimes \mathbf{C} = (\mathbf{A} \otimes \mathbf{B}) \otimes \mathbf{C} = \mathbf{A} \otimes (\mathbf{B} \otimes \mathbf{C})
$$
$$
(\mathbf{A} + \mathbf{B}) \otimes (\mathbf{C} + \mathbf{D}) = \mathbf{A} \otimes \mathbf{C} + \mathbf{A} \otimes \mathbf{D} + \mathbf{B} \otimes \mathbf{C} + \mathbf{B} \otimes \mathbf{D}
\tag{2.4}
$$

if $\mathbf{A} + \mathbf{B}$ and $\mathbf{C} + \mathbf{D}$ exist, and

$$
(\mathbf{A} \otimes \mathbf{B})(\mathbf{C} \otimes \mathbf{D}) = \mathbf{AC} \otimes \mathbf{BD}
\tag{2.5}
$$

if $\mathbf{AC}$ and $\mathbf{BD}$ exist. Some other useful properties of the Kronecker product are [Magnus &

Neudecker 1988, Searle 1982]:

(i) $a \otimes \mathbf{A} = a\mathbf{A} = \mathbf{A}a = \mathbf{A} \otimes a$; where a is a scalar

(ii) $(\mathbf{A} \otimes \mathbf{B})' = \mathbf{A}' \otimes \mathbf{B}'$

(iii) $\mathbf{a}' \otimes \mathbf{b} = \mathbf{ba}' = \mathbf{b} \otimes \mathbf{a}'$

(iv) $\text{tr}(\mathbf{A} \otimes \mathbf{B}) = \text{tr}(\mathbf{A})\,\text{tr}(\mathbf{B})$; for square matrices $\mathbf{A}$ and $\mathbf{B}$ (2.6)

(v) $(\mathbf{A} \otimes \mathbf{B})^{-1} = \mathbf{A}^{-1} \otimes \mathbf{B}^{-1}$; if $\mathbf{A}$ and $\mathbf{B}$ are nonsingular

(vi) $\text{r}(\mathbf{A} \otimes \mathbf{B}) = \text{r}(\mathbf{A})\,\text{r}(\mathbf{B})$

(vii) $(\mathbf{A} \otimes \mathbf{B})^{+} = \mathbf{A}^{+} \otimes \mathbf{B}^{+}$

where the superscript '+' is used to indicate the Moore–Penrose inverse [Campbell & Meyer 1979, Rao & Mitra 1971] and $\text{r}(\mathbf{A})$ is the abbreviation of the rank of $\mathbf{A}$.

The Kronecker product can be used to define dyads. A dyad is the outer product of two vectors [see Equation 2.6 (iii)]. Figure 2.2 shows the outer product of the vectors $\mathbf{a}$ and $\mathbf{b}$. The dyad $\mathbf{ab}'$ can also be written as $\mathbf{a} \otimes \mathbf{b}' = \mathbf{b}' \otimes \mathbf{a}$.

Another product which can be of use in multi-way analysis is the Hadamard or elementwise product, which is defined for matrices $\mathbf{A}$ and $\mathbf{B}$ of equal size ($I \times J$):

$$\mathbf{A} * \mathbf{B} = \begin{bmatrix} a_{11}b_{11} & \cdot & \cdot & \cdot & a_{1J}b_{1J} \\ \cdot & \cdot & & & \cdot \\ \cdot & & \cdot & & \cdot \\ \cdot & & & \cdot & \cdot \\ a_{I1}b_{I1} & \cdot & \cdot & \cdot & a_{IJ}b_{IJ} \end{bmatrix} \qquad (2.7)$$

where a_{ij} and b_{ij} are the elements of $\mathbf{A}$ and $\mathbf{B}$, respectively. Hence, the Hadamard product is an element-by-element product.

EXAMPLE 2.3

Hadamard product

In the example used earlier for the Kronecker product, the Hadamard product of $\mathbf{A}$ and $\mathbf{B}$ becomes:

$$\mathbf{A} * \mathbf{B} = \begin{bmatrix} 1 & 2 \\ 3 & 4 \end{bmatrix} * \begin{bmatrix} 0 & 7 \\ -1 & 8 \end{bmatrix} = \begin{bmatrix} 0 & 14 \\ -3 & 32 \end{bmatrix}$$

Some properties of the Hadamard product are [Magnus & Neudecker 1988]:

(i) $\mathbf{A} * \mathbf{B} = \mathbf{B} * \mathbf{A}$

(ii) $(\mathbf{A} * \mathbf{B})' = \mathbf{A}' * \mathbf{B}'$

(iii) $(\mathbf{A} * \mathbf{B}) * \mathbf{C} = \mathbf{A} * (\mathbf{B} * \mathbf{C})$ (2.8)

(iv) $(\mathbf{A} + \mathbf{B}) * (\mathbf{C} + \mathbf{D}) = \mathbf{A} * \mathbf{C} + \mathbf{A} * \mathbf{D} + \mathbf{B} * \mathbf{C} + \mathbf{B} * \mathbf{D}$

(v) $\mathbf{A} * \mathbf{I} = \text{diag}(a_{11}, \ldots, a_{nn})$

where $\mathbf{I}$ is the identity matrix and $\mathbf{A}$ is a square matrix with diagonal elements $a_{11}, \ldots, a_{nn}$. $\text{Diag}(a_{11}, \ldots, a_{nn})$ is the matrix with diagonal elements equal to $a_{11}, \ldots, a_{nn}$ and all off-diagonal elements equal to zero [Schott 1997].

The third product which is useful in three-way analysis is the Khatri–Rao product [McDonald 1980, Rao & Mitra 1971] which is defined as follows:

$$\mathbf{A} = [\mathbf{A}_1 \ .. \ \mathbf{A}_K] \quad \text{and} \quad \mathbf{B} = [\mathbf{B}_1 \ .. \ \mathbf{B}_K]$$
$$\mathbf{A} \odot \mathbf{B} = [\mathbf{A}_1 \otimes \mathbf{B}_1 \ .. \ \mathbf{A}_K \otimes \mathbf{B}_K] \tag{2.9}$$

where $\mathbf{A}$ and $\mathbf{B}$ are partitioned matrices with an equal number of partitions. Some useful properties of the Khatri–Rao product are

$$(\mathbf{A} \odot \mathbf{B}) \odot \mathbf{C} = \mathbf{A} \odot (\mathbf{B} \odot \mathbf{C})$$
$$(\mathbf{T}_1 \otimes \mathbf{T}_2)(\mathbf{A} \odot \mathbf{B}) = \mathbf{T}_1 \mathbf{A} \odot \mathbf{T}_2 \mathbf{B} \tag{2.10}$$

A special case occurs when $\mathbf{A}$ and $\mathbf{B}$ are partitioned in their K columns. This type of partitioning will be used throughout this book. It then additionally holds that

$$(\mathbf{A} \odot \mathbf{B})'(\mathbf{A} \odot \mathbf{B}) = (\mathbf{A}'\mathbf{A}) * (\mathbf{B}'\mathbf{B}) \tag{2.11}$$

EXAMPLE 2.4

Khatri–Rao product

In the example used earlier for the Kronecker and Hadamard product,

$$\mathbf{A} \odot \mathbf{B} = \left[\begin{pmatrix} 1 \\ 3 \end{pmatrix} \otimes \begin{pmatrix} 0 \\ -1 \end{pmatrix} \begin{pmatrix} 2 \\ 4 \end{pmatrix} \otimes \begin{pmatrix} 7 \\ 8 \end{pmatrix} \right] = \begin{bmatrix} 0 & 14 \\ -1 & 16 \\ 0 & 28 \\ -3 & 32 \end{bmatrix}$$

The natural extension of a dyad is a triad. The triad consists of a multiplication of three vectors $\mathbf{a}\,(I \times 1)$, $\mathbf{b}\,(J \times 1)$ and $\mathbf{c}\,(K \times 1)$, and is visualized in Figure 2.5. The multiplication involved is sometimes referred to as a tensor product ($\triangle$). In order to distinguish between Kronecker and tensor products, different symbols will be used. A Kronecker product is a tensor product, but not every tensor product is a Kronecker product [Burdick 1995]. In

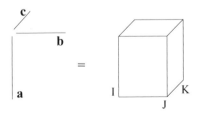

Figure 2.5. Visualization of a triad, making up a rank-one three-way array.

figures such as Figure 2.5 where the context is clear, the tensor product symbol is usually left out.

Tensor product

An example of the use of tensor products is the following.

$$\mathbf{a} = \begin{bmatrix} 1 \\ 2 \\ 3 \end{bmatrix}; \quad \mathbf{b} = \begin{bmatrix} -1 \\ 2 \end{bmatrix}; \quad \mathbf{c} = \begin{bmatrix} 2 \\ 1 \end{bmatrix}$$

$$\mathbf{a} \triangle \mathbf{b} \triangle \mathbf{c} = \underline{\mathbf{X}}$$

$$\mathbf{X}_1 = \begin{bmatrix} -2 & 4 \\ -4 & 8 \\ -6 & 12 \end{bmatrix}; \quad \mathbf{X}_2 = \begin{bmatrix} -1 & 2 \\ -2 & 4 \\ -3 & 6 \end{bmatrix}$$

where $\mathbf{X}_1$ is the first frontal slice of $\underline{\mathbf{X}}$ and $\mathbf{X}_2$ is the back frontal slice of $\underline{\mathbf{X}}$.

Matricizing and Sub-arrays

The principle of matricizing is explained in Figure 2.6 [Kiers 2000]. The three-way array $\underline{\mathbf{X}}$ ($I \times J \times K$) is matricized as a concatenated two-way array (or matrix) $\mathbf{X}_{(I \times JK)}$ of size ($I \times JK$). In Figure 2.6 one way of matricizing $\underline{\mathbf{X}}$ is shown. In the columns of the new matrix $\mathbf{X}_{(I \times JK)}$, the original index j runs the fastest and k the slowest. There is also a configuration possible in which k runs fastest and j slowest. This will be notated as $\mathbf{X}_{(I \times KJ)}$. There exist four other ways of matricizing $\underline{\mathbf{X}}$ in concatenated arrays, that is, arrays of sizes: ($J \times IK$), ($J \times KI$), ($K \times IJ$) and ($K \times JI$).

There are different terms in the literature used for the matricizing as exemplified in Figure 2.6, e.g. unfolding, juxtaposition, concatenation, stringing-out. The term unfolding should be avoided since it has a specific meaning in psychometrics [Davies & Coxon 1982].

The notation for a matricized three-way array $\underline{\mathbf{X}}$ ($I \times J \times K$) was already introduced above, in explaining the principle of matricizing. Sometimes it is clear from the context which matricized matrix is meant in which case the subscript is omitted.

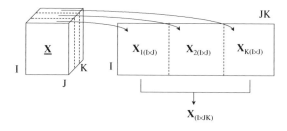

Figure 2.6. Matricizing a three-way array $\underline{\mathbf{X}}$ to a two-way array $\mathbf{X}_{(I \times JK)}$.

EXAMPLE 2.6

Matricizing operation

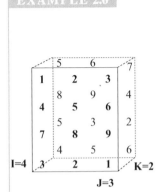

Figure 2.7. A three-way array $\underline{\mathbf{X}}$ ($4 \times 3 \times 2$); boldface characters are in the foremost frontal slice and normal characters are in the back slice.

An example of matricizing a three-way array $\underline{\mathbf{X}}$ is given in Figure 2.7. Three different matricized three-way arrays are as follows:

$$\mathbf{X}_{(I \times JK)} = \begin{bmatrix} 1 & 2 & 3 & & 5 & 6 & 7 \\ 4 & 5 & 6 & & 8 & 9 & 4 \\ 7 & 8 & 9 & & 5 & 3 & 2 \\ 3 & 2 & 1 & & 4 & 5 & 6 \end{bmatrix}$$

$$\mathbf{X}_{(J \times IK)} = \begin{bmatrix} 1 & 4 & 7 & 3 & & 5 & 8 & 5 & 4 \\ 2 & 5 & 8 & 2 & & 6 & 9 & 3 & 5 \\ 3 & 6 & 9 & 1 & & 7 & 4 & 2 & 6 \end{bmatrix}$$

$$\mathbf{X}_{(K \times IJ)} = \begin{bmatrix} 1 & 4 & 7 & 3 & & 2 & 5 & 8 & 2 & & 3 & 6 & 9 & 1 \\ 5 & 8 & 5 & 4 & & 6 & 9 & 3 & 5 & & 7 & 4 & 2 & 6 \end{bmatrix}$$

Figures 2.3 and 2.4 indicates that a three-way array $\underline{\mathbf{X}}$ has several sub-arrays. Examples of sub-arrays are slices, columns, rows and tubes, as discussed earlier.

EXAMPLE 2.7

Sub-arrays

In the example of Figure 2.7, some of the sub-arrays are

$$\mathbf{X}_{i=3} = \begin{bmatrix} 7 & 5 \\ 8 & 3 \\ 9 & 2 \end{bmatrix}$$

$$\mathbf{X}_{j=1} = \begin{bmatrix} 1 & 5 \\ 4 & 8 \\ 7 & 5 \\ 3 & 4 \end{bmatrix}$$

$$\mathbf{X}_{k=2} = \begin{bmatrix} 5 & 6 & 7 \\ 8 & 9 & 4 \\ 5 & 3 & 2 \\ 4 & 5 & 6 \end{bmatrix}$$

2.4 Linearity Concepts

Linearity

There is much confusion about the terms linear and linearity. Suppose a model is needed for relating x_i (the predictor) to y_i (the predictand), where i indexes the objects and runs from 1 to I. Such a model is linear in the *parameters* and in the *variables* if it can be written as

$$y_i = b_0 + b_1 x_i + e_i; \quad i = 1, \ldots, I \tag{2.12}$$

which is the usual linear regression model with an intercept (b_0) and an error term (ε_i). An example of a model linear in the *parameters* but nonlinear in x is

$$y_i = b_0 + b_1 x_i + b_2 x_i^2 + e_i; \quad i = 1, \ldots, I \tag{2.13}$$

It is useful to distinguish these two types of linearity. If in the chemical sciences the term 'linearity' is loosely used, then often linearity in the parameters is meant. Hence, models in Equations (2.12) and (2.13) are linear in the parameters, because when x is fixed, y is a linear function of the parameters b_0, b_1 (and b_2).

Bilinearity

The notion of linearity can be extended to bilinearity. Suppose that a model is needed for $\mathbf{X}(I \times J)$, with elements x_{ij}. A two-component PCA model of $\mathbf{X}$ (see Chapter 1 and for a more detailed explanation Chapter 3) is then:

$$x_{ij} = t_{i1} p_{j1} + t_{i2} p_{j2} + e_{ij}; \quad i = 1, \ldots, I; \ j = 1, \ldots, J \tag{2.14}$$

and it can be seen that the modeled part of $\mathbf{X}$ is both linear in t_{i1}, t_{i2} (with fixed p_{j1}, p_{j2}) and in p_{j1}, p_{j2} (with fixed t_{i1}, t_{i2}). Hence, Equation (2.14) is called a bilinear model of $\mathbf{X}$ [Kruskal 1984].

There is some confusion in the second-order calibration literature [Sanchez & Kowalski 1988] where the term bilinearity is reserved for equations of the form of Equation (2.14) with a one component PCA model for $\mathbf{X}$. However, bilinearity is more general and holds for an arbitrary number of components [Kruskal 1984].

Trilinearity

The notion of bilinearity can be extended further. Consider a model of the three-way array $\underline{\mathbf{X}}(I \times J \times K)$, with elements x_{ijk}:

$$x_{ijk} = a_{i1} b_{j1} c_{k1} + a_{i2} b_{j2} c_{k2} + e_{ijk}; \quad i = 1, \ldots, I; j = 1, \ldots, J; k = 1, \ldots, K \tag{2.15}$$

where $\underline{\mathbf{X}}(I \times J \times K)$ is modeled with two components. Then the modeled part of x_{ijk} is trilinear in the parameter sets a, b and c. Hence, the model in Equation (2.15) can be called a trilinear model of $\underline{\mathbf{X}}$ and is, in fact, the PARAFAC model of $\underline{\mathbf{X}}$ (see Chapter 1 and for a more detailed explanation Chapter 4).

The notion of trilinearity can be extended to quadrilinearity, etc. The basic idea remains the same: fixing the parameters in all sets but one, the model is linear in the set of parameters which are not fixed. This holds for all parameter sets [Kruskal 1984].

2.5 Rank of Two-way Arrays

Rank is a very important mathematical property of two-way arrays (matrices). The mathematical definition of rank is unambiguous, but for chemistry this rank definition is not always the most convenient one. In the first section, different rank definitions are given with examples. The rank of three-way and multi-way arrays is considerably more complicated than for the two-way arrays and is treated separately.

Definitions and properties

Suppose that a matrix A ($I \times J$) is given. The matrix A has J columns. The number of linearly independent columns of A is called the column-rank of A.[1] The matrix A defines a linear transformation from R^J to R^I. The range or column space of A is defined as $\Re(A) = \{y | Ax = y; \forall x \in R^J\}$ or, stated in words, the range of A consists of all the ys for which there is an x such that $Ax = y$. Clearly, $\Re(A) \subseteq R^I$. The rank of A is the dimension of $\Re(A)$ and this equals the column-rank of A.

EXAMPLE 2.8

Rank and column-rank

The idea of rank and column-rank can be illustrated with an example. Suppose that

$$A = \begin{bmatrix} 1 & 4 \\ 2 & 5 \\ 3 & 6 \end{bmatrix}$$

then A has two independent columns. Hence, the column-rank is two. The matrix A defines a linear transformation $Ax = y$, and using

$$Ax = \begin{bmatrix} 1 & 4 \\ 2 & 5 \\ 3 & 6 \end{bmatrix} \begin{bmatrix} x_1 \\ x_2 \end{bmatrix} = x_1 \begin{bmatrix} 1 \\ 2 \\ 3 \end{bmatrix} + x_2 \begin{bmatrix} 4 \\ 5 \\ 6 \end{bmatrix} = y$$

the columns of A form a basis for $\Re(A)$. Hence, $\Re(A)$ is two-dimensional and the rank of A is therefore two. In another example, suppose that

$$A = \begin{bmatrix} 1 & 3 \\ 2 & 6 \end{bmatrix}$$

then A has rank one because it has two dependent columns (the second column is three times the first column). The linear transformation $Ax = y$ can be written as

$$Ax = \begin{bmatrix} 1 & 3 \\ 2 & 6 \end{bmatrix} \begin{bmatrix} x_1 \\ x_2 \end{bmatrix} = x_1 \begin{bmatrix} 1 \\ 2 \end{bmatrix} + x_2 \begin{bmatrix} 3 \\ 6 \end{bmatrix} = (x_1 + 3x_2) \begin{bmatrix} 1 \\ 2 \end{bmatrix} = y$$

and the range of A is spanned by $(1\ 2)'$ and has dimension one, the rank of A.

---

[1] The set of vectors $a_i (i = 1, \ldots, I)$ is linearly independent if $\sum_{i=1}^{I} c_i a_i = 0$ implies that $c_i = 0$, for all i.

Similarly, the number of linearly independent rows of **A** is called the row-rank of **A**. The row-rank of **A** is the column-rank of **A**′. A fundamental theorem in matrix algebra states that the row-rank and the column-rank of a matrix are equal (and equal to *the* rank) [Schott 1997]. Hence, it follows that the rank $r(\mathbf{A}) \leq \min(I,J)$. The matrix **A** has full rank if and only if $r(\mathbf{A}) = \min(I,J)$. Sometimes the term full column-rank is used. This means that $r(\mathbf{A}) = \min(I,J) = J$, implying that $J \leq I$. The term full row-rank is defined analogously.

If the rank of **A** is R, then there are R columns of **A** spanning $\Re(\mathbf{A})$; let $\mathbf{t}_1, \ldots, \mathbf{t}_R$ be a basis of $\Re(\mathbf{A})$. Likewise, there are R rows in **A** spanning $\Re(\mathbf{A}')$ and let $\mathbf{p}_1, \ldots, \mathbf{p}_R$ be a basis of $\Re(\mathbf{A}')$. Then there exist **t** and **p** such that **A** can be written as [Schott 1997]

$$\mathbf{A} = \mathbf{t}_1\mathbf{p}_1' + \cdots + \mathbf{t}_R\mathbf{p}_R' \text{ or}$$
$$a_{ij} = t_{i1}p_{j1} + \cdots + t_{iR}p_{jR}; \quad i = 1, \ldots, I; j = 1, \ldots, J \tag{2.16}$$

where a_{ij} is an element of **A**, t_{ir} is an element of $\mathbf{t}_r$, and p_{jr} is an element of $\mathbf{p}_r$. Each component $\mathbf{t}_r\mathbf{p}_r'$ is a rank one matrix and it is important to realize that Equation (2.16) is an *exact* relationship; it is not an approximation. The choice of the bases $\mathbf{t}_1, \ldots, \mathbf{t}_R$ and $\mathbf{p}_1, \ldots, \mathbf{p}_R$ is not unique: there is rotational freedom and intensity (or scale)/sign-indeterminacy. In Chapter 3, some more aspects of this are outlined. Equation (2.16) shows that each matrix of rank R can be written as a summation of R rank one matrices.

Some important equalities follow. Suppose **A** is $(I \times J)$ and the rank $r(\mathbf{A}) = R$, then:

(i) $r(\mathbf{A}) = r(\mathbf{AB}) = r(\mathbf{CA}) = r(\mathbf{CAB})$; for square nonsingular **C** and **B** of the proper size

(ii) $r(\mathbf{A}) = r(\mathbf{A}') = r(\mathbf{A}'\mathbf{A}) = r(\mathbf{AA}')$

(iii) **A** can always be written as $\mathbf{A} = \mathbf{FG}$, with $\mathbf{F}(I \times R)$, $\mathbf{G}(R \times J)$, both of full rank, which is equivalent to Equation (2.16)   (2.17)

(iv) $\Re(\mathbf{A}) = \Re(\mathbf{AA}')$

(v) $r(\mathbf{A}) = r(\mathbf{S})$, **S** being the singular values of **A** arranged in a diagonal matrix (see Chapter 3)

Proofs of these equalities can be found in standard matrix algebra books [Magnus & Neudecker 1988, Schott 1997, Searle 1982].

Pseudo-rank, rank deficiency and chemical sources of variation

It is a property of two-way arrays that, if a matrix **X** is generated of size $(I \times J)$ with $I \times J$ random numbers (assuming a perfect random number generator), then **X** has less than full rank with probability zero. This has consequences for measured data sets in chemistry: if a matrix **X** is the result of a measurement it will always be of full rank, because measurement noise cannot be avoided in experiments (apart from problems with, e.g., rounded off noise and limited AD conversion which can lower the rank of **X**).

The mathematical concept of rank is not very convenient in chemical modeling. Take, for example, ultraviolet spectra (100 wavelengths) measured on ten different samples, each of which contains the same absorbing species at different concentrations. The resulting data matrix **X** has size (10×100) and, if the Lambert–Beer law holds, is essentially of rank one.

Due to measurement errors, however, the mathematical rank is ten. The model of $\mathbf{X}$ is

$$\mathbf{X} = \mathbf{cs}' + \mathbf{E} \tag{2.18}$$

where the vector $\mathbf{c}$ contains the concentrations of the absorbing species in the ten samples, $\mathbf{s}$ represents the pure ultraviolet spectrum of the absorbing species and $\mathbf{E}$ is the noise part.

The important variation in $\mathbf{X}$ has contribution only from a rank one matrix ($\mathbf{cs}'$). The pseudo-rank of this system is therefore one. A formal definition of pseudo-rank of $\mathbf{X}$ can be given by assuming that the variation in $\mathbf{X}$ can be separated into two parts:

$$\mathbf{X} = \hat{\mathbf{X}} + \mathbf{E} \tag{2.19}$$

where $\mathbf{E}$ contains the measurement noise (and other undesirable phenomena) and $\hat{\mathbf{X}}$ represents the systematic variation in $\mathbf{X}$. The pseudo-rank of $\mathbf{X}$ is defined as the (mathematical) rank of $\hat{\mathbf{X}}$. In common practice, the pseudo-rank is considerably lower than the mathematical rank of a matrix of measured data.

In the example above on ultraviolet spectroscopy there is one single chemical source of variation in $\mathbf{X}$: the concentration of the analyte. Due to the linearity of the system this single source of variation generates an $\hat{\mathbf{X}}$ matrix of rank one. Sometimes the term 'chemical rank' is used to indicate the number of chemical sources of variation in a data matrix.

It is not always true that K sources of chemical variation generate an $\hat{\mathbf{X}}$ matrix of rank K. Consider short-wave near-infrared (800–1000 nm) spectra of mixtures of water and ethanol at different concentrations of water and ethanol and collect these spectra in $\mathbf{X}$. If the Lambert–Beer law holds, $\mathbf{X}$ can be modeled as

$$\mathbf{X} = \mathbf{c}_w \mathbf{s}_w' + \mathbf{c}_e \mathbf{s}_e' + \mathbf{E} \tag{2.20}$$

where $\mathbf{c}_w$ and $\mathbf{c}_e$ contain the concentrations of water and ethanol, respectively, and $\mathbf{s}_w$, $\mathbf{s}_e$ are the pure spectra of water and ethanol, respectively. However, the pseudo-rank of $\mathbf{X}$ is higher than two (typically, four) [Wülfert *et al.* 1998]. This is due to the hydrogen bonding activity of water and ethanol, which changes the spectra of water and ethanol depending on their relative concentrations in the mixture. The systematic part of the data matrix $\mathbf{X}$ has become more complicated due to these nonlinearities. Hence, only two chemical sources of variation (chemical rank is two) generate a pseudo-rank four measurement matrix.

Rank deficiency is a phenomenon opposite to the above. In cases with rank deficiencies the pseudo-rank is less than the chemical rank. Rank deficiencies in a measurement matrix can arise due to linear relations in or restrictions on the data. Suppose that the underlying model of $\mathbf{X}$ is

$$\mathbf{X} = \mathbf{c}_1 \mathbf{s}_1' + \mathbf{c}_2 \mathbf{s}_2' + \mathbf{c}_3 \mathbf{s}_3' + \mathbf{E} \tag{2.21}$$

and assume that $\mathbf{s}_1 = \mathbf{s}_2$. Then the pseudo-rank of $\mathbf{X}$ is not three but two, due to the linear relationship $\mathbf{s}_1 - \mathbf{s}_2 = \mathbf{0}$ (see Elaboration 2.1)

ELABORATION **2.1**

Rank-deficiency

In a flow-injection analysis experiment three analytes were studied, namely 2-hydroxybenzaldehyde (2-HBA), 3-hydroxybenzaldehyde (3-HBA) and 4-hydroxybenzaldehyde (4-HBA). A pH gradient between 4.5 and 11.4 was imposed on the segment in the channel containing the analytes and UV–Vis spectra were measured continuously. The pK_a values of the analytes are in the range of the pH gradient, hence, the analytes were present in the protonated and deprotonated form in the flow injection channel. For a detailed description and analysis, see. [Nørgaard & Ridder 1994, Reis *et al.* 2000].

In this example only the analytes 2-hydroxybenzaldehyde (2-HBA) and 3-hydro-xybenzaldehyde (3-HBA) are discussed. The pK_a values of these analytes are 8.37 and 8.98, respectively. Hence, both the acidic and basic form of the analytes will be present along the pH gradient. Therefore, the ultraviolet spectrum of the acidic and basic form of each solute produces the signal together with the concentration profiles of the acidic and basic forms. If the spectra of the acidic forms are called $s_{a,2}$ and $s_{a,3}$ for 2-HBA and 3-HBA, respectively, and $c_{a,2}$, $c_{a,3}$ are the concentration profiles of the acidic forms of 2-HBA and 3-HBA, and if analogous definitions are used for the basic forms ($s_{b,2}$, $s_{b,3}$, $c_{b,2}$, $c_{b,3}$) then the responses of the pure solutes 2-HBA and 3-HBA can be written as

$$\mathbf{X}_{2\text{-HBA}} = \mathbf{c}_{a,2}\mathbf{s}'_{a,2} + \mathbf{c}_{b,2}\mathbf{s}'_{b,2} + \mathbf{E}_{2\text{-HBA}}$$

$$\mathbf{X}_{3\text{-HBA}} = \mathbf{c}_{a,3}\mathbf{s}'_{a,3} + \mathbf{c}_{b,3}\mathbf{s}'_{b,3} + \mathbf{E}_{3\text{-HBA}}$$

where the symbol $\mathbf{E}$ is used to indicate a matrix containing measurement error.

The pseudo-rank of both analyte responses equals two. Taking a mixture $\mathbf{X}_M$ of both analytes and measuring that mixture in the same system gives

$$\mathbf{X}_M = \mathbf{c}_{a,2}\mathbf{s}'_{a,2} + \mathbf{c}_{b,2}\mathbf{s}'_{b,2} + \mathbf{c}_{a,3}\mathbf{s}'_{a,3} + \mathbf{c}_{b,3}\mathbf{s}'_{b,3} + \mathbf{E}_M$$

which gives a matrix of apparently pseudo-rank four. There is, however, a restriction on the data due to properties of the FIA measuring system. The shapes of the total concentration profiles of the separate analytes are the same because these shapes are largely determined by the dispersion in the FIA channel. Mathematically, this means that the following holds

$$\mathbf{c}_{a,2} + \mathbf{c}_{b,2} = \alpha(\mathbf{c}_{a,3} + \mathbf{c}_{b,3})$$

where α is an arbitrary positive constant. Using this constraint in the equation for $\mathbf{X}_M$ gives

$$\mathbf{X}_M = (\alpha\mathbf{c}_{a,3} + \alpha\mathbf{c}_{b,3} - \mathbf{c}_{b,2})\mathbf{s}'_{a,2} + \mathbf{c}_{b,2}\mathbf{s}'_{b,2} + \mathbf{c}_{a,3}\mathbf{s}'_{a,3} + \mathbf{c}_{b,3}\mathbf{s}'_{b,3} + \mathbf{E}_M$$

$$= \mathbf{c}_{a,3}(\alpha\mathbf{s}'_{a,2} + \mathbf{s}'_{a,3}) + \mathbf{c}_{b,3}(\alpha\mathbf{s}'_{a,2} + \mathbf{s}'_{b,3}) + \mathbf{c}_{b,2}(\mathbf{s}'_{b,2} - \mathbf{s}'_{a,2}) + \mathbf{E}_M$$

which gives a pseudo-rank of three for matrix $\mathbf{X}_M$.

There is another way of representing the restriction using an extra matrix $\mathbf{H}$. Collect the concentration profiles in the matrix $\mathbf{C} = [\mathbf{c}_{a,2} \ \mathbf{c}_{b,2} \ \mathbf{c}_{a,3} \ \mathbf{c}_{b,3}]$ and the spectra in the matrix $\mathbf{S} = [\mathbf{s}_{a,2} \ \mathbf{s}_{b,2} \ \mathbf{s}_{a,3} \ \mathbf{s}_{b,3}]$. Then $\mathbf{X}_M$ can be written as

$$\mathbf{X}_M = \mathbf{C}\mathbf{S}' + \mathbf{E}_M$$

The restriction on the concentration profiles can be accounted for by a special matrix $\mathbf{H}$

$$\mathbf{X}_M = \mathbf{C}\mathbf{H}\mathbf{S}' + \mathbf{E}_M$$

$$\mathbf{H} = \begin{bmatrix} 0 & 0 & 0 & 0 \\ -1 & 1 & 0 & 0 \\ \alpha & 0 & 1 & 0 \\ \alpha & 0 & 0 & 1 \end{bmatrix}$$

which can be inferred by carefully reading the elaborate expression for $\mathbf{X}_M$ with the constraints included. The rank of $\mathbf{H}$ is only three and, hence, the pseudo-rank of $\mathbf{X}_M$ is also three [see Equation (2.17)]. Now, $\mathbf{X}_M$ can be written as

$$\mathbf{X}_M = \tilde{\mathbf{C}}\tilde{\mathbf{H}}\mathbf{S}' + \mathbf{E}_M$$

$$\tilde{\mathbf{C}} = [\mathbf{c}_{b,2} \quad \mathbf{c}_{a,3} \quad \mathbf{c}_{b,3}]$$

$$\tilde{\mathbf{H}} = \begin{bmatrix} -1 & 1 & 0 & 0 \\ \alpha & 0 & 1 & 0 \\ \alpha & 0 & 0 & 1 \end{bmatrix}$$

where both $\tilde{\mathbf{H}}$ and $\tilde{\mathbf{C}}$ are now of rank 3. The last equation shows a special type of a two-mode component model (see Chapter 3).

Not every restriction on the data gives a rank reduction. Consider again an example where spectroscopic measurements are performed on different samples. Suppose that there are three absorbing species in the samples with pure spectra $\mathbf{s}_1$, $\mathbf{s}_2$ and $\mathbf{s}_3$, and concentration $\mathbf{c}_1$, $\mathbf{c}_2$ and $\mathbf{c}_3$, indicating the concentrations of the different absorbing species in the samples. A restriction often encountered is closure, which means that $\mathbf{c}_1 + \mathbf{c}_2 + \mathbf{c}_3 = \mathbf{1}$, where $\mathbf{1}$ is a vector of ones.[2] Closure occurs, for instance, if the three species are the absorbing species of a batch process because closure is then forced by the law of preservation of mass.

Assuming that the Lambert–Beer law holds, then the resulting measurement matrix $\mathbf{X}_M$ can be written as

$$\mathbf{X}_M = [\mathbf{c}_1 \quad \mathbf{c}_2 \quad \mathbf{c}_3] \begin{bmatrix} \mathbf{s}_1' \\ \mathbf{s}_2' \\ \mathbf{s}_3' \end{bmatrix} + \mathbf{E} = \mathbf{C}\mathbf{S}' + \mathbf{E} \tag{2.22}$$

where the symbol $\mathbf{E}$ is again used to indicate measurement noise and $\mathbf{C}$, $\mathbf{S}$ are implicitly defined. Note that despite the fact that $\mathbf{c}_1 + \mathbf{c}_2 + \mathbf{c}_3 = \mathbf{1}$, $\mathbf{C}$ is still of full rank. Upon assuming that also $\mathbf{S}$ is of full rank, then the pseudo-rank of $\mathbf{X}_M$ is three and not affected by the restriction (see Elaboration 2.2).

ELABORATION **2.2**

Closure and rank

The difference between the closure restriction and the restriction of equal total concentration profiles in the flow injection analysis example (see Elaboration 2.1) is that the latter introduces a linear dependency between the columns of $\mathbf{C}$, whereas the closure restriction does not. For a linear dependency between vectors $\mathbf{a}$, $\mathbf{b}$ and $\mathbf{c}$ it must hold that $\lambda_1 \mathbf{a} + \lambda_2 \mathbf{b} + \lambda_3 \mathbf{c} = 0$, for certain nonzero $\lambda_1, \lambda_2, \lambda_3$. This is the case for the equal total concentration profiles restrictions in the FIA example but not for the closure restriction.

An example of a matrix with closure is

$$\mathbf{A} = \begin{bmatrix} 0.3 & 0.7 \\ 0.5 & 0.5 \\ 0.1 & 0.9 \end{bmatrix}$$

---

[2] Other constants than one also impose closure, the $\mathbf{1}$ is just an example.

and the columns of $\mathbf{A}$ are linearly independent, making the rank of $\mathbf{A}$ equal to two. If $\mathbf{A}$ is column-centered, then the situation changes:

$$\mathbf{A}_c = \begin{bmatrix} 0 & 0 \\ 0.2 & -0.2 \\ -0.2 & 0.2 \end{bmatrix}$$

where $\mathbf{A}_c$ is the column-centered $\mathbf{A}$. The columns of $\mathbf{A}_c$ are linearly dependent, hence the rank of $\mathbf{A}_c$ is one (see Chapter 9).

After column centering of $\mathbf{X}_M$ in Equation (2.22), the obtained matrix $\tilde{\mathbf{X}}_M$ has pseudo-rank two and becomes rank-deficient. This holds in certain cases for the combination of closure and column centering (see Appendix 2.A). Discussions on closure, centering, rank-deficiency and its relationships can be found in the literature [Amrhein *et al.* 1996, Pell *et al.* 1992].

Determination of pseudo-rank

Efficient algorithms exist for establishing the rank of a matrix $\mathbf{X}$ ($I \times J$). Using the singular value decomposition of $\mathbf{X}$ ($\mathbf{X} = \mathbf{USV}'$, see Chapter 3) and the property that $r(\mathbf{X}) = r(\mathbf{S})$ any efficient singular value decomposition algorithm can be used. Counting the number of nonzero entries on the diagonal of $\mathbf{S}$ is then sufficient for establishing the rank. Likewise, use can be made of the fact that $r(\mathbf{X}) = r(\mathbf{X}'\mathbf{X})$ and an eigenvalue decomposition of $\mathbf{X}'\mathbf{X}$ ($\mathbf{X}'\mathbf{X} = \mathbf{V}\mathbf{\Lambda}\mathbf{V}'$) can be used. Again, only checking the values on the diagonal of $\mathbf{\Lambda}$ is sufficient. Efficient algorithms exist for both the SVD and the eigenvalue decomposition of the symmetric matrix $\mathbf{X}'\mathbf{X}$ [Golub & van Loan 1989, Wilkinson 1965].

Establishing the pseudo-rank of $\mathbf{X}$ is considerably more complicated since the noise in $\mathbf{X}$ has to be separated from the systematic part of $\mathbf{X}$ [see Equation (2.19)]. One approach is to use the singular values of $\mathbf{X}$ to judge what the pseudo-rank is. A cutoff value for the singular values has to be established and having this cutoff value the pseudo-rank is easily established as the number of singular values higher than this cutoff value. Cutoff values can be based on scree plots [Cattell 1966], or F-type statistics [Malinowski 1991], or on the statistical distribution of singular values [Faber *et al.* 1994, Malinowski 1987].

Another conceptually different approach is cross-validation. In Equation (2.19), $\hat{\mathbf{X}}$ is regarded as a model for $\mathbf{X}$, and as such the model should be able to predict the values of $\mathbf{X}$. This can be checked by performing a cross-validation scheme in which parts of $\mathbf{X}$ are left out of the calculations and kept apart, the model is built and used to predict the left out entries. The sum of squared differences between the predicted and the real entries serves as a measure of discrepancy. All data in $\mathbf{X}$ are left out once, and the squared differences are summed in a so called PRESS statistics (PRediction Error Sum of Squares). The model that gives the lowest PRESS is selected and the pseudo-rank of $\mathbf{X}$ is defined as the number of components in that model.

There are different ways to implement a cross-validation scheme for two-way arrays [Eastment & Krzanowski 1982, Wold 1978]. An overview of methods to establish the pseudo-rank of a matrix is given by Faber [Faber *et al.* 1994].

k-Rank

The concept of k-rank was introduced by Kruskal [1976], see also Harshman and Lundy [1984]. It is a useful concept in deciding whether a PARAFAC decomposition of a given array is unique (see Chapters 4 and 6).

Consider a matrix $\mathbf{X}\,(I \times J)$ of rank R. Then $\mathbf{X}$ has some set of R independent columns. However, some other set of R columns of $\mathbf{X}$ might not be independent. Assume that every set of R columns is independent and call this property universal R-column independence. Obviously, if a matrix $\mathbf{X}$ is universal R-column independent then $\mathrm{r}(\mathbf{X}) \geq R$. The largest integer k for which a matrix $\mathbf{X}$ is universal k-column independent is called the k-rank of $\mathbf{X}$ and is denoted by $k_{\mathbf{X}}$. Stated otherwise, the k-rank of $\mathbf{X}$ is the largest subset size for all subsets of columns of $\mathbf{X}$ that always has full rank.

EXAMPLE 2.9

k-Rank

Examples of the concept of k-rank are:

$$\mathbf{A} = \begin{bmatrix} 1 & 1 & 2 & 5 \\ 1 & 1 & 3 & 3 \\ 1 & 1 & 4 & 2 \end{bmatrix} ; \quad \mathrm{r}(\mathbf{A}) = 3, k_{\mathrm{A}} = 1$$

$$\mathbf{B} = \begin{bmatrix} 1 & 2 & 3 \\ 2 & 3 & 4 \\ 3 & 4 & 5 \end{bmatrix} ; \quad \mathrm{r}(\mathbf{B}) = 3, k_{\mathrm{B}} = 3$$

$$\mathbf{C} = \begin{bmatrix} 1 & 2 & 3 \\ 2 & 3 & 5 \\ 3 & 4 & 7 \end{bmatrix} ; \quad \mathrm{r}(\mathbf{C}) = 2, k_{\mathrm{C}} = 2$$

Matrix $\mathbf{A}$ has rank three. Not all possible subsets of two columns have rank two. Therefore, only all subsets of one column have full rank. Hence, k_{A} is one. Matrix $\mathbf{B}$ has rank three. All subsets of two columns have rank two. There is only one subset of three columns: the matrix itself. Therefore, the k-rank of $\mathbf{B}$ is three. Matrix $\mathbf{C}$ has rank two. The third column is the sum of the first and second column. All subsets of two columns have rank two (full rank). The subset of three columns does not have full rank. Hence, k_{C} is two.

2.6 Rank of Three-way Arrays

Although there is a clear definition of the rank of three-way arrays, the properties of two-way rank do not automatically generalize to three-way rank. This will be shown in the following sections.

Definitions and properties

The rank of a three-way array $\underline{\mathbf{X}}$ is the smallest number of trilinear (PARAFAC) components needed to fit $\underline{\mathbf{X}}$ exactly. This coincides with the definition of two-way rank; it is a direct generalization. Hence, a three-way array $\underline{\mathbf{X}}$ has rank R if R is the smallest value for which there exist vectors $\mathbf{a}_1, \ldots, \mathbf{a}_R; \mathbf{b}_1, \ldots, \mathbf{b}_R$ and $\mathbf{c}_1, \ldots, \mathbf{c}_R$ such that

$$x_{ijk} = \sum_{r=1}^{R} a_{ir} b_{jr} c_{kr}; \quad i = 1, \ldots, I; j = 1, \ldots, J; k = 1, \ldots, K \tag{2.23}$$

where x_{ijk} is the typical element of $\underline{\mathbf{X}}(I \times J \times K)$ and a_{ir}, b_{jr}, c_{kr} are typical elements of $\mathbf{a}_r, \mathbf{b}_r$ and $\mathbf{c}_r$, respectively. Equation (2.23) is a direct generalization of Equation (2.16) and each $I \times J \times K$ three-way array with typical elements $a_{ir} b_{jr} c_{kr}$ has (three-way) rank one.

Consider all arrays of size $2 \times 2 \times 2$. Then it can be shown that the maximum possible rank of such an array is 3 [Kruskal 1989]. This is in remarkable contrast with two-way arrays: a 2×2 array (matrix) has maximum rank 2. Likewise, it has been shown that the maximum rank of a $3 \times 3 \times 3$ array is 5 [Kruskal 1989]. Results are also available for the maximum rank of a $2 \times J \times K$ array [Ja'Ja' 1979] which is $m + \min(m,n)$, where $m = \min(J,K), n = [\max(J,K)/2]$ and $[x]$ means the largest integer $\leq x$.

Kruskal has shown that there is a positive probability[3] for a randomly generated $2 \times 2 \times 2$ array to have a rank lower than three [Kruskal 1989]. In a simulation study he found that of all $2 \times 2 \times 2$ arrays, there is a probability of 0.79 of obtaining a rank two array and a probability of 0.21 of obtaining a rank three array. The probability of obtaining an array of rank one or lower is zero. These results have been generalized to the case of $2 \times n \times n$ arrays [Ten Berge 1991]. This is in sharp contrast with results for two-way arrays, where the full rank case has probability one.

Due to the special properties of three-way rank, it is important to distinguish different types of rank for three-way arrays [Kruskal 1989, Lickteig 1985]:

- *individual rank*: given a three-way array $\underline{\mathbf{X}}(I \times J \times K)$, what is the rank of $\underline{\mathbf{X}}$?
- *maximum rank*: given the set of all arrays of size $I \times J \times K$, what is the maximum rank of any array of this set?
- *typical rank*: given the set of all arrays of size $I \times J \times K$, what is the rank of 'almost all' arrays of this size?
- *border rank*: an array $\underline{\mathbf{X}}$ has border rank at most R if and only if $\underline{\mathbf{X}}$ can be approximated arbitrarily well by an array of rank at most R.

The individual rank of a three-way array concerns an *exact* fit, whereas the border rank concerns an *approximate* fit. Some upper bounds are available for maximum and typical ranks [Ja'Ja' 1979, Kruskal 1989, Lickteig 1985, Ten Berge 1991, Ten Berge 2000, Ten Berge & Kiers 1999].[4] In the example of the $2 \times 2 \times 2$ array above, the maximum rank is three and the typical rank is {2, 3}, because almost all arrays have rank two or three. In practice the individual rank of an array is very hard to establish.

At first sight the treatment and concepts of rank might seem very theoretical. However, the concept of rank plays an important role in second-order calibration and curve resolution. Three-way rank may also have repercussions for degrees of freedom and hence for significance testing.

---

[3] The actual results are in terms of volumes of the 8-dimensional space generated by the $2 \times 2 \times 2$ arrays, but for clarity the results are stated in terms of probability.

[4] Intriguing, but yet unanswered, questions are: (i) what are the maximum and typical ranks of *all* array sizes, and (ii) how do three-way ranks generalize to four-way arrays and higher?

Row-rank, column-rank, tube-rank, dimensionality vector, properties

For a two-way matrix **X** the dimensionality of the space generated by its columns (column-rank) equals the dimensionality generated by its rows (row-rank). This does not hold for three-way arrays.

In Section 2.3 the principle of matricizing was explained. A three-way array $\underline{\mathbf{X}}(I \times J \times K)$ can be matricized in three different ways (and three more with the same ranks, so they are not mentioned here): (i) row-wise, giving a $J \times IK$ two-way array $\mathbf{X}_{(J \times IK)}$, (ii) column-wise, giving an $I \times JK$ two-way array $\mathbf{X}_{(I \times JK)}$ and (iii) tube-wise, giving a $K \times IJ$ two-way array $\mathbf{X}_{(K \times IJ)}$. Let P, Q and R be the ranks of the arrays $\mathbf{X}_{(J \times IK)}$, $\mathbf{X}_{(I \times JK)}$ and $\mathbf{X}_{(K \times IJ)}$, respectively, then (P, Q, R) is called the *dimensionality vector* of $\underline{\mathbf{X}}$. P is the dimension of the space spanned by the row-vectors in $\underline{\mathbf{X}}$ and is consequently called the row-rank. Similar definitions apply for the column and tube rank.

P, Q and R are not necessarily equal. This is in contrast with two-way arrays, because for a two-way array **X** also a dimensionality vector (P, Q) can be defined, but it always holds that $P = Q = r(\mathbf{X})$.

The dimensionality vector (P, Q, R) of a three-way array $\underline{\mathbf{X}}$ with rank S obeys certain inequalities [Kruskal 1989]:

$$
\begin{aligned}
&\text{(i)} \quad P \le QR; Q \le PR; R \le PQ \\
&\text{(ii)} \quad \max(P, Q, R) \le S \le \min(PQ, QR, PR)
\end{aligned}
\tag{2.24}
$$

and these inequalities will be important for the size of the core-array in Tucker3 models (see Chapter 6). Specifically, the dimensionality vector of $\underline{\mathbf{X}}$ is the size of the smallest core-array in a Tucker3 model that will fit $\underline{\mathbf{X}}$ exactly [Kruskal 1989].

EXAMPLE 2.10

Dimensionality vector of an array

The $(4 \times 3 \times 2)$ three-way array of Figure 2.7 has three associated matricized matrices (see Example 2.6). These arrays have rank 4, 3, 2, respectively, showing that P, Q and R can be unequal.

Pseudo-rank, rank deficiency and chemical sources of variation

The pseudo-rank of three-way arrays is a straightforward generalization of the two-way definition. Suppose that a three-way array $\underline{\mathbf{X}}$ is modeled with a three-way model and this results in

$$
\underline{\mathbf{X}} = \hat{\underline{\mathbf{X}}} + \underline{\mathbf{E}}
\tag{2.25}
$$

where $\underline{\mathbf{E}}$ is a three-way array of residuals. The three-way rank of $\hat{\underline{\mathbf{X}}}$ is the pseudo-rank of $\underline{\mathbf{X}}$. The three-way method used to model $\underline{\mathbf{X}}$ does not necessarily have to be a PARAFAC model, but the pseudo-rank of $\underline{\mathbf{X}}$ is nevertheless equivalent to the minimum number of PARAFAC components necessary to exactly fit $\hat{\underline{\mathbf{X}}}$.

The notion of chemical sources of variation is still useful in the context of three-way arrays, but the relationship between the number of chemical sources of variation and rank of the three-way array is not trivial (see Elaborations 2.3 and 2.4). This relationship is dependent on how the three-way array originated from the experiment, whether nonlinearities are present, etc. This is an area of present and future research. [Amrhein *et al.* 1996].

ELABORATION **2.3**

Chemical sources of variation and three-way rank

As an example of the relationship between chemical sources of variation and three-way rank consider second-order calibration. In that type of calibration, instruments are used that give a matrix response for measuring a single sample. The data can, for example, come from fluorescence (emission–excitation) spectroscopy or liquid chromatography–ultraviolet spectroscopy. A standard $\mathbf{X}_1(J \times K)$ in which certain analytes are present in known concentrations is used to quantify for those analytes in a mixture $\mathbf{X}_2(J \times K)$, in which unknown interferents might be present. This results in a three-way array $\underline{\mathbf{X}}$ where $\mathbf{X}_1$ and $\mathbf{X}_2$ are the two individual slices. Second-order calibration usually comes down to building a PARAFAC model for that $\underline{\mathbf{X}}$.

Suppose that in second-order calibration a standard $\mathbf{X}_1$ contains one analyte (hence, there is one chemical source of variation) and this standard is measured in such a way that the pseudo-rank of $\mathbf{X}_1$ equals one. The mixture $\mathbf{X}_2$, measured under the same experimental circumstances, contains the analyte and one unknown interferent. If the instrumental profiles (e.g. spectra and chromatograms) of the analyte and interferent are different, then the three-way array $\underline{\mathbf{X}}$ having $\mathbf{X}_1$ and $\mathbf{X}_2$ as its two individual slices has two chemical sources of variation. This equals the number of PARAFAC components needed to model the systematic part of the data, which is the three-way rank of $\hat{\underline{\mathbf{X}}}$, the systematic part of $\underline{\mathbf{X}}$.

Rank-deficiencies may perhaps occur in three-way matrices. The situation is more complicated than in two-way analysis, since rank deficiencies in one loading matrix of a three-way array are not the same as a three-way rank deficiency. Elaboration 2.4 explains this.

ELABORATION **2.4**

Rank-deficiency in three-way arrays

In Elaboration 2.1 the flow injection analysis example is used to illustrate two-way rank-deficiency. The same example is expanded to illustrate three-way rank-deficiency. The models of the three analytes 2-hydroxybenzaldehyde (2-HBA), 3- hydroxybenzaldehyde (3-HBA) and 4- hydroxybenzaldehyde (4-HBA) are

$$\mathbf{X}_{2\text{-HBA}} = \mathbf{c}_{a,2}\mathbf{s}'_{a,2} + \mathbf{c}_{b,2}\mathbf{s}'_{b,2} + \mathbf{E}_{2\text{-HBA}}$$

$$\mathbf{X}_{3\text{-HBA}} = \mathbf{c}_{a,3}\mathbf{s}'_{a,3} + \mathbf{c}_{b,3}\mathbf{s}'_{b,3} + \mathbf{E}_{3\text{-HBA}}$$

$$\mathbf{X}_{4\text{-HBA}} = \mathbf{c}_{a,4}\mathbf{s}'_{a,4} + \mathbf{c}_{b,4}\mathbf{s}'_{b,4} + \mathbf{E}_{4\text{-HBA}}$$

where the symbols are the same as in Elaboration 2.1. If the three different matrices $\underline{\mathbf{X}}_{2\text{-HBA}}$, $\underline{\mathbf{X}}_{3\text{-HBA}}$ and $\underline{\mathbf{X}}_{4\text{-HBA}}$ are stacked on top of each other to form $\underline{\mathbf{X}}(3 \times J \times K)$, then a model of $\underline{\mathbf{X}}$ can be found

$$\underline{\mathbf{X}} = \hat{\underline{\mathbf{X}}} + \underline{\mathbf{E}}$$

where $\underline{\mathbf{E}}$ is again the residual. Due to the restrictions on the concentration profiles

$$\mathbf{c}_{a,2} + \mathbf{c}_{b,2} = \alpha(\mathbf{c}_{a,3} + \mathbf{c}_{b,3}) = \beta(\mathbf{c}_{a,4} + \mathbf{c}_{b,4})$$

there are only four independently varying concentration profiles. Hence, when $\underline{\mathbf{X}}$ is matricized such that the concentration profiles mode (the time mode) is kept intact, the pseudo-rank of the resulting matrix is four. If $\mathbf{X}$ is matricized to a $3 \times JK$ matrix where each row is a strung out version of a single measured standard, then a matrix of rank three is the result. Because in general the six different ultraviolet spectra form a linear independent set, a Tucker3 (6,4,3) model is needed to fit $\underline{\mathbf{X}}$. Yet, the three-way rank of $\underline{\mathbf{X}}$ is six because six PARAFAC components are necessary for fitting the data perfectly in the noiseless case. Hence, there is no three-way rank deficiency.

Determination of rank

The rank of a three-way array is defined by the number of PARAFAC components needed to fit the data exactly. Hence, the determination of rank comes down to establishing this number of components. Establishing the rank of the systematic part of an array $\underline{\mathbf{X}}$ is a part of the validation of three-way models and is treated in Chapter 7.

2.7 Algebra of Multi-way Analysis

The algebra of multi-way arrays is described in a field of mathematics called tensor analysis, which is an extension and generalization of matrix algebra. A zero-order tensor is a scalar; a first-order tensor is a vector; a second-order tensor is a matrix; a third-order tensor is a three-way array; a fourth-order tensor is a four-way array and so on. The notions of addition, subtraction and multiplication of matrices can be generalized to multi-way arrays. This is shown in the following sections [Borisenko & Tarapov 1968, Budiansky 1974].

Addition and subtraction of multi-way arrays

Addition and subtraction of multi-way arrays is a direct generalization of that of two-way arrays. Suppose that $\underline{\mathbf{X}}$ and $\underline{\mathbf{Y}}$ have the same dimensions ($I \times J \times K$) then $\underline{\mathbf{Z}} = \underline{\mathbf{X}} + \underline{\mathbf{Y}}$ is defined by

$$z_{ijk} = x_{ijk} + y_{ijk}; \quad i = 1, \ldots, I; \; j = 1, \ldots, J; \; k = 1, \ldots, K \quad\quad (2.26)$$

where x_{ijk}, y_{ijk} and z_{ijk} are the typical elements of $\underline{\mathbf{X}}$, $\underline{\mathbf{Y}}$ and $\underline{\mathbf{Z}}$, respectively. Subtraction is defined analogously. Addition as defined by Equation (2.26) has the properties of commutativity and associativity.

Addition and subtraction are only defined for arrays of the same orders and dimensions. Generalizations to sums of more than two multi-way arrays are straightforward.

Outer products of multi-way arrays

The outer product $\underline{\mathbf{Z}}$ of two three-way arrays $\underline{\mathbf{X}}(I \times J \times K)$ and $\underline{\mathbf{Y}}(L \times M \times N)$ is defined as

$$z_{ijklmn} = x_{ijk} y_{lmn}; \quad i = 1, \ldots, I; \; j = 1, \ldots, J; \; k = 1, \ldots, K$$
$$l = 1, \ldots, L; \; m = 1, \ldots, M; \; n = 1, \ldots, N \tag{2.27}$$

where z_{ijklmn}, x_{ijk} and y_{lmn} are typical elements of $\underline{\mathbf{Z}}$, $\underline{\mathbf{X}}$ and $\underline{\mathbf{Y}}$, respectively. The six-way array $\underline{\mathbf{Z}}$ has dimensions $(I \times J \times K \times L \times M \times N)$. This outer product should not be confused with the normal matrix product. The matrix product of $\mathbf{A}(I \times K)$ and $\mathbf{B}(K \times J)$ is $\mathbf{Z}(I \times J)$, while the outer product becomes $\underline{\mathbf{Z}}(I \times K \times K \times J)$. The outer product is noncommutative, that is, $z_{ijklmn} \neq z_{lmnijk}$. The law of associativity still holds [Borisenko & Tarapov 1968].

To give a simple example, the outer product $\mathbf{Z}(I \times J)$ of two vectors $\mathbf{x}(I \times 1)$ and $\mathbf{y}(J \times 1)$ is

$$z_{ij} = x_i y_j; \quad i = 1, \ldots, I; \; j = 1, \ldots, J \tag{2.28}$$

or

$$\mathbf{Z} = \mathbf{xy}'$$

where z_{ij}, x_i and y_j are typical elements of $\mathbf{Z}$, $\mathbf{x}$ and $\mathbf{y}$, respectively.

It is also possible to have an outer product of more than two arrays. Consider taking the outer product of $\mathbf{x}$ $(I \times 1)$, $\mathbf{y}$ $(J \times 1)$ and $\mathbf{z}$ $(K \times 1)$. Then using the definition in Equation (2.28):

$$u_{ij} = x_i y_j; \quad i = 1, \ldots, I; j = 1, \ldots, J$$
$$w_{ijk} = u_{ij} z_k; \quad i = 1, \ldots, I; j = 1, \ldots, J; k = 1, \ldots, K \tag{2.29}$$

and the result is a triple product of all elements of $\mathbf{x}$, $\mathbf{y}$ and $\mathbf{z}$ making up w_{ijk} which are the typical elements of the $(I \times J \times K)$ three-way array $\underline{\mathbf{W}}$. This kind of products appear often in later chapters.

EXAMPLE 2.11

Outer products of arrays

The same example as used before in Example 2.5 is considered.

$$\mathbf{a} = \begin{bmatrix} 1 \\ 2 \\ 3 \end{bmatrix}; \quad \mathbf{b} = \begin{bmatrix} -1 \\ 2 \end{bmatrix}; \quad \mathbf{c} = \begin{bmatrix} 2 \\ 1 \end{bmatrix}$$

$$\mathbf{X}_1 = \begin{bmatrix} -2 & 4 \\ -4 & 8 \\ -6 & 12 \end{bmatrix}; \quad \mathbf{X}_2 = \begin{bmatrix} -1 & 2 \\ -2 & 4 \\ -3 & 6 \end{bmatrix}$$

where $\mathbf{X}_1$ is the frontal slice of $\underline{\mathbf{X}}$ and $\mathbf{X}_2$ is the back slice of $\underline{\mathbf{X}}$, where $\underline{\mathbf{X}}$ is the three-way array resulting from the outer product of $\mathbf{a}$, $\mathbf{b}$ and $\mathbf{c}$.

2.8 Summary

This chapter introduces concepts of two-way and multi-way algebra. This includes definition of arrays and their subparts, a number of useful matrix products and the different concepts of two-way and three-way rank. Also linearity, bilinearity and trilinearity are defined. It is important to remember about rank that the problem-defined rank (chemical rank, pseudo-rank) has to be chosen by the data analyst using certain criteria where the mathematically defined ranks are extremes that are not useful for modeling of noisy data.

Appendix 2.A

In this appendix it is shown that the combination of closure and column centering gives rank reduction in certain cases.

Suppose that a measurement matrix $\mathbf{X}$ is generated in an experiment and it has the following structure

$$\mathbf{X} = \mathbf{CS'} + \mathbf{E} \tag{2.30}$$

where $\mathbf{C}$ is the $(J \times R)$ matrix of concentrations of R absorbing species in J samples; $\mathbf{S}$ is the $(K \times R)$ matrix of pure spectra of these species and $\mathbf{E}$ is a matrix of residuals. Suppose that $R < K < J$, and closure exists in the measurements, which can be written as

$$\mathbf{C1}_R = \mathbf{1}_J \tag{2.31}$$

where the symbol $\mathbf{1}_L$ is used to indicate a column vector of L values of 1. The column centering of a matrix $\mathbf{A}$ $(J \times K)$ can be written formally by premultiplying $\mathbf{A}$ with $[\mathbf{I} - \mathbf{1}_J\mathbf{1}'_J/J]$ (see Chapter 9). If the matrix $\mathbf{X}$ is premultipied by this centering matrix, then so is $\mathbf{C}$. This gives

$$[\mathbf{I} - \mathbf{1}_J\mathbf{1}'_J/J]\mathbf{C} = \tilde{\mathbf{C}}$$
$$[\mathbf{I} - \mathbf{1}_J\mathbf{1}'_J/J]\mathbf{C1}_R = [\mathbf{I} - \mathbf{1}_J\mathbf{1}'_J/J]\mathbf{1}_J = [\mathbf{1}_J - \mathbf{1}_J\mathbf{1}'_J\mathbf{1}_J/J] = \tag{2.32}$$
$$[\mathbf{1}_J - \mathbf{1}_J] = \mathbf{0} \Rightarrow \tilde{\mathbf{C}}\mathbf{1}_R = \mathbf{0}$$

and the last line in Equation (2.32) shows that $r(\tilde{\mathbf{C}}) = r(\mathbf{C}) - 1$, because one of the R columns of $\tilde{\mathbf{C}}$ can be written as a combination of the other columns. Hence, also the pseudo-rank of $\mathbf{X}$ is reduced by one.

3

TWO-WAY COMPONENT AND REGRESSION MODELS

3.1 Models for Two-way One-block Data Analysis: Component Models

General idea of component models

A two-way array or data matrix usually has rows representing objects and columns representing variables. This array has $I \times J$ entries, e.g. I objects and J variables measured on each object.

For real measurements, the array can be thought of as having a structural part and a noise part. This is shown in Figure 3.1. The structural part can be further decomposed into rank 1 matrices, each consisting of outer products of vectors (see also Figure 1.6). The rows in a two-way array can be seen as points in a multivariate space spanned by the variables. If there is correlation between variables, the spread of the points is not random. These points allow a description using fewer coordinates than the number of original variables. An example with three variables where the objects form a linear structure in space is shown in Figure 3.2. In this case a line describes this structure well. Deviations from the line may be measurement noise.

The variables x_1, x_2 and x_3 in Figure 3.2 could for example represent measurements of iron, red color and red blood cell volume in the blood of patients (the objects). Because all the three variables are correlated, the straight line representation gives a sufficient summary of the data. The three-dimensional coordinate system may be exchanged with a new one-dimensional coordinate system where the new *latent* variable is a combination of the original three *manifest (measured)* variables. For a more complex chemical example, suppose chemical composition is measured in a number of lakes. Then the lakes are the I objects and the results of the chemical analyses (e.g., Cl^-, NO_3^-, SO_4^{2-}, Na^+, Fe^{2+}, PO_4^{3-},

Multi-way Analysis With Applications in the Chemical Sciences.   A. Smilde, R. Bro and P. Geladi
© 2004 John Wiley & Sons, Ltd   ISBN: 0-471-98691-7

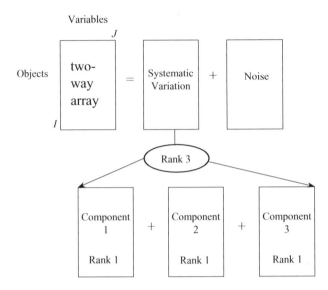

Figure 3.1. A two-way array can be decomposed into a systematic and a noise part. The systematic part is favorably represented as a principal component model. It can be shown as a sum of R rank one matrices (in this case $R = 3$).

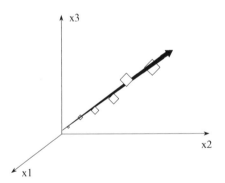

Figure 3.2. For the three variables, all objects can be represented by their positions in variable space. With correlated variables, it may suffice with, e.g., one new coordinate in a direction defined as a weighted sum of the original variables to describe the major strucure in the data. The distances of the objects to the new coordinate are small and can be thought of as measurement noise.

organic matter, NH_4^+) are the J variables which are measured with noise. It can be expected that the concentrations of the ions are correlated because no ion exists independently of others. Hence, the original J variables can be replaced with fewer latent variables. If the manifest variables are independent, then it is not possible to summarize these manifest variables in a few latent variables.

The advantages of a reduction in the number of variables are numerous, e.g., the interpretation of the data becomes easier and the influence of noise is reduced because the latent variables are weighted averages of the original manifest variables.

The mathematics of what is described above is equivalent to principal component analysis. The ideas of principal component analysis (PCA) go back to Beltrami [1873] and Pearson [1901]. They tried to describe the structural part of data sets by lines and planes of best fit in multivariate space. The PCA method was introduced in a rudimentary version by Fisher and Mackenzie [1923]. The name 'principal component analysis' was introduced by Hotelling [1933]. An early calculation algorithm is given by Müntz [1913]. More details can be found in the literature [Jackson 1991, Jolliffe 1986, Stewart 1993].

Principal component analysis (PCA)

Before discussing PCA it is useful to introduce two standard tools from matrix algebra [Golub & van Loan 1989, Schott 1997].

SINGULAR VALUE DECOMPOSITION (SVD) AND EIGENDECOMPOSITION

Suppose that a matrix $\mathbf{X}$ ($I \times J$) is available. Assuming that $J \leq I$, the singular value decomposition (SVD) of $\mathbf{X}$ is

$$\mathbf{X} = \mathbf{USV}' \tag{3.1}$$

with $\mathbf{U}$ ($I \times J$), $\mathbf{S}$ ($J \times J$) and $\mathbf{V}$ ($J \times J$) with $\mathbf{U}'\mathbf{U} = \mathbf{I}$, $\mathbf{V}'\mathbf{V} = \mathbf{VV}' = \mathbf{I}$ and $\mathbf{S}$ is diagonal with the singular values of $\mathbf{X}$ in decreasing order on its diagonal. The singular values are all nonnegative (by convention) and when they are distinct the decomposition is unique up to joint reflection of columns of $\mathbf{U}$ and $\mathbf{V}$. For the case that $J > I$, the SVD of $\mathbf{X}$ can be found by using Equation (3.1) on the transpose of $\mathbf{X}$

$$\mathbf{X}' = \mathbf{VSU}' \Rightarrow \mathbf{X} = \mathbf{USV}' \tag{3.2}$$

with $\mathbf{V}$ ($J \times I$), $\mathbf{V}'\mathbf{V} = \mathbf{I}$, $\mathbf{U}$($I \times I$), $\mathbf{U}'\mathbf{U} = \mathbf{UU}' = \mathbf{I}$ and $\mathbf{S}$ is diagonal containing the singular values of $\mathbf{X}$ on its diagonal again.

The matrix $\mathbf{X}'\mathbf{X}$ ($J \times J$) is Gramian (symmetric and positive (semi-)definite) and can be decomposed as [Schott 1997]

$$\mathbf{X}'\mathbf{X} = \mathbf{K\Lambda K}' \tag{3.3}$$

where $\mathbf{K}'\mathbf{K} = \mathbf{I}$ and $\mathbf{\Lambda}$ ($J \times J$) is a diagonal matrix with nonnegative values (now an intrinsic property!) for convenience arranged in decreasing order. The columns of $\mathbf{K}$ ($J \times J$) are the eigenvectors of $\mathbf{X}'\mathbf{X}$ and the elements on the diagonal of $\mathbf{\Lambda}$ are the corresponding eigenvalues. Multipying Equation (3.3) to the right by $\mathbf{K}$ gives the familiar eigenvalue/eigenvector equation for $\mathbf{X}'\mathbf{X}$:

$$\mathbf{X}'\mathbf{XK} = \mathbf{K\Lambda} \tag{3.4}$$

where use is made of the orthogonality of $\mathbf{K}$. Equation (3.3) is the eigendecomposition of $\mathbf{X}'\mathbf{X}$ and if the values on the diagonal of $\mathbf{\Lambda}$ are distinct, this decomposition is unique upto reflection of the columns of $\mathbf{K}$.

By writing $\mathbf{X'X} = \mathbf{VS'U'USV'} = \mathbf{VS^2V'}$ and assuming that all singular values are distinct, $\mathbf{VS^2V'}$ is equal to $\mathbf{K\Lambda K'}$. Hence, in that case $\mathbf{V} = \mathbf{K}$ (upto reflection) and $\mathbf{\Lambda} = \mathbf{S^2}$. This shows the close relationship between the SVD of $\mathbf{X}$ and the eigendecomposition of $\mathbf{X'X}$.

Equation (3.1) can be rewritten as

$$\begin{array}{ll}(1) & \mathbf{U'X} = \mathbf{SV'} \\ (2) & \mathbf{XV} = \mathbf{US}\end{array} \qquad (3.5)$$

by using the orthogonality properties of $\mathbf{U}$ and $\mathbf{V}$. The vectors $\mathbf{u}_1,\ldots,\mathbf{u}_J$ (collected in $\mathbf{U}$) are called the left singular vectors of $\mathbf{X}$, and likewise the vectors $\mathbf{v}_1,\ldots,\mathbf{v}_J$ (collected in $\mathbf{V}$) are called the right singular vectors.

FORMAL TREATMENT OF PCA

There are two ways of viewing and introducing PCA. Both approaches give similar results and are closely related. They are illustrated in Figure 3.3. A line L is sought that describes the original data as well as possible. That is, the orthogonally projected data (open circles) are as close as possible to the original data (closed circles). This is the Pearson approach: a line of closest fit [Eckart & Young 1936, Pearson 1901]. The emphasis is on the *variance explained by* the principal components.

On the other hand, line L represents a new variable (a linear combination of the manifest variables $\mathbf{x}_1$ and $\mathbf{x}_2$) and the scores on this new variables are the open circles. In this approach, the line L is chosen such that the scores corresponding to the open circles have the highest possible variance. Hence, the emphasis is on the *variance of* the principal components and this approach is usually taken in the statistics literature. The figure implies that both approaches are related. They will both be treated in some detail below. Additional details are given in Appendix 3.A and a thorough description of the differences and similarities of the approaches is given elsewhere [Cadima & Jolliffe 1997, Ten Berge & Kiers 1996, Ten Berge & Kiers 1997].

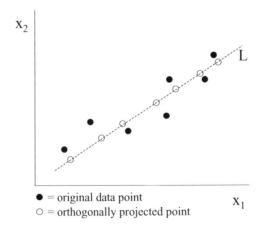

Figure 3.3. Both views of principal components explained. Closed circles are the original data points and open circles are the projections of these points on the first principal component.

'Variance of the principal components' approach.   Suppose that the data matrix $\mathbf{X}$ ($I \times J$) contains measured or calculated values of J variables measured on I objects. Because J can be large and often contains redundant information it is convenient to have a few linear combinations of these variables serve as the variation or information carriers of $\mathbf{X}$. Selecting a linear combination of the variables of $\mathbf{X}$ can be formalized in selecting $\mathbf{t}_1 = \mathbf{Xw}_1$ where $\mathbf{t}_1(I \times 1)$ contains the scores on the new variable and $\mathbf{w}_1(J \times 1)$ is a vector of weights defining the linear combination. Assume that $\mathbf{X}$ is column-centered and possibly scaled. The problem comes down to finding a $\mathbf{w}_1$ that maximizes the variance of $\mathbf{t}_1$. Obviously, the size of $\mathbf{w}_1$ has to be restricted, otherwise the variance of $\mathbf{t}_1$ can be made arbitrarily large. This is done by restricting $\mathbf{w}_1$ to have length one. Hence, the problem is

$$\max_{\|\mathbf{w}_1\|=1} \mathrm{var}(\mathbf{t}_1) \quad \text{or} \quad \max_{\|\mathbf{w}_1\|=1} (\mathbf{t}_1'\mathbf{t}_1) \tag{3.6}$$

and because $\mathbf{X}$ was assumed to be centered, $\mathbf{t}_1$ is centered too, hence both problems in Equation (3.6) give the same $\mathbf{w}_1$. Equation (3.6) can be rewritten as

$$\max_{\|\mathbf{w}_1\|=1} (\mathbf{t}_1'\mathbf{t}_1) = \max_{\|\mathbf{w}_1\|=1} (\mathbf{w}_1'\mathbf{X}'\mathbf{Xw}_1) \tag{3.7}$$

which is a standard problem in optimization. The solution of this problem is $\mathbf{w}_1 = \mathbf{v}_1$, where $\mathbf{v}_1$ is the first eigenvector of $\mathbf{X}'\mathbf{X}$, or, equivalently, the first right singular vector of $\mathbf{X}$ [Ten Berge 1993]. The corresponding $\mathbf{t}_1$ equals $\mathbf{Xv}_1$, and this equals $s_1\mathbf{u}_1$, where $\mathbf{u}_1$ is the first left singular vector of $\mathbf{X}$ and s_1 is the largest (=first) singular value of $\mathbf{X}$. The next step is to find a second component $\mathbf{t}_2 = \mathbf{Xw}_2$ under the constraints that $\mathbf{w}_2'\mathbf{w}_2 = 1$ and $\mathbf{w}_1'\mathbf{w}_2 = 0$. The second constraint is imposed to get an orthogonal direction ensuring that the first and second components do not coincide. The solution is $\mathbf{w}_2 = \mathbf{v}_2$ and $\mathbf{t}_2 = s_2\mathbf{u}_2$, where $\mathbf{u}_2$ and $\mathbf{v}_2$ are the second left and right singular vectors, respectively, of $\mathbf{X}$ corresponding to the second largest singular value s_2 of $\mathbf{X}$. The resulting $\mathbf{t}_1$ and $\mathbf{t}_2$ are orthogonal ($\mathbf{t}_1'\mathbf{t}_2 = 0$) although this constraint is not specifically imposed.

This procedure can be repeated for the third, fourth and further components. This approach gives a unique solution of PCA: no rotations are possible without destroying the properties of the principal components unless some eigenvalues are identical. Of course, there is still a sign indeterminacy: $\mathbf{t}_1 = \mathbf{Xv}_1$ is equivalent to $-\mathbf{t}_1 = \mathbf{X}(-\mathbf{v}_1)$.

'Variance explained by the principal components' or Pearson's approach.   Consider $\mathbf{t}_1 = s_1\mathbf{u}_1$ from above as a summarizer of $\mathbf{X}$, then $\mathbf{X}$ can be regressed on $\mathbf{t}_1$ and regression coefficients $\mathbf{p}_1(J \times 1)$, called loadings, can be calculated indicating to what extend each column of $\mathbf{X}$ is described by $\mathbf{t}_1$. This is equivalent to minimizing $\|\mathbf{X} - \mathbf{t}_1\mathbf{p}_1'\|^2$ over $\mathbf{p}_1$, which has the solution $\mathbf{p}_1 = \mathbf{X}'\mathbf{t}_1(\mathbf{t}_1'\mathbf{t}_1)^{-1}$. It can be shown that $\mathbf{w}_1 = \mathbf{p}_1 = \mathbf{v}_1$ (see Appendix 3.A). Hence, the loadings are directly available.

This reflects exactly the approach of Pearson: defining a line of closest fit. The vector $\mathbf{p}_1$ gives a direction in the J-dimensional space (defining a line) and $\mathbf{t}_1$ represents the scores (orthogonal projections) on that line. The outer product $\mathbf{t}_1\mathbf{p}_1'$ is a rank one matrix and is the best rank-one approximation of $\mathbf{X}$ in a least squares sense. This approach can be generalized for more than one component. Then the problem becomes one of finding the subspace of closest fit.

Formally, the problem of finding a subspace of closest fit is

$$\min_{\mathbf{T},\mathbf{P}} \|\mathbf{X} - \mathbf{TP}'\|^2 = \min_{\mathbf{W}} \|\mathbf{X} - \mathbf{XWP}'\|^2 \tag{3.8}$$

Once $\mathbf{W}$ is known, $\mathbf{P}$ can always be found. Hence, the minimization in Equation (3.8) is only over $\mathbf{W}$. The components $\mathbf{T}$ are restricted to be in the column-space of $\mathbf{X}$. This restriction is inactive (see Appendix 3.A), that is, solving for $\mathbf{T}$ directly in Equation (3.8) automatically gives components in the column-space of $\mathbf{X}$. Problem (3.8) is also known as the 'one-mode component problem' [Magnus & Neudecker 1988] because components are sought to summarize the columns of $\mathbf{X}$ (which is only one of the modes of $\mathbf{X}$). Hence, the purpose is to reduce one mode of $\mathbf{X}$. A more detailed treatment of the Pearson approach is given in Appendix 3.A.

It is important to discuss the concept of uniqueness at this point. The principal components are unique but are not unique in providing a basis for the plane of closest fit. This plane can also be described using another basis, e.g., $\mathbf{P}$ can be rotated by $\mathbf{Z}$ (where $\mathbf{Z}$ is an $R \times R$ nonsingular matrix). Then upon using $(\mathbf{Z}')^{-1}$ to counterrotate $\mathbf{T}$, the solution $\mathbf{TP}'$ does not change: $\mathbf{TP}' = \mathbf{T}(\mathbf{Z}')^{-1}\mathbf{Z}'\mathbf{P}' = \tilde{\mathbf{T}}(\mathbf{PZ})' = \tilde{\mathbf{T}}\tilde{\mathbf{P}}'$, where $\tilde{\mathbf{P}}$ is the new basis and $\tilde{\mathbf{T}}$ are the scores with respect to the new basis. This property is known as 'rotational freedom' [Harman 1967]. Summarizing, the plane found is unique, but not its basis vectors.[1]

In the special case that (i) $\mathbf{W}'\mathbf{W} = \mathbf{I}$, (ii) $\mathbf{T}'\mathbf{T}$ is diagonal, and (iii) the columns of $\mathbf{T}$ are ordered in decreasing variance, the two approaches (*variance of* and *variance by*) give identical results (see Appendix 3.A). This shows the intimate relationship between the methods. If the vectors $\mathbf{t}$ and $\mathbf{p}$ are restricted to be orthogonal then it is possible to estimate the vectors $\mathbf{t}$ and $\mathbf{p}$ one at a time. First, using $\mathbf{X}$, $\mathbf{t}_1$ and $\mathbf{p}_1$ are calculated. Next, the contribution of the first principal component $\mathbf{t}_1\mathbf{p}_1'$ is subtracted from $\mathbf{X}$, giving $\mathbf{E}_1$ (this is called deflation). Then, $\mathbf{t}_2$ and $\mathbf{p}_2$ are calculated from $\mathbf{E}_1$. This is the basis of the NIPALS algorithm for calculating principal components [Wold 1975].

Notation of PCA models

In this section the use of different types of notation in two-way analysis is illustrated with the PCA model or the SVD of a two-way array $\mathbf{X}$. This will facilitate understanding the notation for three-way component models (Chapter 4).

NOTATION WITH TYPICAL ELEMENTS AND SUMMATION

The singular value decomposition of $\mathbf{X}$ ($I \times J$) was given by Equation (3.1) and is repeated here

$$\mathbf{X} = \mathbf{USV}' \tag{3.9}$$

where it is assumed that $J \leq I$, for convenience. The SVD of $\mathbf{X}$ can be truncated to any arbitrary number of components. Usually the number of components is taken to be the pseudo-rank of $\mathbf{X}$, which can be established by different techniques (see Chapter 2). Suppose that R is the number of components used in approximating $\mathbf{X}$, then the truncated rank-R

---

[1] A rank-one principal component solution is always unique.

SVD of **X** is

$$\hat{\mathbf{X}} = \mathbf{U}_R \mathbf{S}_R \mathbf{V}_R' \tag{3.10}$$

where $\mathbf{U}_R$ $(I \times R)$ contains the first R left singular vectors of $\mathbf{X}$ ($\mathbf{U}_R'\mathbf{U}_R = \mathbf{I}$); $\mathbf{S}_R$ is an $R \times R$ diagonal matrix with the first and largest R singular values of $\mathbf{X}$ on its diagonal; $\mathbf{V}_R$ $(J \times R)$ contains the first R right singular vectors of $\mathbf{X}$ ($\mathbf{V}_R'\mathbf{V}_R = \mathbf{I}$). In the previous section it was shown that Equation (3.10) is the best rank-R decomposition of $\mathbf{X}$.

Equation (3.10) is sometimes expressed using the symbols $\mathbf{T}$ (scores) and $\mathbf{P}$ (loadings), where $\mathbf{T} = \mathbf{U}_R \mathbf{S}_R$ and $\mathbf{P} = \mathbf{V}_R$. However, this is a matter of choice (see below). If t_{ir}, p_{jr}, and x_{ij} are the typical elements of $\mathbf{T}$ $(I \times R)$, $\mathbf{P}$ $(J \times R)$ and $\mathbf{X}$, respectively, then Equation (3.10) can be rewritten as

$$\hat{x}_{ij} = \sum_{r=1}^{R} t_{ir} p_{jr}; \quad i = 1, \ldots I; \; j = 1, \ldots, J \tag{3.11}$$

Equation (3.11) shows that the elements of $\mathbf{T}$ and $\mathbf{P}$ give a model of $\mathbf{X}$. This is one way of writing a PCA model of $\mathbf{X}$ and this type of notation can also be extended easily to the three-way case, as will be shown in Chapter 4.

EXAMPLE 3.1

PCA in typical elements and summation

Suppose that a PCA on a matrix $\mathbf{X}$ (3×2) is performed and this results in

$$\mathbf{t}_1 = \begin{bmatrix} t_{11} \\ t_{21} \\ t_{31} \end{bmatrix} = \begin{bmatrix} 1 \\ 2 \\ 3 \end{bmatrix}; \qquad \mathbf{t}_2 = \begin{bmatrix} t_{12} \\ t_{22} \\ t_{32} \end{bmatrix} = \begin{bmatrix} -2 \\ -0.5 \\ 1 \end{bmatrix}$$

$$\mathbf{p}_1 = \begin{bmatrix} p_{11} \\ p_{21} \end{bmatrix} = \begin{bmatrix} 0.71 \\ 0.71 \end{bmatrix}; \qquad \mathbf{p}_2 = \begin{bmatrix} p_{12} \\ p_{22} \end{bmatrix} = \begin{bmatrix} 0.71 \\ -0.71 \end{bmatrix}$$

where for simplicity the values in **p** have been truncated to two digits. Then the approximation of the (3,2)-th element of **X** is

$$\hat{x}_{32} = \sum_{r=1}^{2} t_{3r} p_{2r} = 3 \times 0.71 + 1 \times (-0.71) = 1.42$$

In this way all six elements of **X** can be approximated. The **p** vectors form an orthonormal set and the **t** vectors form an orthogonal set.

NOTATION WITH MATRICES AND VECTORS

The rank-R principal component decomposition of $\mathbf{X}$ can also be written as

$$\hat{\mathbf{X}} = \mathbf{T}\mathbf{P}' \tag{3.12}$$

where $\hat{\mathbf{X}}$, $\mathbf{T}$ and $\mathbf{P}$ are defined as before. Equation (3.12) is the matrix notation of a principal component model of $\mathbf{X}$ and this can be expanded to three-way analysis using matricized three-way arrays. Suppose that a three-way array $\underline{\mathbf{X}}$ has two frontal slices $\mathbf{X}_1(I \times J)$ and $\mathbf{X}_2$ $(I \times J)$ and these two slices are modeled with the same scores $\mathbf{T}$ $(I \times R)$ and the same loadings $\mathbf{P}$ ($J \times R$), where $R \leq \min(I, J)$. Then the following holds:

$$\text{(i)} \quad \hat{\mathbf{X}}_1 = \mathbf{TP}'$$
$$\text{(ii)} \quad \hat{\mathbf{X}}_2 = \mathbf{TP}' \tag{3.13}$$
$$\text{(iii)} \quad [\hat{\mathbf{X}}_1 \quad \hat{\mathbf{X}}_2] = [\mathbf{T} \quad \mathbf{T}]\mathbf{P}'$$

and Equation (3.13) follows by evaluating this expression using the rules of partitioned matrices. Equation (3.13) makes use of matricized three-way arrays $\underline{\hat{\mathbf{X}}}$ (containing the two slices $\hat{\mathbf{X}}_1$ and $\hat{\mathbf{X}}_2$) and $\underline{\mathbf{T}}$ (containing the two slices $\mathbf{T}$ and $\mathbf{T}$). Equation (3.13) can be expanded to more than two matrices and this allows for a concise way of expressing three-way models. Suppose, e.g., that the two slices $\mathbf{X}_1$ and $\mathbf{X}_2$ are modeled with the same loading $\mathbf{P}$, but with their own $\mathbf{T}_1$ and $\mathbf{T}_2$, then this can be expressed as

$$[\hat{\mathbf{X}}_1 \quad \hat{\mathbf{X}}_2] = [\mathbf{T}_1 \quad \mathbf{T}_2]\mathbf{P}' \tag{3.14}$$

In the following chapters this type of notation will be used frequently.

There is another way to express the principal component decomposition of $\mathbf{X}$ by using the column vectors $\mathbf{t}_1, \ldots, \mathbf{t}_R$ of $\mathbf{T}$ and the column vectors $\mathbf{p}_1, \ldots, \mathbf{p}_R$ of $\mathbf{P}$:

$$\hat{\mathbf{X}} = \mathbf{t}_1\mathbf{p}_1' + \mathbf{t}_2\mathbf{p}_2' + \cdots + \mathbf{t}_R\mathbf{p}_R' = \sum_{r=1}^{R} \mathbf{t}_r\mathbf{p}_r' \tag{3.15}$$

where the convention is used that all vectors are column vectors. This is the outer product notation of a principal component model (see Chapter 2).

EXAMPLE 3.2

Outer product notation of PCA

For the previous small numerical example, the outer product notation of the PCA model of $\mathbf{X}$ becomes

$$\hat{\mathbf{X}} = \mathbf{t}_1\mathbf{p}_1' + \mathbf{t}_2\mathbf{p}_2' =$$

$$\begin{bmatrix} 1 \\ 2 \\ 3 \end{bmatrix} [0.71 \quad 0.71] + \begin{bmatrix} -2 \\ -0.5 \\ 1 \end{bmatrix} [0.71 \quad -0.71] =$$

$$\begin{bmatrix} 0.71 & 0.71 \\ 1.42 & 1.42 \\ 2.13 & 2.13 \end{bmatrix} + \begin{bmatrix} -1.42 & 1.42 \\ -0.355 & 0.355 \\ 0.71 & -0.71 \end{bmatrix} =$$

$$\begin{bmatrix} -0.71 & 2.13 \\ 1.065 & 0.355 \\ 2.84 & -0.71 \end{bmatrix}$$

and note that both matrices are added together to give the PCA model of $\mathbf{X}$ have rank one.

Written in terms of the original SVD, Equation (3.15) becomes

$$\hat{\mathbf{X}} = s_{11}\mathbf{u}_1\mathbf{v}_1' + s_{22}\mathbf{u}_2\mathbf{v}_2' + \cdots + s_{RR}\mathbf{u}_R\mathbf{v}_R' = \sum_{r=1}^{R} s_{rr}\mathbf{u}_r\mathbf{v}_r' \tag{3.16}$$

There is not an equivalent generalization of the notation in Equations (3.15) and (3.16) for three-way models.

NOTATION USING KRONECKER PRODUCT, VEC-OPERATOR AND KHATRI–RAO PRODUCT

While the following formulations may seem unnecessarily complicated, they are helpful because they will ease the understanding of how three-way models are expressed. The equation for the principal component model of $\mathbf{X}$ can also be expressed in Kronecker products, using Equation (2.6.iii):

$$\hat{\mathbf{X}} = \mathbf{t}_1 \otimes \mathbf{p}_1' + \cdots + \mathbf{t}_R \otimes \mathbf{p}_R' = \sum_{r=1}^{R} \mathbf{t}_r \otimes \mathbf{p}_r' \tag{3.17}$$

and by using Equation (2.2.ii), it holds that

$$\text{vec } \hat{\mathbf{X}} = \mathbf{p}_1 \otimes \mathbf{t}_1 + \cdots + \mathbf{p}_R \otimes \mathbf{t}_R = \sum_{r=1}^{R} \mathbf{p}_r \otimes \mathbf{t}_r \tag{3.18}$$

and using Equation (2.2.iv)

$$\text{vec } \hat{\mathbf{X}} = \mathbf{P} \otimes \mathbf{T} \text{ vec } \mathbf{I} \tag{3.19}$$

showing that there are many alternative ways to write the principal component model of $\mathbf{X}$.

In Chapter 2 the Khatri–Rao product was introduced. Using Equations (2.15–2.17), it is also possible to write the PCA of $\mathbf{X}$ as

$$\text{vec } \hat{\mathbf{X}} = (\mathbf{P} \odot \mathbf{T})\mathbf{1} \tag{3.20}$$

where $\mathbf{1}$ is a $(R \times 1)$ vector of ones. In three-way analysis frequent use will be made of the Kronecker product, vec-operation and the Khatri–Rao product.

EXAMPLE 3.3

PCA in Kronecker product-, vec- and Khatri-Rao product notation

Using the same example as in the previous two frames, the PCA model of $\mathbf{X}$ according to Equation (3.17) is

$$\hat{\mathbf{X}} = \mathbf{t}_1 \otimes \mathbf{p}_1' + \mathbf{t}_2 \otimes \mathbf{p}_2' =$$

$$\begin{bmatrix} 1 \\ 2 \\ 3 \end{bmatrix} \otimes [\,0.71 \quad 0.71\,] + \begin{bmatrix} -2 \\ -0.5 \\ 1 \end{bmatrix} \otimes [\,0.71 \quad -0.71\,] =$$

$$\begin{bmatrix} 0.71 & 0.71 \\ 1.42 & 1.42 \\ 2.13 & 2.13 \end{bmatrix} + \begin{bmatrix} -1.42 & 1.42 \\ -0.355 & 0.355 \\ 0.71 & -0.71 \end{bmatrix} = \begin{bmatrix} -0.71 & 2.13 \\ 1.065 & 1.775 \\ 2.84 & 1.42 \end{bmatrix}$$

which is similar to the outer product notation. In vec-notation (Equation (3.18)) the PCA model of $\mathbf{X}$ becomes:

$$\text{vec } \hat{\mathbf{X}} = \text{vec}(\mathbf{p}_1 \otimes \mathbf{t}_1) + \text{vec}(\mathbf{p}_2 \otimes \mathbf{t}_2) =$$

$$\text{vec} \begin{bmatrix} 0.71 \\ 1.42 \\ 2.13 \\ 0.71 \\ 1.42 \\ 2.13 \end{bmatrix} + \text{vec} \begin{bmatrix} -1.42 \\ -0.355 \\ 0.71 \\ 1.42 \\ 0.355 \\ -0.71 \end{bmatrix} = \begin{bmatrix} -0.71 \\ 1.065 \\ 2.84 \\ 2.13 \\ 1.775 \\ 1.42 \end{bmatrix}$$

which is exactly the vec version of the PCA model of $\mathbf{X}$. By defining $\mathbf{P} = [\mathbf{p}_1\ \mathbf{p}_2]$ and $\mathbf{T} = [\mathbf{t}_1\ \mathbf{t}_2]$, the Kronecker product notation (Equation (3.19)) of the PCA model of $\mathbf{X}$ becomes

$$\text{vec } \hat{\mathbf{X}} = \mathbf{P} \otimes \mathbf{T} \text{ vec } \mathbf{I} = \begin{bmatrix} 0.71 & -1.42 & 0.71 & -1.42 \\ 1.42 & -0.355 & 1.42 & -0.355 \\ 2.13 & 0.71 & 2.13 & 0.71 \\ 0.71 & -1.42 & -0.71 & 1.42 \\ 1.42 & -0.355 & -1.42 & 0.355 \\ 2.13 & 0.71 & -2.13 & -0.71 \end{bmatrix} \begin{bmatrix} 1 \\ 0 \\ 0 \\ 1 \end{bmatrix} = \begin{bmatrix} -0.71 \\ 1.065 \\ 2.84 \\ 2.13 \\ 1.775 \\ 1.42 \end{bmatrix}$$

which is again the vec version of the PCA model of $\mathbf{X}$. Writing out Equation (3.19) in general terms for the simple example gives

$$\text{vec } \hat{\mathbf{X}} = \mathbf{P} \otimes \mathbf{T} \text{ vec } \mathbf{I} = [\,\mathbf{p}_1 \otimes \mathbf{t}_1 \quad \mathbf{p}_1 \otimes \mathbf{t}_2 \quad \mathbf{p}_2 \otimes \mathbf{t}_1 \quad \mathbf{p}_2 \otimes \mathbf{t}_2\,] \begin{bmatrix} 1 \\ 0 \\ 0 \\ 1 \end{bmatrix} = \mathbf{p}_1 \otimes \mathbf{t}_1 + \mathbf{p}_2 \otimes \mathbf{t}_2$$

and this shows that the vec $\mathbf{I}$ term selects the proper Kronecker products of $\mathbf{p}$s and $\mathbf{t}$s to form the vec version of the PCA model of $\mathbf{X}$. Finally, the working of the Khatri–Rao product (Equation (3.20)) can be illustrated by

$$(\mathbf{P} \odot \mathbf{T})\mathbf{1} = [\mathbf{p}_1 \quad \mathbf{p}_2] \odot [\mathbf{t}_1 \quad \mathbf{t}_2]\mathbf{1} = [\mathbf{p}_1 \otimes \mathbf{t}_1 \quad \mathbf{p}_2 \otimes \mathbf{t}_2] \begin{bmatrix} 1 \\ 1 \end{bmatrix}$$

$$= \mathbf{p}_1 \otimes \mathbf{t}_1 + \mathbf{p}_2 \otimes \mathbf{t}_2 = \text{vec } \hat{\mathbf{X}}$$

and this again equals the vec version of the PCA model of $\mathbf{X}$.

ILLUSTRATION WITH PICTURES

Pictorial representations of vectors and matrices illustrating the principal component model of $\mathbf{X}$ are shown in Figures 3.4 and 3.5. Figure 3.4 is the pictorial illustration of Equations (3.15) and (3.17) for $R = 2$. Figure 3.5 is the representation of Equation (3.16) for $R = 2$ and the size of the circles represent the magnitudes of s_{11} and s_{22}. Note that the norm of all vectors in Figure 3.5 is equal to one, by definition. Such pictures can be generalized to three-way models which is shown in the following chapters.

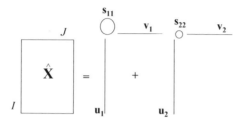

Figure 3.4. A pictorial representation of the principal component model of **X** (Equation (3.15) with $R = 2$). For reasons of simplicity, the prime $(')$ on P_1 is omitted here and in similar figures.

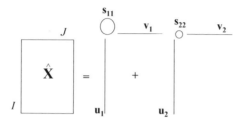

Figure 3.5. A pictorial representation of the principal component model of **X** in SVD terms (Equation (3.16) with $R = 2$).

Two-mode component analysis

The one-mode component analysis problem can be extended to the two-mode component analysis problem in which reduction in both modes is achieved [Levin 1965, Magnus & Neudecker 1988]. Suppose that $\mathbf{X}$ $(I \times J)$ is available, then the two-mode component model of $\mathbf{X}$ is found by solving

$$\min_{\mathbf{A},\mathbf{B},\mathbf{G}} \|\mathbf{X} - \mathbf{A}\mathbf{G}\mathbf{B}'\|^2 \qquad (3.21)$$

where $\mathbf{A}$ is $(I \times P)$ with $\mathbf{A}'\mathbf{A} = \mathbf{I}$; $\mathbf{B}$ is $(J \times Q)$ with $\mathbf{B}'\mathbf{B} = \mathbf{I}$ and $\mathbf{G}$ is $(P \times Q)$ which is called the core matrix. The number of components in $\mathbf{A}$ and $\mathbf{B}$ can be different. A solution to this problem can be found by using the SVD of $\mathbf{X}$ $(= \mathbf{U}\mathbf{S}\mathbf{V}')$:

$$
\begin{aligned}
\mathbf{A} &= \mathbf{U}_P \\
\mathbf{B} &= \mathbf{V}_Q \qquad (3.22)\\
\mathbf{G} &= \mathbf{S}_{P,Q}
\end{aligned}
$$

where $\mathbf{U}_P$ contains the first P columns of $\mathbf{U}$, $\mathbf{V}_Q$ contains the first Q columns of $\mathbf{V}$ and $\mathbf{S}_{P,Q}$ is the $P \times Q$ upper left part of $\mathbf{S}$. There is rotational freedom in solutions of Equation (3.22), because $\mathbf{A}$ and $\mathbf{B}$ can be postmultiplied with an orthogonal $\mathbf{Z}_1$ $(P \times P)$ and an orthogonal $\mathbf{Z}_2$ $(Q \times Q)$, respectively, and this can be compensated by premultiplying and postmultiplying $\mathbf{G}$ by $\mathbf{Z}_1'$ and $\mathbf{Z}_2'$, respectively.

Although Equation (3.22) seems to be a reasonable thing to do, it is not used often in chemistry. The reason is that in chemistry the interest is usually in approximating $\mathbf{X}$ with a matrix of low rank. The singular value decomposition of $\mathbf{X}$ truncated to R components gives the best rank R approximation of $\mathbf{X}$. Problem (3.22) is presented here, because it naturally leads to a three-way generalization (Tucker3 models) which is very useful. In Example 3.4 a chemical example of a two-mode component model is given to show that it is sometimes useful.

EXAMPLE 3.4

Two-mode component model

In Elaboration 2.1, an example was given of a two-mode component model. A model of a special matrix $\mathbf{X}_M$ was written as

$$\hat{\mathbf{X}}_M = \tilde{\mathbf{C}} \tilde{\mathbf{H}} \mathbf{S}'$$

and by defining $\mathbf{A} = \tilde{\mathbf{C}}; \mathbf{G} = \tilde{\mathbf{H}}; \mathbf{B} = \mathbf{S}$ a two-mode component model of $\mathbf{X}_M$ is obtained. Although a principal component model of $\mathbf{X}_M$ is the most efficient way of summarizing the systematic variation in $\mathbf{X}_M$, such special two-mode component models can be useful.

3.2   Models for Two-way Two-block Data Analysis: Regression Models

General idea of regression models

Calibration is one of the oldest scientific activities. In order to measure something, a proper standard to measure against is needed. Even the ancient land surveyors in Babylonia and Egypt calibrated. Calibration is the use of empirical data and prior knowledge for determining how to predict unknown quantitative information Y from available measurements X, via some mathematical transfer function [Martens & Næs 1989].

In chemistry, calibration has a special place because direct measurement of concentrations is rarely possible. The only direct concentration determination used is the cumbersome technique of gravimetry. For simple binary mixtures, density, refractive index, conductivity, polarization angle etc. may be used for measuring concentration. This is almost a direct chemical measurement because the concentration can be read out of a table. This principle is used, for instance, in measuring the H_2SO_4 concentration of battery acid by density measurement.

Most advanced, sensitive and selective methods of measuring concentrations are based on secondary measures. In spectrophotometry, color intensity is used as a measure of a concentration and the calibration is done with the color intensity of prepared standards of known concentration. An example of this is given in Figure 3.6, where x is concentration and y is color intensity.

Other chemical calibrations depend on, for example, chromatographic peak area, emission intensity, current, voltage, position on a thin layer chromatography or electrophoresis plate (expressing voltage gradients).

The relationship between concentration and secondary measurement is not always linear. Whether it is a straight line or a curved line that describes the calibration, an equation is needed to be able to predict concentrations for future samples. A quantitative calibration curve can also be used to calculate a number of important analytical properties (sensitivity, linearity, offset or baseline, detection limit [Currie 1995]). The calibration line would be

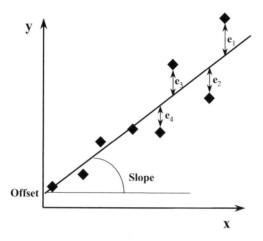

Figure 3.6. The relationship between concentration (horizontal axis) and a secondary variable, e.g. absorbance, (vertical axis) can often be expressed as a straight line with a slope b_1 and an offset b_0. The *es* are residuals.

easily described by an exact function if there was no measurement or model error, but this is rarely the case. Finding a proper formula to express the relationship between two variables can be done by using the principle of regression. This principle was studied, for example, by the anthropologist Francis Galton in his studies of the relation between the heights of fathers and sons [Galton 1885].

An expression for the relationship between color intensity (y) and concentration (x) could be:

$$y_i = b_1 x_i + b_0 + e_i \qquad (3.23)$$

where b_1 is the slope of the line and b_0 the offset; e_i is the error in y_i.[2] See Figure 3.6.

Some nomenclature is in order here: b_0 is called the constant or offset term or intercept; b_1 is the regression coefficient; x is the independent or predictor variable and y is the dependent or response variable; e_i is the residual. More regression nomenclature can be found, e.g., in Draper & Smith [Draper & Smith 1998].

The previous equation leads to a more general expression:

$$y_i = b_0 + b_1 x_i + b_{11} x_i^2 + \cdots + e_i \qquad (3.24)$$

This is nothing but fitting a polynomial throught the points; b_{11} is the coefficient of the quadratic term. Forcing b_{11} to zero gives an equation linear in x. The dots indicate the possibility of adding higher exponents. Another possible extension is:

$$y_i = b_0 + b_1 x_{1i} + b_2 x_{2i} + b_3 x_{3i} + \cdots + e_i \qquad (3.25)$$

where there is more than one variable involved; b_1, b_2, b_3 etc. are the coefficients of the different variables x_1, x_2, x_3 etc. Both of the above problems are called multiple linear regression problems.

---

[2] The usual distinction made in statistics between population parameters (Greek symbols) and estimated parameters (Arabic symbols) is omitted for notational simplicity.

The method to estimate the coefficients of these equations by least squares regression, is essentially the same for all these models. The idea of minimizing the sum (or weighted sum) of the e_i^2 comes from Gauss and Legendre. Least squares minimization penalizes large deviations between measured and fitted values heavily.

Values of y_i are measured for a sufficient number of values of x_i or of the x_is and then the sum of squares of the e_is is minimized [Box & Draper 1987, Draper & Smith 1998].

Regression equations such as Equations (3.23)–(3.25) assume that the xs are error-free and that the error is in y. Dealing with possible error in the xs is important and some of the regression techniques presented below can handle this problem.

Multiple linear regression

In formulas the multiple linear regression equation (3.25) is

$$y = b_1 x_1 + \cdots + b_J x_J + e$$
$$y = Xb + e \tag{3.26}$$

$$\text{with } \mathbf{y} = \begin{bmatrix} y_1 \\ \cdot \\ \cdot \\ \cdot \\ y_I \end{bmatrix}; \quad \mathbf{X} = \begin{bmatrix} x_{11} & \cdot & \cdot & x_{1J} \\ \cdot & \cdot & \cdot & \cdot \\ \cdot & \cdot & \cdot & \cdot \\ x_{I1} & \cdot & \cdot & x_{IJ} \end{bmatrix}; \quad \mathbf{e} = \begin{bmatrix} e_1 \\ \cdot \\ \cdot \\ e_I \end{bmatrix}$$

The associated least squares problem is then

$$\min_{\mathbf{b}} \| \mathbf{y} - \mathbf{X}\mathbf{b} \|^2 \tag{3.27}$$

where $\mathbf{X}$ is $(I \times J)$, $\mathbf{b}$ $(J \times 1)$ and $\mathbf{y}$ $(I \times 1)$ (the case of multivariate $\mathbf{y}$ will be treated later). The column vector $\mathbf{Xb}$ is a vector in $\Re(\mathbf{X})$, hence, multiple linear regression equals finding a vector in the range of $\mathbf{X}$ that is as close as possible to $\mathbf{y}$. For regression problems where an intercept is assumed, the $\mathbf{X}$ and $\mathbf{y}$ have to be mean-centered or $\mathbf{X}$ has to be augmented with a column of ones to represent the intercept.

The solution of Equation (3.27) is $\mathbf{b} = \mathbf{X}^+\mathbf{y}$; where the symbol '+' stands for the Moore–Penrose inverse of $\mathbf{X}$ [Rao & Mitra 1971]. Using the SVD of $\mathbf{X}$ ($\mathbf{X} = \mathbf{USV}'$, where s_{ii} is the ith diagonal element of $\mathbf{S}$), an explicit formula for the Moore–Penrose inverse is

$$\mathbf{X}^+ = \mathbf{VS}^+\mathbf{U}' \tag{3.28}$$

$$s_{ii}^+ = \frac{1}{s_{ii}} \quad \text{if } s_{ii} \neq 0 \quad \text{and} \quad 0 \text{ if } s_{ii} = 0$$

where s_{ii}^+ is the ith diagonal element of $\mathbf{S}^+$. The properties of the solution of Equation (3.28) depend on the structure of $\mathbf{X}$ and $\mathbf{y}$. These properties are summarized in Table 1. The Moore–Penrose inverse is always unique, but any generalized inverse[3] will solve Equation (3.27) [Schott 1997]. The requirement that $\| \mathbf{b} \|$ is minimal gives the unique Moore–Penrose solution [Rao & Mitra 1971].

---

[3] To be more precise, only the {1,3} generalized inverses solve the least squares problem, where 1 and 3 refer to the four Moore–Penrose conditions.

Table 3.1. Properties of the regression problem (LS means least squares, that is, the sum of squared residuals is minimized)

	X of Full Rank	**X** Not of Full Rank
$I = J$	$\mathbf{b} = \mathbf{X}^+\mathbf{y} = \mathbf{X}^{-1}\mathbf{y}$ exact fit	$\mathbf{b} = \mathbf{X}^+\mathbf{y}$ $\mathbf{y} \in \Re(\mathbf{X})$: exact fit, $\|\mathbf{b}\|$ minimal $\mathbf{y} \notin \Re(\mathbf{X})$: LS, $\|\mathbf{b}\|$ minimal
$I > J$	$\mathbf{b} = \mathbf{X}^+\mathbf{y} = (\mathbf{X}'\mathbf{X})^{-1}\mathbf{X}'\mathbf{y}$ $\mathbf{y} \in \Re(\mathbf{X})$: exact fit $\mathbf{y} \notin \Re(\mathbf{X})$: LS	$\mathbf{b} = \mathbf{X}^+\mathbf{y}$ $\mathbf{y} \in \Re(\mathbf{X})$: exact fit, $\|\mathbf{b}\|$ minimal $\mathbf{y} \notin \Re(\mathbf{X})$: LS, $\|\mathbf{b}\|$ minimal
$I < J$	$\mathbf{b} = \mathbf{X}^+\mathbf{y}$ exact fit, $\|\mathbf{b}\|$ minimal	$\mathbf{b} = \mathbf{X}^+\mathbf{y}$ $\mathbf{y} \in \Re(\mathbf{X})$: exact fit, $\|\mathbf{b}\|$ minimal $\mathbf{y} \notin \Re(\mathbf{X})$: LS, $\|\mathbf{b}\|$ minimal

For multivariate **Y**, Equation (3.27) translates to

$$\min_{\mathbf{B}} \|\mathbf{Y} - \mathbf{XB}\|^2 = \min_{\mathbf{B}} \|[\mathbf{y}_1|\dots|\mathbf{y}_M] - \mathbf{X}[\mathbf{b}_1|\dots|\mathbf{b}_M]\|^2$$
$$= \min_{\mathbf{B}} (\|\mathbf{y}_1 - \mathbf{Xb}_1\|^2 + \dots + \|\mathbf{y}_M - \mathbf{Xb}_M\|^2) \tag{3.29}$$

where **Y** is $(I \times M)$ and $\mathbf{y}_m$ is the mth column of **Y**. As can be seen in Equation (3.29), this problem can be described as M separate univariate (multiple) regression problems. If **X** has full column rank, then $\mathbf{B} = (\mathbf{X}'\mathbf{X})^{-1}\mathbf{X}'\mathbf{Y}$.

For the cases where the inverse of $\mathbf{X}'\mathbf{X}$ does not exist or if $\mathbf{X}'\mathbf{X}$ is ill-conditioned (that is, $\mathbf{X}'\mathbf{X}$ is nearly singular), there is always a *numerical* solution to Equations (3.27) and (3.29). However, this does not mean that this solution is always desirable from a *statistical or practical* point of view. Specifically, the estimated regression vector **b** tends to be uncertain because the solution is mostly governed by the noise part of the data. This can lead to high variances of predicted y values for new samples or objects. [Belsley *et al.* 1980].

Many alternatives exists for dealing with problems (3.27) and (3.29) for **X** matrices which are ill-conditioned. Several of such alternatives will be discussed in the next sections. These are important because for multivariate data it is mostly the case that some variables are correlated by nature or by sampling. Correlated variables lead to ill-conditioned data matrices.

Principal component regression

The problem dealt with by principal component regression is regressing **y** $(I \times 1)$ on a possibly ill-conditioned **X** $(I \times J)$. Hence, principal component regression tries to solve Equation (3.27) and Equation (3.29) for ill-conditioned **X**. Principal component regression approximates **X** by a few, say R, components (its principal components) and regresses **y** on these R components. Principal component regression can be written as

$$\mathbf{X} = \mathbf{TP}' + \mathbf{E_X}; \quad \text{with } \mathbf{T} = \mathbf{U}_R\mathbf{S}_R \text{ and } \mathbf{P} = \mathbf{V}_R \text{ from } \mathbf{X} = \mathbf{USV}'$$
$$\mathbf{y} = \mathbf{Tb} + \mathbf{E_Y}; \quad \text{with } \mathbf{b} = (\mathbf{T}'\mathbf{T})^{-1}\mathbf{T}'\mathbf{y} \tag{3.30}$$

where $\mathbf{USV}'$ is the SVD of $\mathbf{X}$, and the subscript 'R' means R-truncation of the respective matrices. The matrix $\mathbf{T}'\mathbf{T}$ is well conditioned, due to its diagonality and the fact that R is usually (much) smaller than J. Equation (3.30) allows errors to be present in the X-variables, which is often the case in chemistry. The regression model, however, does not assume errors present in the scores $\mathbf{T}$. The Moore–Penrose solution (see Equation (3.28)) equals principal component regression if all the principal components of $\mathbf{X}$ associated to nonzero singular values are used.

There is some debate in the literature about which principal components should be taken in $\mathbf{T}$. In Equation (3.30) the first R, associated with the R largest singular values of $\mathbf{X}$ are selected, but alternatives exist. It may, e.g, be advantageous to use the components that correlate maximally with $\mathbf{y}$ [Mason & Gunst 1985], but common practice is to use the ones as in Equation (3.30). Especially for data where most of the variation is known to be unrelated to the dependent variable, it may be better to select components. Note that there is a pay-off: too few components do not fit $\mathbf{X}$ and do not predict $\mathbf{Y}$ well, whereas too many components overfit $\mathbf{Y}$ and $\mathbf{X}$, leading to unstable new predicted $\mathbf{Y}$ values. Validation methods can be used to estimate the optimal number of components [Martens & Næs 1989].

For multivariate $\mathbf{y}$ Equation (3.30) becomes

$$\mathbf{X} = \mathbf{TP}' + \mathbf{E}_X; \quad \text{with } \mathbf{T} = \mathbf{U}_R\mathbf{S}_R \text{ and } \mathbf{P} = \mathbf{V}_R \text{ from } \mathbf{X} = \mathbf{USV}'$$
$$\mathbf{Y} = \mathbf{TB} + \mathbf{E}_Y; \quad \text{with } \mathbf{B} = (\mathbf{T}'\mathbf{T})^{-1}\mathbf{T}'\mathbf{Y} \tag{3.31}$$

and the same comments as under Equation (3.30) hold for this case.

Partial least squares regression

GENERAL IDEA OF PARTIAL LEAST SQUARES REGRESSION

In discussing principal component regression it was mentioned that the principal components do not necessarily correlate maximally with $\mathbf{Y}$. Partial least squares regression finds components that compromise between fitting of $\mathbf{X}$ and predicting $\mathbf{Y}$. The general problem dealt with by partial least squares regression is regressing $\mathbf{Y}$ $(I \times M)$ on an ill-conditioned $\mathbf{X}$ $(I \times J)$. Hence, partial least squares regression also tries to solve problems (3.27) and (3.29) for ill-conditioned $\mathbf{X}$. The central idea of partial least squares regression is to approximate $\mathbf{X}$ by a few, say R, specifically constructed components (the partial least squares regression components) and to regress $\mathbf{Y}$ on the R components. Hence, partial least squares regression tries to model $\mathbf{X}$ and $\mathbf{Y}$ using the common components $\mathbf{T}$:

$$\mathbf{X} = \mathbf{TP}' + \mathbf{E}_X$$
$$\mathbf{Y} = \mathbf{TQ}' + \mathbf{E}_Y \tag{3.32}$$

where $\mathbf{T}$ is an $I \times R$ matrix of scores; $\mathbf{P}$ is a $J \times R$ matrix of X-loadings; $\mathbf{Q}$ is a $M \times R$ matrix of Y loadings; $\mathbf{E}_X$ and $\mathbf{E}_Y$ are residual matrices [Martens & Næs 1989]. Again, errors in $\mathbf{X}$ are allowed for explicitly.

The R components are constructed as a compromise between summarizing $\mathbf{X}$ and predicting $\mathbf{Y}$. Although Equation (3.32) resembles Equation (3.31), the calculated components, loadings and weights are different in the two equations. Stated otherwise, the structural model of partial least squares regression and principal component regression is the same, but the estimated components are different.

In order to illustrate how the partial least squares regression components are constructed, regression with a univariate $\mathbf{y}$ ($I \times 1$) is considered. Suppose $\mathbf{X}$ ($I \times J$) and $\mathbf{y}$ ($I \times 1$) are available and centered columnwise. The first partial least squares regression component $\mathbf{t}_1$ which is calculated to predict $\mathbf{y}$ from $\mathbf{X}$ solves

$$\max_{\mathbf{w}}[\mathbf{cov}(\mathbf{t}, \mathbf{y})|\mathbf{X}\mathbf{w} = \mathbf{t} \text{ and } \|\mathbf{w}\| = 1] \tag{3.33}$$

Equation (3.33) can be rewritten as:

$$\max_{\mathbf{w}}[\mathbf{y}'\mathbf{t}|\mathbf{t} = \mathbf{X}\mathbf{w} \text{ and } \|\mathbf{w}\| = 1] \tag{3.34}$$

where for convenience, without changing the problem, the covariance is expressed without correction for the degrees of freedom. This can be rewritten as

$$\max_{\mathbf{w}}[\mathbf{y}'\mathbf{X}\mathbf{w}|\|\mathbf{w}\| = 1] \tag{3.35}$$

and by defining $\mathbf{z} = \mathbf{X}'\mathbf{y}$, this is equivalent to

$$\max_{\mathbf{w}}[\mathbf{z}'\mathbf{w}|\|\mathbf{w}\| = 1] \tag{3.36}$$

This expression is maximized[4] if and only if

$$\mathbf{w} = \frac{\mathbf{z}}{\|\mathbf{z}\|} = \frac{\mathbf{X}'\mathbf{y}}{\|\mathbf{X}'\mathbf{y}\|} \tag{3.37}$$

which is a well-known result [Höskuldsson 1988].

After finding this first component $\mathbf{t} = \mathbf{X}\mathbf{w}$, $\mathbf{y}$ and $\mathbf{X}$ are 'deflated' and the second component is found. There are two ways to do this: one way generates orthogonal $\mathbf{t}$ vectors and the second way generates nonorthogonal $\mathbf{t}$ vectors [Martens & Næs 1989]. Both ways are discussed briefly, but they give essentially the same results, e.g., the same predictions [Helland 1988]. In Chapter 4 it is shown that both versions can be generalized for three-way analysis giving also the same prediction results.

ORTHOGONAL $\mathbf{t}$ VECTORS VERSION

In this version, the matrix $\mathbf{X}$ is regressed on the first score vector $\mathbf{t}_1$ generating regression coefficients (loadings) $\mathbf{p}_1$. In computing the next component the independent data, $\mathbf{X}$, are deflated by exchanging them with the residuals $\mathbf{E}^{(1)} = \mathbf{X} - \mathbf{t}_1\mathbf{p}_1'$ holding the part of the data not yet used. The $\mathbf{y}$ vector can also be deflated, but this is not mandatory since deflating $\mathbf{X}$ makes the independent data orthogonal to the part of $\mathbf{y}$ already described[5] [Burnham *et al.* 1996, De Jong & Ter Braak 1994, Helland 1988].

The next component is found by solving a similar problem as in Equation (3.33) with the proper matrices involved ($\mathbf{E}^{(1)}$ instead of $\mathbf{X}$). This results in the solution $\mathbf{w}_2$ and $\mathbf{t}_2$. The process continues with the next deflation step giving $\mathbf{p}_2$ and stops when the number

---

[4] This can be understood by applying the cosine rule to the vectors $\mathbf{z}$ and $\mathbf{w}$. Maximizing $\mathbf{z}'\mathbf{w}$ for $\mathbf{z}$ and $\mathbf{w}$ of length one is equivalent to maximizing the cosine of the angle between $\mathbf{z}$ and $\mathbf{w}$. This cosine is maximal for an angle of zero degrees. Hence, $\mathbf{w}$ is in the direction of $\mathbf{z}$.

[5] Deflation of $\mathbf{y}$, however, provides residuals which are important for diagnostic purposes.

of components is sufficiently large. The appropriate number of components can be determined, e.g., by cross-validation. [Wold *et al.* 1984]. In this implementation of partial least squares regression the score vectors, **t**, are orthogonal, due to the deflation step (see Appendix 3.B).

After having found R components and by defining $\mathbf{T} = [\mathbf{t}_1, \ldots, \mathbf{t}_R]$, $\mathbf{W} = [\mathbf{w}_1, \ldots, \mathbf{w}_R]$ and $\mathbf{P} = [\mathbf{p}_1, \ldots, \mathbf{p}_R]$ this results in the following model

$$\mathbf{T} = \mathbf{XW}(\mathbf{P'W})^{-1}$$
$$\mathbf{X} = \mathbf{TP'} + \mathbf{E}_X \qquad (3.38)$$
$$\mathbf{Y} = \mathbf{Tq} + \mathbf{e}_y$$

where $\mathbf{t}_r = \mathbf{X}^{(r-1)}\mathbf{w}_r$ is found in the rth dimension of the model by maximizing the covariance of $\mathbf{t}_r$ with $\mathbf{y}$ under the constraint that $\|\mathbf{w}_r\| = 1$ and $\mathbf{X}^{(r-1)}$ is the $\mathbf{X}$ matrix after the $(r-1)$th deflation step. For this algorithm, $\mathbf{W'W} = \mathbf{I}$; $\mathbf{T'T} = $ diagonal and $\mathbf{w}'_i\mathbf{p}_j = 0\,(i < j)$. It can be shown that $\mathbf{P}$ solves the problem of minimizing $\|\mathbf{X} - \mathbf{TP'}\|^2$ for given $\mathbf{T}$, using the orthogonality of the $\mathbf{t}$ vectors [Höskuldsson 1988]. However, $\mathbf{P}$ and $\mathbf{T}$ do not solve the problem of minimizing $\|\mathbf{X} - \mathbf{TP'}\|^2$ for general $\mathbf{P}$ and $\mathbf{T}$.

The essential part of partial least squares regression is the criterion of maximizing the covariance of $\mathbf{t}$ and $\mathbf{y}$. Because the covariance is a product of the variance of $\mathbf{t}$ *as well as* the correlation between $\mathbf{t}$ and $\mathbf{y}$ it follows that each component is characterized by (i) the variation in $\mathbf{t}$ is high 'ensuring' that noise is not being modeled, and (ii) the correlation is high meaning that $\mathbf{t}$ has predictive relevance for $\mathbf{y}$. Given that valid systematic variation is generally larger than noise, and that a linear relationship can express the relation between $\mathbf{X}$ and $\mathbf{y}$, the criterion in partial least squares regression is thus sensible.

The full partial least squares regression algorithm is given in Appendix 3.B.

NONORTHOGONAL **t** VECTORS VERSION

After having found the first component $\mathbf{t}_1 = \mathbf{Xw}_1$, it is also possible to deflate $\mathbf{X}$ with $\mathbf{t}_1\mathbf{w}'_1$; hence, $\mathbf{E}^{(1)} = \mathbf{X} - \mathbf{t}_1\mathbf{w}'_1$. The dependent variable, $\mathbf{y}$, can be deflated and the whole process can be started again by solving Equation (3.33) with the proper matrices and vectors involved. This implementation does not give orthogonal score vectors $\mathbf{t}_1, \ldots, \mathbf{t}_R$. This might be the reason for its unpopularity, but the solution in terms of fitted $\mathbf{y}$ and predicted $\mathbf{y}$s for new samples is identical to the orthogonal t-vector version [Helland 1988].

The final partial least squares regression model after R components in the nonorthogonal t-vector version reads (by defining again $\mathbf{T} = [\mathbf{t}_1, \ldots, \mathbf{t}_R]$ and $\mathbf{W} = [\mathbf{w}_1, \ldots, \mathbf{w}_R]$)

$$\mathbf{T} = \mathbf{XW}$$
$$\mathbf{X} = \mathbf{TW'} + \mathbf{E}_X \qquad (3.39)$$
$$\mathbf{Y} = \mathbf{Tq} + \mathbf{e}_y$$

where, again, in each dimension of the PLS model the covariance of $\mathbf{t}$ and $\mathbf{y}$ is maximized. It holds that $\mathbf{W'W} = \mathbf{I}$ and $\mathbf{T'T}$ is tridiagonal [Burnham *et al.* 1996]. For convenience the same symbols $\mathbf{T}$ and $\mathbf{q}$ are used in Equations (3.38) and (3.39), but their values are not the same. The matrix $\mathbf{W}$ is the same in both Equations (3.38) and (3.39) [Helland 1988].

$\mathbf{T} = \mathbf{XW}$ solves the problem of minimizing $\|\mathbf{X} - \mathbf{TW}'\|^2$ for given $\mathbf{W}$ due to the property $\mathbf{W}'\mathbf{W} = \mathbf{I}$. Nevertheless, $\mathbf{T} = \mathbf{XW}$ does not solve the problem of minimizing $\|\mathbf{X} - \mathbf{TW}'\|^2$ for general $\mathbf{T}$ and $\mathbf{W}$ because $\mathbf{W}$ is constrained to define the direction of maximal covariance with $\mathbf{y}$.

The algorithm for estimating the parameters in Equation (3.39) is given in Appendix 3.B.

GENERAL SCHEME OF PARTIAL LEAST SQUARES REGRESSION

Both versions can be summarized in the following general scheme

$$\mathbf{T} = \mathbf{XV}$$
$$\mathbf{X} = \mathbf{TP}' + \mathbf{E_X} \qquad (3.40)$$
$$\mathbf{y} = \mathbf{Tq} + \mathbf{e_y}$$

where $\mathbf{t}_r = \mathbf{E}^{(r-1)}\mathbf{w}_r$ and $\mathbf{E}^{(r-1)}$ is the deflated $\mathbf{X}$ after the $(r-1)$th component. With different restrictions put on $\mathbf{T}$, $\mathbf{V}$, $\mathbf{P}$ and $\mathbf{W}$ the result is one of the two above mentioned models.

Both principal component regression and partial least squares regression solve the problem of ill-conditioned $\mathbf{X}$ (and $\mathbf{X}'\mathbf{X}$) by taking a few vectors $\mathbf{t}$ in $\Re(\mathbf{X})$. In general, principal component regression and partial least squares regression do not give the same solution. There is some debate about the relative merits of both methods. Practical experience shows that the performance of PLS and principal component regression does not differ much in many situations [Frank & Friedman 1993]. Recent theoretical results give a basis for this observation [Stoica & Söderström 1998]. An important property of principal component regression and PLS is that latent variables (scores, loadings) are calculated. These can be used for diagnostic purposes and for visualization.

3.3 Summary

This chapter discusses two-way models and serves as a introduction for the chapters to come. A distinction is made between component models and regression models.

Component models deal with data that can be meaningfully arranged in a single block. A common method to deal with this type of data is principal component analysis and this technique is explained in some detail. A less common method which will prove useful in the following, called two-mode component analysis, is also discussed briefly.

Regression models deal with data that can be meaningfully arranged in two related blocks. Examples are given for problems that generate such data, e.g., calibration problems. Common methods to deal with such data are introduced and discussed briefly: multiple linear regression, partial least squares regression and principal component regression.

Appendix 3.A: Some PCA Results

For PCA the weights equal the loadings

In PCA the weights equal the loadings. This can be shown as follows. Let $\mathbf{USV}'$ be the SVD of $\mathbf{X}$. Then $\mathbf{w}_1 = \mathbf{v}_1$ and $\mathbf{t}_1 = s_1\mathbf{u}_1$. The loadings $\mathbf{p}_1$ can be found by solving the regression problem $\|\mathbf{X} - \mathbf{t}_1\mathbf{p}_1'\|^2$. This results in

$$\mathbf{p}_1' = (\mathbf{t}_1'\mathbf{t}_1)^{-1}\mathbf{t}_1'\mathbf{X} \quad \text{or} \quad \mathbf{p}_1 = \mathbf{X}'\mathbf{t}_1(\mathbf{t}_1'\mathbf{t}_1)^{-1} \tag{3.41}$$

and using the SVD of $\mathbf{X}$, and the fact that $\mathbf{t}_1 = s_1\mathbf{u}_1$ gives

$$\mathbf{p}_1 = \mathbf{V}'\mathbf{SU}'\mathbf{u}_1 s_1 (s_1\mathbf{u}_1'\mathbf{u}_1 s_1)^{-1} \Rightarrow \mathbf{p}_1 = \mathbf{v}_1 = \mathbf{w}_1 \tag{3.42}$$

using the orthogonality of $\mathbf{U}$ and $\mathbf{V}$. Similar derivations can be made for $\mathbf{p}_2$, $\mathbf{p}_3$ etc.

The formal PCA Problem in the Pearson approach

The formal problem of PCA can be stated as follows [Ten Berge 1993]

$$\min_{\mathbf{P},\mathbf{W}} \|\mathbf{X} - \mathbf{XWP}'\|^2 = \min_{\mathbf{T},\mathbf{P}} \|\mathbf{X} - \mathbf{TP}'\|^2$$
$$\text{s.t. } \mathbf{W}'\mathbf{W} = \mathbf{I} \tag{3.43}$$

where the abbreviation s.t. means 'subject to'. The constraint $\mathbf{W}'\mathbf{W} = \mathbf{I}$ has been added for convenience and without loss of generality. The weights used to form the linear combinations of the columns of $\mathbf{X}$ are collected in $\mathbf{W}$ ($J \times R$), where $R \leq J$ equals the number of linear combinations or new variables carrying the information in $\mathbf{X}$. Usually, R is taken smaller than J, otherwise no dimension reduction takes place. The new values of the objects on the formed linear combinations are collected in $\mathbf{T}$ ($I \times R$), where $\mathbf{T} = \mathbf{XW}$. Having found these linear combinations, a regression is carried out of $\mathbf{X}$ on $\mathbf{T}$, in order to obtain the best fit of $\mathbf{X}$. The resulting regression weights are collected in $\mathbf{P}$ ($J \times R$).

The solution of problem (3.43), [Ten Berge 1993] is

$$\mathbf{W} = \mathbf{K}_R\mathbf{Q}$$
$$\mathbf{T} = \mathbf{XW} = \mathbf{XK}_R\mathbf{Q} \tag{3.44}$$
$$\mathbf{P} = \mathbf{K}_R\mathbf{Q} = \mathbf{W}$$

where $\mathbf{Q}$ is an arbitrary orthogonal ($R \times R$) matrix and $\mathbf{K}_R$ is the ($J \times R$) matrix that contains the first R eigenvectors of $\mathbf{X}'\mathbf{X}$, corresponding to the R largest eigenvalues of $\mathbf{X}'\mathbf{X}$.

Several comments are appropriate:

1. If the additional constraint is added that $\mathbf{Q} = \mathbf{I}$ then the columns of $\mathbf{T}$ are orthogonal to each other. The first column in $\mathbf{T}$ has the highest possible variance; the second column has the highest variance under the additional constraint of being orthogonal to the first column, and so on. This gives the principal components of the '*variance of*' approach.

2. Instead of using the constraint $\mathbf{W}'\mathbf{W} = \mathbf{I}$, it is also possible to apply the constraint that $\mathbf{T}'\mathbf{T}$ is diagonal in Equation (3.43). This results in both $\mathbf{W}'\mathbf{W}$ and $\mathbf{P}'\mathbf{P}$ being diagonal. If

in addition all the variance is captured in $\mathbf{T}$, then $\mathbf{W'W} = \mathbf{P'P} = \mathbf{I}$. This is the customary way in chemometrics to formulate the PCA model and this gives again the principal components which have maximum variance.

3. For $\mathbf{Q} = \mathbf{I}$, the solutions of Equation (3.43) with increasing R are nested: the rank R solution contains the rank $R - 1$ solution with an extra component (the Rth component) added. This is easily seen from Equation (3.44): $\mathbf{K}_R = [\mathbf{K}_{R-1} | \mathbf{k}_R]$, where $\mathbf{k}_R$ is the rth eigenvector of $\mathbf{X'X}$.

4. The scores $\mathbf{T}$ and loadings $\mathbf{P}$ can always be transformed to $\tilde{\mathbf{T}} = \mathbf{TZ}$ and $\tilde{\mathbf{P}} = \mathbf{P}(\mathbf{Z}^{-1})'$ without changing the fit of $\mathbf{X}$ in Equation (3.43). If $\mathbf{T}$ and $\mathbf{P}$ resulted from Equation (3.44) then column orthogonality of $\mathbf{P}$ still holds after orthogonal transformation ($\mathbf{Z'Z} = \mathbf{ZZ'} = \mathbf{I}$ and $\tilde{\mathbf{P}}'\tilde{\mathbf{P}} = \mathbf{I}$), whereas the column orthogonality of $\mathbf{T}$ is destroyed (see later). This also shows that the constraint $\mathbf{W'W} = \mathbf{I}$ is not active with respect to the fit.

5. The problem in Equation (3.43) can of course also be solved for a scaled and/or centered matrix $\mathbf{X}$. Scaling and/or centering affect the solution of the problem in Equation (3.43). There is no simple relationship between solutions of centered/uncentered or scaled/unscaled versions of the same original $\mathbf{X}$.

The SVD can be used to calculate the solution of problem (3.43) for the case of $\mathbf{Q} = \mathbf{I}$, because then $\mathbf{W} = \mathbf{K}_R = \mathbf{V}_R$ and $\mathbf{T} = \mathbf{XW} = \mathbf{XV}_R = \mathbf{U}_R\mathbf{S}_R$, where $\mathbf{V}_R$ contains the first R columns of $\mathbf{V}$, $\mathbf{U}_R$ contains the first R columns in $\mathbf{U}$ and $\mathbf{S}_R$ contains the $R \times R$ upper left part of $\mathbf{S}$. This shows that there is an intimate relationship between SVD and PCA, the two methods are, however, not equivalent: PCA solves the problem of Equation (3.43) whereas the SVD of $\mathbf{X}$ is simply its decomposition into $\mathbf{USV'}$.

Orthogonality and rotation

Rewriting $\mathbf{T'T}$, it follows that $\mathbf{T'T} = (\mathbf{XK}_R\mathbf{Q})'(\mathbf{XK}_R\mathbf{Q}) = \mathbf{Q'K}'_R\mathbf{X'XK}_R\mathbf{Q} = \mathbf{Q'\Lambda}_R\mathbf{Q}$, where $\mathbf{\Lambda}_R$ is the $R \times R$ upper left part of $\mathbf{\Lambda}$. Hence, $(\mathbf{T'T})^{-1} = \mathbf{Q'\Lambda}_R^{-1}\mathbf{Q}$. Note that $\mathbf{T'T} = \mathbf{\Lambda}_R$ for $\mathbf{Q} = \mathbf{I}$ showing that the score vectors are indeed orthogonal to each other if $\mathbf{Q} = \mathbf{I}$.

When the scores $\mathbf{T}$ and the loadings $\mathbf{P}$ are transformed with an orthogonal $\mathbf{Z}$ ($R \times R$) then $\tilde{\mathbf{T}} = \mathbf{TZ}$ and $\tilde{\mathbf{P}} = \mathbf{PZ}$. Clearly, $\tilde{\mathbf{P}}'\tilde{\mathbf{P}} = \mathbf{Z'P'PZ} = \mathbf{I}$, whereas $\tilde{\mathbf{T}}'\tilde{\mathbf{T}} = \mathbf{Z'T'TZ} = \mathbf{Z'\Lambda}_R\mathbf{Z}$, which is not diagonal. Hence, the columns of $\tilde{\mathbf{T}}$ are not orthogonal any more.

Restriction on T

The restriction on the columns of $\mathbf{T}$ to be in the column-space of $\mathbf{X}$ is not active. Regardless of $\mathbf{P}$, $\mathbf{T}$ should minimize $\|\mathbf{X} - \mathbf{TP'}\|^2$. This is a simple regression step, which results in $\mathbf{T} = \mathbf{XP}(\mathbf{P'P})^{-1}$. Hence, each column of $\mathbf{T}$ is automatically in the column space of $\mathbf{X}$.

Appendix 3.B: PLS Algorithms

The partial least squares regression algorithms.

Orthogonal t vectors version

1. $\mathbf{w}_1 = \mathbf{X}^T\mathbf{y}/\|\mathbf{X}^T\mathbf{y}\|$
2. $\mathbf{t}_1 = \mathbf{X}\mathbf{w}_1$
3. $\mathbf{p}_1 = \mathbf{X}^T\mathbf{t}_1/(\mathbf{t}_1^T\mathbf{t}_1)$
4. $\mathbf{E}^{(1)} = \mathbf{X} - \mathbf{t}_1\mathbf{p}_1^T$
5. $\mathbf{w}_2 = \mathbf{E}^{(1)^T}\mathbf{y}^{(1)}/\|\mathbf{E}^{(1)^T}\mathbf{y}^{(1)}\|$ (3.45)
6. $\mathbf{t}_2 = \mathbf{E}^{(1)}\mathbf{w}_2$
7. $\mathbf{p}_2 = \mathbf{E}^{(1)^T}\mathbf{t}_2/(\mathbf{t}_2^T\mathbf{t}_2)$
8. $\mathbf{E}^{(2)} = \mathbf{E}^{(1)} - \mathbf{t}_2\mathbf{p}_2^T$

$\ldots$

until R components

Step 3 in Equation (3.45) (the deflation of $\mathbf{X}$) is a regression step where $\mathbf{X}$ is regressed on $\mathbf{t}_1$. Hence, the residuals of that regression ($\mathbf{E}^{(1)}$) are orthogonal to $\mathbf{t}_1$. Because $\mathbf{t}_2$ is a linear combination of the columns of $\mathbf{E}^{(1)}$, it follows that $\mathbf{t}_2$ is orthogonal to $\mathbf{t}_1$.

Nonorthogonal t vectors version

1. $\mathbf{w}_1 = \mathbf{X}^T\mathbf{y}/\|\mathbf{X}^T\mathbf{y}\|$
2. $\mathbf{t}_1 = \mathbf{X}\mathbf{w}_1$
3. $\mathbf{E}^{(1)} = \mathbf{X} - \mathbf{t}_1\mathbf{w}_1^T$
4. $\mathbf{w}_2 = \mathbf{E}^{(1)^T}\mathbf{y}^{(1)}/\|\mathbf{E}^{(1)^T}\mathbf{y}^{(1)}\|$ (3.46)
5. $\mathbf{t}_2 = \mathbf{E}^{(1)}\mathbf{w}_2$
6. $\mathbf{E}^{(2)} = \mathbf{E}^{(1)} - \mathbf{t}_2\mathbf{w}_2^T$

$\ldots$

until R components

Note that the algorithms in Equations (3.45) and (3.46) are noniterative, because the $\mathbf{y}$ is univariate. For multivariate $\mathbf{Y}$, the PLS algorithms become iterative. There are alternative algorithms for partial least squares regression, depending on the size of $\mathbf{X}$ [De Jong & Ter Braak 1994, Lindgren *et al.* 1993]. Moreover, it is a matter of choice whether to deflate $\mathbf{X}$ and/or $\mathbf{y}$ [Burnham *et al.* 1996].

4

THREE-WAY COMPONENT AND REGRESSION MODELS

4.1 Historical Introduction to Multi-way Models

A short historical introduction of multi-way analysis is given here. Most of this historical work comes from psychometrics and the most important references related to the models presented in the following sections are given. The pioneering work began in the middle of the twentieth century and ended around 1980, when the most important multi-way models and their algorithms had been introduced.

Some of the first ideas on multi-way analysis were published by Raymond Cattell [1944,1952]. Thurstone's principle of parsimony states that a simple structure should be found to describe a data matrix or its correlation matrix with the help of factors [Thurstone 1935]. For the simultaneous analysis of several matrices together, Cattell proposed to use the principle of 'parallel proportional profiles' [Cattell 1944]. The principle of 'parallel proportional profiles' states that a set of common factors should be found that can be fitted with different dimension weights to many data matrices at the same time. This is the same as finding a common set of factors for a stack of matrices, a three-way array. To quote Cattell:

> *The principle of parsimony, it seems should not demand 'Which is the simplest set of factors reproducing this particular correlation matrix?' but rather 'Which set of factors will be most parsimonious at once to this and other matrices considered together?'*

Another quote states:

> *The criterion is then no longer that the rotation shall offer fewest factor loadings for any matrix; but that it shall offer fewest dissimilar (and therefore fewest total) loadings in all matrices together.*

This is one of the goals of three-way analysis.

Multi-way Analysis With Applications in the Chemical Sciences.   A. Smilde, R. Bro and P. Geladi
© 2004 John Wiley & Sons, Ltd   ISBN: 0-471-98691-7

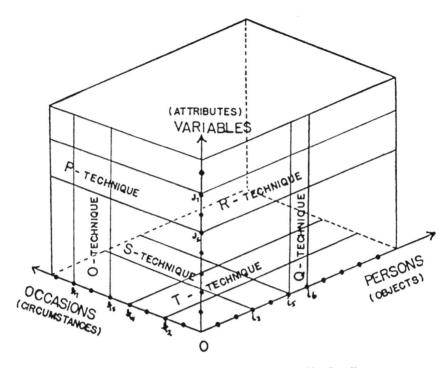

Figure 4.1. The first three-way array proposed by Cattell.

Cattell's most important historical multi-way paper is that in which he defines multi-way arrays [Cattell 1952]. He defines object, circumstance/time, attribute, scale and observer as the five ways for an idealized psychological multi-way array and quickly reduces this to person–attribute–circumstance three-way arrays for practical reasons. Figure 4.1 gives a simplified view of Cattell's first three-way array. Discussion of T, S, O, P, R and Q techniques is outside the scope of this book; the figure is only given as an illustration. A real multi-way solution for analyzing such arrays was not given in the paper.

A famous paper by Ledyard Tucker, based on a technical report from 1963, appeared in 1964 [Tucker 1964]. In this paper, decomposition of a three-way array into loading matrices **A**, **B** and **C** and a three-way core **G** is introduced with a hypothetical numerical example of 12 individuals, 9 traits and 5 raters. The nomenclature introduced by Tucker is still in use today. For example, the term 'mode' is introduced to avoid confusion with the term 'dimension'. A simplified representation of the arrays is given in Figure 4.2. An appendix of the paper contains an introduction to three-way array algebra. The mathematics of the first Tucker paper was not complete and a new paper in 1966 [Tucker 1966] was an attempt to correct this by introducing the Kronecker product. A Ph.D. thesis of Joseph Levin resulted in a paper in 1965 [Levin 1965] in which the similarity between the core of a Tucker decomposition and the singular values in a singular value decomposition is shown. The method first presented by Tucker is shown in a modified form as the 'Tucker3 analysis' in a later section of this chapter.

Another three-way model was introduced independently by Carroll & Chang [Carroll & Chang 1970] and by Harshman [Harshman 1970] in 1970. Carroll & Chang call their model CANDECOMP (Canonical Decomposition) while Harshman uses the name

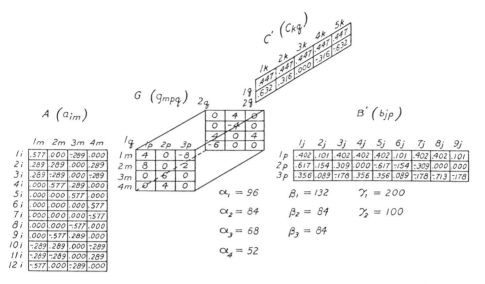

Figure 4.2. A graphical representation of the first three-way analysis as proposed by Tucker. Excerpt from *Contributions to Mathematical Psychology* by Norman Frederiksen and Harold Gulliksen, copyright ©1964 by Holt, Rinhart and Winston, reprinted by permission of the publisher.

PARAFAC (Parallel factor analysis). However, the model behind the two is the same. The basic idea of the model is to use the same factors to describe the variation in several matrices simultaneously albeit with different weighting coefficients for each matrix. This is exactly the idea behind 'parallel proportional profiles' of Cattell. In this book, the method is termed PARAFAC and it is presented in the following section. A three-way PARAFAC model consists of loading matrices **A**, **B** and **C** which have the same number of factors (Figure 4.3). The notation used in Figure 4.3 differs from the notation in this book, but Figure 4.3 is given only as an illustration. A PARAFAC model of a multi-way array usually yields *unique axes* (meaning that there is no rotational freedom in the orientation of the loading vectors), whereas a Tucker model only provides unique subspaces.

The principle of parallel proportional profiles has a natural chemical interpretation for curve resolution. Similar ideas evolved independently in chemistry initiated by Ho *et al.* in the late 1970s and early 1980s [Ho *et al.* 1978, Ho *et al.* 1980, Ho *et al.* 1981]. They developed the method of rank annihilation, which is close to the idea of a PARAFAC decomposition.

4.2 Models for Three-way One-block Data: Three-way Component Models

Parallel factor analysis (PARAFAC)

The PARAFAC model and the CANDECOMP model are closely related and will be abbreviated as PARAFAC models. The PARAFAC model is also known under the name trilinear

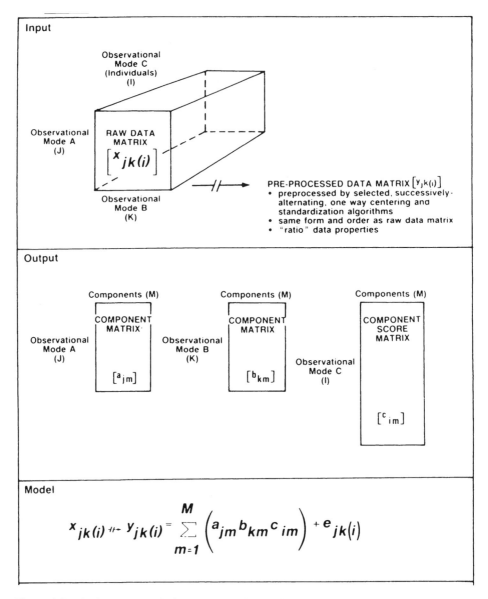

Figure 4.3. The three-way analysis as proposed by Harshman and Carroll & Chang. Compared to Figure 4.2 the core array is absent (actually, it is a unit superdiagonal core-array) and all matrices have the same number of factors. Reproduced by permission of Praeger Publishers, an imprint of GPG, Inc., W, CT, USA.

decomposition [Sanchez & Kowalski 1990]. The PARAFAC model is introduced below using different types of notation. The more usual types of notation (using summation and simultaneous components) are presented in this section, whereas the less usual types of notation (using Kronecker products, Khatri–Rao products and tensor products) are dealt with in Appendix 4.A.

The PARAFAC model is introduced here by generalizing the singular value decomposition. A two-way model of a matrix $\mathbf{X}$ ($I \times J$), with typical elements x_{ij}, based on a singular value decomposition truncated to R components reads in summation notation

$$x_{ij} = \sum_{r=1}^{R} a_{ir} g_{rr} b_{jr} + e_{ij}; i = 1, \ldots, I; j = 1, \ldots, J \tag{4.1}$$

where $\mathbf{A}$ ($I \times R$), with elements a_{ir}, and $\mathbf{B}$ ($J \times R$), with elements b_{jr}, are both orthogonal matrices, and $\mathbf{G} = \text{diag}(g_{11}, \ldots, g_{RR})$ is a diagonal matrix called the singular value or *core-matrix* containing the R largest singular values of $\mathbf{X}$ in decreasing order and e_{ij} are the residuals. Written in matrix notation, Equation (4.1) becomes

$$\mathbf{X} = \mathbf{AGB'} + \mathbf{E} \tag{4.2}$$

where $\mathbf{E}$ contains the elements e_{ij}. A change of notation has been made, in comparison with Section 3.1, to make the generalization easier. The values of g_{rr} can be absorbed in a_{ir}, b_{jr} or in both, that is a matter of choice. This leads to the model

$$\mathbf{X} = \mathbf{AB'} + \mathbf{E} \tag{4.3}$$

where the names have been maintained although $\mathbf{A}$ and $\mathbf{B}$ in Equations (4.2) and (4.3) will not be the same. This model may also be written

$$x_{ij} = \sum_{r=1}^{R} a_{ir} b_{jr} + e_{ij} \tag{4.4}$$

For a three-way array $\mathbf{X}$ ($I \times J \times K$), with elements x_{ijk}, Equation (4.4) generalizes to the PARAFAC model

$$x_{ijk} = \sum_{r=1}^{R} a_{ir} b_{jr} c_{kr} + e_{ijk} \tag{4.5}$$

where R is the number of components used in the PARAFAC model and e_{ijk} is a residual term containing all the unexplained variation. A pictorial description of this model is given in Figure 4.4. The model in Equation (4.5) is a trilinear model: fixing two sets of parameters (e.g. as and bs), x_{ijk} is expressed as a linear function of the remaining parameters (e.g. the cs). The model in Equation (4.5) can also be re-expressed in line with Equation (4.1) by introducing a scaling parameter g_{rrr}:

$$x_{ijk} = \sum_{r=1}^{R} g_{rrr} a_{ir} b_{jr} c_{kr} + e_{ijk} \tag{4.6}$$

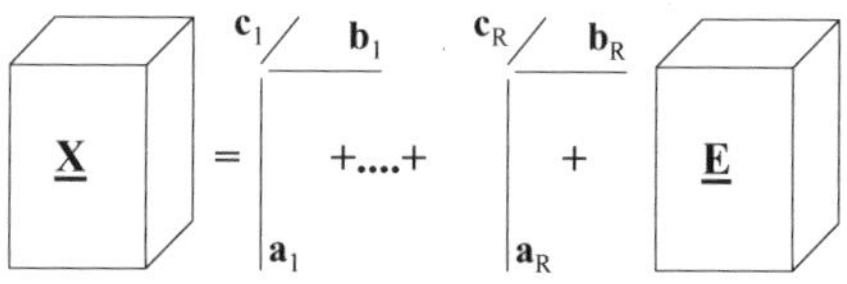

Figure 4.4. The PARAFAC model with R components.

EXAMPLE 4.1

Example of a PARAFAC model

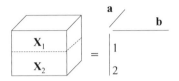

Figure 4.5. A PARAFAC model of $\underline{\mathbf{X}}$.

Data that ideally behave according to a PARAFAC model are, e.g., hyphenated data such as chromatography–ultraviolet data. A sample containing 1 mol/l of an analyte absorbing in the ultraviolet (UV) range is injected into a chromatograph and detected using diode-array UV detection. The result is a measurement matrix $\mathbf{X}_1$ ($I \times J$), where I is the number of time points at which a UV spectrum is taken and J is the number of UV wavelengths. A model for $\mathbf{X}_1$ is $\mathbf{X}_1 = \mathbf{ab}'$, where $\mathbf{a}$ ($I \times 1$) is the chromatogram of the absorbing analyte and $\mathbf{b}$ ($J \times 1$) is its UV spectrum. For simplicity, the residual term is dropped in the equation. If the same analyte is measured under the same conditions at a concentration of 2 mol/l, this results in $\mathbf{X}_2 = 2\mathbf{ab}'$. The matrices $\mathbf{X}_1$ and $\mathbf{X}_2$ can be stacked on top of each other thereby obtaining the three-way array $\underline{\mathbf{X}}$ ($I \times J \times 2$). A model for the variation in $\underline{\mathbf{X}}$ is given in Figure 4.5. The factor underlying the variation in the third mode of $\underline{\mathbf{X}}$ is the concentration of the absorbing analyte. The properties of the analyte (that is, its $\mathbf{a}$ and $\mathbf{b}$) affect both objects in a synchronous way, the only difference being a proportional difference. This leads to a PARAFAC model. For the above to hold, there must be no significant retention time shifts. If the retention time shifts, then it means that the elution profiles will change from experiment to experiment. The PARAFAC model cannot handle such deviations from the underlying model.

Collect the elements a_{ir} in $\mathbf{A}$ ($I \times R$); b_{jr} in $\mathbf{B}$ ($J \times R$) and c_{kr} in $\mathbf{C}$ ($K \times R$). Then Equation (4.5) can be written in terms of matrices $\mathbf{X}_k$, where $\mathbf{X}_k$ is the kth ($I \times J$) slice of the three-way array $\underline{\mathbf{X}}$ ($I \times J \times K$), i.e. $\mathbf{X}_k$ has typical elements x_{ijk} ($i = 1, \ldots, I; j = 1, \ldots, J$ for each k):

$$\mathbf{X}_k = \mathbf{A}\mathbf{D}_k\mathbf{B}' + \mathbf{E}_k = c_{k1}\mathbf{a}_1\mathbf{b}_1' + \cdots + c_{kR}\mathbf{a}_R\mathbf{b}_R' + \mathbf{E}_k \qquad (4.7)$$

where $\mathbf{D}_k$ is a diagonal matrix with the kth row of $\mathbf{C}$ on its diagonal (elements $c_{k1}, \ldots, c_{kR}$); $\mathbf{a}_r$, $\mathbf{b}_r$ are the rth columns of $\mathbf{A}$ and $\mathbf{B}$, respectively and the residual term $\mathbf{E}_k$ ($I \times J$) is defined similarly to $\mathbf{X}_k$. Hence, each $\mathbf{X}_k$ is modeled using the same components $\mathbf{A}$ and $\mathbf{B}$, but with different weights, represented by $\mathbf{D}_k$. A pictorial representation of Equation (4.7) is shown in Figure 4.6.

Across all slices $\mathbf{X}_k$, the components $\mathbf{a}_r$ and $\mathbf{b}_r$ remain the same, only their weights $d_{k1}, \ldots, d_{kR}$ are different. Hence, all slices $\mathbf{X}_k$ are modeled with parallel and proportional profiles $d_{k1}\mathbf{a}_1\mathbf{b}_1', \ldots, d_{kR}\mathbf{a}_R\mathbf{b}_R'$. This allows for writing the model using a simultaneous components notation. There are three fully equivalent ways of writing the PARAFAC model

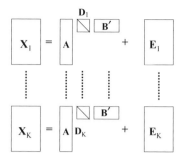

Figure 4.6. Simultaneous component notation of a PARAFAC model.

in the simultaneous components notation due to the symmetry of model (4.7). These are written in terms of frontal, horizontal, and vertical slabs respectively:

$$\mathbf{X}_k = \mathbf{A}\mathbf{D}_k(\mathbf{C})\mathbf{B}',\ k = 1, \ldots, K$$
$$\mathbf{X}_i = \mathbf{B}\mathbf{D}_i(\mathbf{A})\mathbf{C}',\ i = 1, \ldots, I \qquad (4.8)$$
$$\mathbf{X}_j = \mathbf{A}\mathbf{D}_j(\mathbf{B})\mathbf{C}',\ j = 1, \ldots, J$$

where the diagonal matrices $\mathbf{D}_k(\mathbf{C})$, $\mathbf{D}_i(\mathbf{A})$ or $\mathbf{D}_j(\mathbf{B})$ are to be interpreted as operators extracting the kth or ith or jth row of the appropriate loading matrix ($\mathbf{C}$, $\mathbf{A}$, and $\mathbf{B}$ respectively).

EXAMPLE 4.2

Hyphenated data in different PARAFAC notations

Following the example earlier on hyphenated data, assume that two analytes are present in samples $\mathbf{X}_1$ and $\mathbf{X}_2$, both of size $I \times J$. Let $\mathbf{a}_1$ and $\mathbf{a}_2$ represent the chromatograms of the analytes 1 and 2, respectively, and $\mathbf{b}_1$, $\mathbf{b}_2$ their respective UV spectra. Assume further that analyte 1 is present at 1 mol/l and 2 mol/l in $\mathbf{X}_1$ and $\mathbf{X}_2$, respectively and analyte 2 is present at 3 mol/l and 4 mol/l in $\mathbf{X}_1$ and $\mathbf{X}_2$, respectively. The data set $\underline{\mathbf{X}}$, consisting of the slices $\mathbf{X}_1$ and $\mathbf{X}_2$ can be modeled according to Figure 4.7 (for convenience the residual term is dropped). It is also possible to write the PARAFAC model of the hyphenated data in a simultaneous component form. This is shown in Figure 4.8 (again the error term is dropped). Figures 4.7 and 4.8 show that the profiles $\mathbf{a}$ and $\mathbf{b}$ remain the same in both samples $\mathbf{X}_1$ and $\mathbf{X}_2$, which is due to the properties of the hyphenated data (the shapes of the chromatograms and UV spectra are independent of the concentrations of the analytes).

The different expressions for the PARAFAC model of Equation (4.5) are given because in different situations some are more convenient than others. In Appendix 4.A additional ways of writing a PARAFAC model are given.

Under mild conditions the PARAFAC model gives unique estimates (up to permutation, sign and scaling indeterminacy); that is, the calculated $\mathbf{A, B}$ and $\mathbf{C}$ cannot be changed without changing the residuals (no rotational freedom). The details of this property are treated in Section 5.2. This property is also called the intrinsic axes property because with

PARAFAC not only unique subspaces (as in principal component analysis) but also unique basis vector orientations are found.

The parameters in **A**, **B** and **C** can be estimated with different algorithms; these will be treated in Chapter 6. The factors have to be estimated simultaneously, as opposed to e.g. PCA, where the components can be estimated one at a time. This is so because the components in PARAFAC are not orthogonal and hence not independent of each other. Estimating PARAFAC components sequentially (using a deflation step as in some PCA algorithms; see Section 3.1) gives different results compared to estimating the components simultaneously, and the sequential approach does not give the least squares solution (see Example 4.3).

EXAMPLE 4.3

The difference between sequential and simultaneous fitting

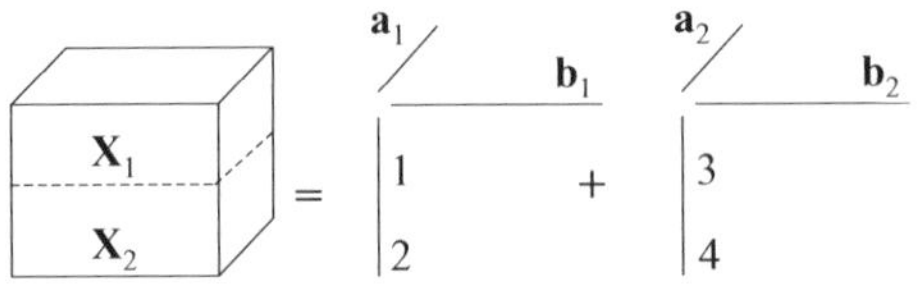

Figure 4.7. A PARAFAC model of hyphenated data.

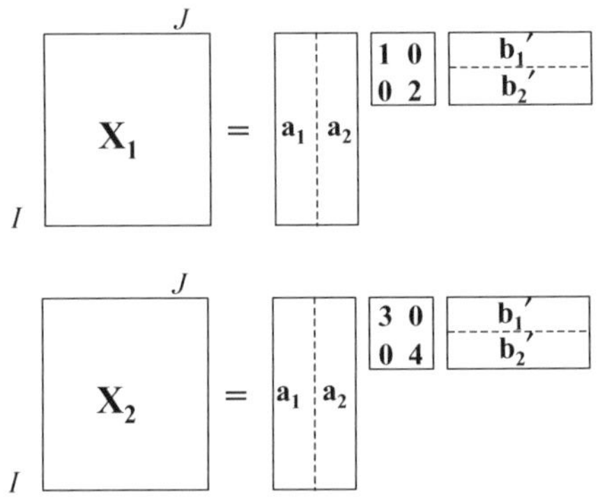

Figure 4.8. Simultaneous components notation of the PARAFAC model of the hyphenated data.

Let

$$A = \begin{bmatrix} 1 & 2 \\ 1 & 1 \end{bmatrix}, \quad B = \begin{bmatrix} 2 & 3 \\ 2 & 1 \end{bmatrix}, \quad \text{and} \quad C = \begin{bmatrix} 4 & 1 \\ 1 & 2 \end{bmatrix}$$

and let the $2 \times 2 \times 2$ three-way array $\underline{X}$ be defined as $\underline{X} = A(C \odot B)'$. Hence

$$X = \begin{bmatrix} 14 & 10 & 14 & 6 \\ 11 & 9 & 8 & 4 \end{bmatrix}$$

Fitting a two-component PARAFAC model using a least squares simultaneous algorithm provides (unique) estimates of **A**, **B**, and **C** that give a perfect model of **X**

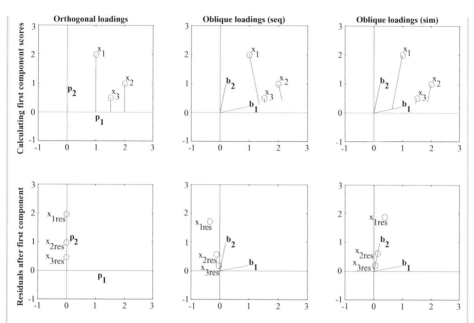

Figure 4.9. A data matrix **X** represented in the two-dimensional variable space. Top plots show how the scores of the first component are calculated using different loading vectors. The bottom plots show the residuals after subtracting the first component from the data.

(zero residuals). Calculating one component at a time; each time from the residuals of the data not yet explained; a two-component model is obtained with the following residuals:

$$\mathbf{X} - \hat{\mathbf{X}} = \begin{bmatrix} -0.36 & -0.82 & 0.99 & 0.19 \\ 0.52 & 1.18 & -1.43 & -0.28 \end{bmatrix}$$

Clearly, these residuals are far from perfect, and thus, even though the data are known to be perfectly trilinear with two components, the sequential PARAFAC algorithm fails to find a reasonable estimate of the parameters. However, this difference between sequential and simultaneous fitting is not related to the three-way nature of the PARAFAC model. Rather it is the orthogonality of the components in principal component analysis that enables the components to be calculated sequentially. A simple two-way example will help in illustrating this.

Consider a 3×2 two-way matrix **X** with elements:

$$\mathbf{X} = \begin{bmatrix} 1 & 2 \\ 2 & 1 \\ 1.5 & 0.5 \end{bmatrix}$$

This matrix can be considered as three points (rows) in a two-dimensional space. This two-dimensional space can be represented by the two vectors $\mathbf{p}_1 = [1\ 0]$ and $\mathbf{p}_2 = [0\ 1]$ (Figure 4.9 top left). Using such orthogonal loadings, it is immaterial whether the scores are calculated one at a time or simultaneously. In the top left plot, the scores of the first component are shown. After subtracting the first component the residual variation

$(\mathbf{X}_{res} = \mathbf{X} - \mathbf{t}_1 \mathbf{p}_1' = \mathbf{X} - \mathbf{X}\mathbf{p}_1 \mathbf{p}_1')$ is only within the variation of the second component $(\mathbf{p}_2)$ (see bottom left plot). However, if the loadings are correlated as is the case for $\mathbf{b}_1 = [.98 \,.20]$ and $\mathbf{b}_2 = [.20 \,.98]$, then it makes a difference whether the scores are calculated one at a time. This is shown in the top and bottom middle plots of Figure 4.9: first regressing only onto $\mathbf{b}_1$ (top middle) the scores obtained are not correct, and the residuals after subtracting the first component (bottom middle) are not within the space spanned by the second component $(\mathbf{b}_2)$. This aspect of having oblique components is what causes the sequential fitting of the PARAFAC model to fail. If the scores are calculated correctly from both loading vectors simultaneously as $[\mathbf{a}_1 \, \mathbf{a}_2] = \mathbf{X}([\mathbf{b}_1 \, \mathbf{b}_2])^+$ then the subsequent residual from subtracting the first component $(\mathbf{X}_{res} = \mathbf{X} - \mathbf{a}_1 \mathbf{b}_1')$ is only within the space spanned by $\mathbf{b}_2$ (right-most plots). Therefore, unless the components are orthogonal in at least two of the modes, sequential fitting of the PARAFAC model will lead to less well-fitting models than the simultaneously fitted model [Leurgans & Ross 1992].

Tucker models

Ledyard Tucker was one of the pioneers in multi-way analysis. He proposed [Tucker 1964, Tucker 1966] a series of models nowadays called N-mode principal component analysis or Tucker models. An extensive treatment of Tucker models is given by Kroonenberg and de Leeuw [1980] and Kroonenberg [1983]. In the following, three different Tucker models will be treated.

TUCKER3 MODELS

Starting from Equation (4.1), a possible generalization of the two-way two-mode component model is to use a non-diagonal core-matrix $\tilde{\mathbf{G}}$, which results in the following model:

$$x_{ij} = \sum_{p=1}^{P} \sum_{q=1}^{Q} a_{ip} \tilde{g}_{pq} b_{iq} + e_{ij} \qquad (4.9)$$

or, in matrix notation,

$$\mathbf{X} = \mathbf{A}\tilde{\mathbf{G}}\mathbf{B}' + \mathbf{E} \qquad (4.10)$$

where the '$\sim$' on top of the $\mathbf{G}$ is used to indicate the different core matrices of Equations (4.10) and (4.2), and $\tilde{g}_{pq}$ is an element of $\tilde{\mathbf{G}}$. Unlike the model of Equation (4.1), there is no requirement in model (4.9) to have the same number of components in $\mathbf{A}$ and $\mathbf{B}$. By letting p and q run to different P and Q and letting $\tilde{\mathbf{G}}$ be of size $P \times Q$, the number of components can differ in the two modes. A nondiagonal core matrix explicitly means that in the model there are *interactions* between factors (see Elaboration 4.1 for the significance of this in three-way modeling). This is an important property of the Tucker models in general. In traditional principal component analysis, loading vectors only interact factorwise. For example, the second score vector interacts with the second loading vector by a magnitude defined be the second singular value. In model (4.9) all vectors can interact. For example, the first score vector interacts with the third loading vector with a magnitude defined by the element $\tilde{g}_{13}$. For a physical interpretation of the model, the concept of interaction of

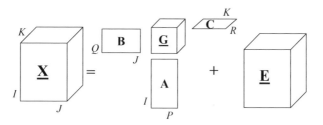

Figure 4.10. Pictorial representation of the Tucker3 model.

factors may be difficult to grasp (see Elaboration 4.3 for an idealized interpretation), but on a mathematical level, such a model has several attractive features with respect to three-way modeling.

Equation (4.9) can be generalized to a three-way array $\underline{\mathbf{X}}$, with elements x_{ijk},

$$x_{ijk} = \sum_{p=1}^{P}\sum_{q=1}^{Q}\sum_{r=1}^{R} a_{ip}b_{jq}c_{kr}g_{pqr} + e_{ijk} \qquad (4.11)$$

where e_{ijk} is an element of $\underline{\mathbf{E}}\,(I \times J \times K)$; a_{ip}, b_{jq} and c_{kr} are the typical elements of the loading matrices $\mathbf{A}\,(I \times P)$, $\mathbf{B}\,(J \times Q)$ and $\mathbf{C}\,(K \times R)$; and g_{pqr} is the typical element of the core-array $\underline{\mathbf{G}}\,(P \times Q \times R)$. This is the (P,Q,R) Tucker3 model of $\underline{\mathbf{X}}$, where the notation (P,Q,R) is used to indicate that the model has P, Q and R factors in the three different modes. A pictorial representation of the Tucker3 model is given in Figure 4.10.

ELABORATION **4.1**

Tucker3 models cannot be rotated to PARAFAC models

Consider a two-way array $\mathbf{X}\,(I \times J)$ for which the following model is available

$$\mathbf{X} = \mathbf{AGB}' + \mathbf{E}$$

where $\mathbf{A}$ is an $I \times R$ loading matrix for the first mode; $\mathbf{B}$ is a $J \times R$ loading matrix of the second mode; and $\mathbf{G}$ is an $R \times R$ core-matrix, not necessarily diagonal. Assume without loss of generality that $\mathbf{A}'\mathbf{A} = \mathbf{B}'\mathbf{B} = \mathbf{I}$. This model might seem more general than a PCA model of $\mathbf{X}$. However, $\mathbf{G}$ can be decomposed as $\mathbf{G} = \mathbf{USV}'$, with $\mathbf{U}\,(R \times R)$, $\mathbf{U}'\mathbf{U} = \mathbf{UU}' = \mathbf{I}$; $\mathbf{V}\,(R \times R)$, $\mathbf{V}'\mathbf{V} = \mathbf{VV}' = \mathbf{I}$; and $\mathbf{S}$ is diagonal. Hence, the model becomes

$$\mathbf{X} = \mathbf{AUSV}'\mathbf{B}' + \mathbf{E} = \tilde{\mathbf{A}}\mathbf{S}\tilde{\mathbf{B}}' + \mathbf{E}$$

where $\tilde{\mathbf{A}}$ and $\tilde{\mathbf{B}}$ are defined implicitly and both matrices are again orthogonal. Therefore, the model reduces to the ordinary rank-reduced SVD of $\mathbf{X}$ and nothing is gained in terms of fit by using a nonzero off-diagonal $\mathbf{G}$ because this $\mathbf{G}$ can always be rotated to a diagonal form without loosing fit.[1]

The PARAFAC model is identical to a Tucker3 model where the core array has been fixed such that all off-superdiagonal elements are zero (see Equation (4.6)). The (R,R,R)

---

[1] The case of a general $\mathbf{G}$ of size $(P \times Q)$ can be treated in a similar way.

Tucker3 model has nonzero off-diagonal elements in its core. Although the Tucker3 model has rotational freedom, it is not, in general, possible to rotate the Tucker3 core-array to a superdiagonal form and thereby obtain a PARAFAC model. Hence, models with and without diagonal core generalize to two different three-way models. This is one of the remarkable differences between two-way analysis and three-way analysis.

In Kronecker product notation the Tucker3 model reads

$$\mathbf{X} = \mathbf{AG(C \otimes B)'} + \mathbf{E} \tag{4.12}$$

where $\mathbf{X} = [\mathbf{X}_1\, \mathbf{X}_2 \cdots \mathbf{X}_K]$ is an $(I \times JK)$ matrix with $\mathbf{X}_k$ defined as above Equation (4.7), $\mathbf{E}$ defined similarly and $\mathbf{G} = [\mathbf{G}_1\, \mathbf{G}_2 \ldots \mathbf{G}_R]$ is the matricized core-array $\underline{\mathbf{G}}$ of size $(P \times QR)$ where $\mathbf{G}_r$ is the rth slab of size $(P \times Q)$ of $\underline{\mathbf{G}}$. This notation makes use of the concept of matricizing three-way arrays, as explained in Section 2.3 (see also Elaboration 4.2).

ELABORATION **4.2**

Using Kronecker multiplication for Tucker3 models

To explain the use of the Kronecker multiplication in the Tucker3 model, first observe that a two-way two-mode component model with nondiagonal core can be written

$$\mathbf{X} = \mathbf{AGB'} + \mathbf{E}$$

In this model, there are interactions between the components. For example, the first column of $\mathbf{A}$ called $\mathbf{a}_1$ will interact with the second column of $\mathbf{B}$ called $\mathbf{b}_2$ with the level of interaction given by g_{12}. If this model is vectorized, i.e. simply rearranged without changing the model, it can be written as

$$\text{vec}\,\mathbf{X} = \text{vec}(\mathbf{AGB'}) + \text{vec}\mathbf{E} = (\mathbf{B} \otimes \mathbf{A})(\text{vec}\,\mathbf{G}) + \text{vec}\mathbf{E}$$

Thus, by using the Kronecker product, it is possible to express the structural model within these two modes in one combined mode. This is exactly what is used in the three-way Tucker3 model. The second and third modes are vectorized for all rows, leading to the model in Equation (4.12).

Due to the symmetry of the Tucker3 model there are three alternative ways of writing the same model in Kronecker notation in terms of either frontal, horizontal or vertical slabs. These are given in Appendix 4.B.

ELABORATION **4.3**

Deriving a Tucker model from an assumed latent structure

There are different ways to derive the Tucker3 model. One way is nicely illustrated by Kroonenberg [1984]. Assume that a three-way array $\underline{\mathbf{X}}\,(I \times J \times K)$ is available with elements x_{ijk} representing the measurement of the jth variable on the ith object at the kth occasion. Suppose that two idealized objects are present, $\mathbf{p}_1$ and $\mathbf{p}_2$; two latent variables are present, $\mathbf{q}_1$ and $\mathbf{q}_2$; and two prototype occasions are present, $\mathbf{r}_1$ and $\mathbf{r}_2$. The scores of all the idealized objects on the latent variables at the prototype occasions are also available: g_{111} is the score of idealized object 1 on latent variable 1 at prototype occasion 1 and g_{211}, g_{121}, g_{221}, g_{112}, g_{212}, g_{122}, g_{222} are similarly defined and available. All the vectors $\mathbf{p}$, $\mathbf{q}$, $\mathbf{r}$ and $\mathbf{g}$ represent a latent structure.

A central idea of latent variable models is that the measurements (often termed scores in psychological literature) are linear combinations of the latent variables (see also Chapter 2). Using this idea, the measurement of real object i on the real variable j at the real occasion k can be constructed. This proceeds in steps. The measurement of real object i on the latent variable $\mathbf{q}_1$ under prototype occasion $\mathbf{r}_1$ is a linear combination of the measurements on the idealized objects $\mathbf{p}_1$ and $\mathbf{p}_2$, using weights a_{ip_1} and a_{ip_2}, then

$$s_{iq_1r_1} = a_{ip_1}g_{111} + a_{ip_2}g_{211}$$

or, in shorthand notation (which will be used from now on),

$$s_{i11} = a_{i1}g_{111} + a_{i2}g_{211}$$

and s_{i11} represents the measurement of real object i on latent variable 1 at prototype occasion 1. Similarly,

$$s_{i21} = a_{i1}g_{121} + a_{i2}g_{221}$$
$$s_{i12} = a_{i1}g_{112} + a_{i2}g_{212}$$
$$s_{i22} = a_{i1}g_{122} + a_{i2}g_{222}$$

where it is important to note that the weights a_{i1} and a_{i2} are independent of the latent variables and prototype occasions. All interrelationships between objects, variables and occasions are reflected in the gs.

The next step is to construct the measurements on real object i on real variable j, instead of on the latent variables $\mathbf{q}_1$ and $\mathbf{q}_2$. The idea is again to use linear combinations,

$$v_{ij1} = b_{j1}s_{i11} + b_{j2}s_{i21}$$
$$v_{ij2} = b_{j1}s_{i12} + b_{j2}s_{i22}$$

where again the weights b_{j1} and b_{j2} determine to what extent the latent variables determine the real variable j.

The final step is to combine idealized occasions using weights c_{k1} and c_{k2},

$$x_{ijk} = c_{k1}v_{ij1} + c_{k2}v_{ij2}$$

and further substituting and rearrangements gives

$$x_{ijk} = \sum_{r=1}^{2} c_{kr} \left[\sum_{q=1}^{2} b_{jq} \left(\sum_{p=1}^{2} a_{ip}g_{pqr} \right) \right]$$

or

$$x_{ijk} = \sum_{p=1}^{2}\sum_{q=1}^{2}\sum_{r=1}^{2} a_{ip}b_{jq}c_{kr}g_{pqr}$$

Hence, ideas of latent variables that express themselves through linear combinations in the real three-way array $\underline{\mathbf{X}}$ lead naturally to a Tucker3 model.

EXAMPLE 4.4

Environmental Tucker3 example

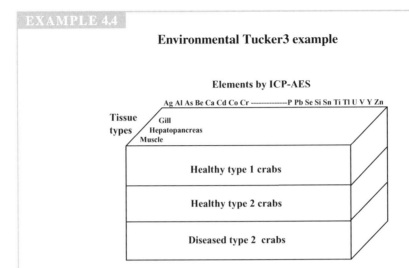

Figure 4.11. ICP–AES study of crabs.

Environmental studies are very complex and often no hard model is known. Studying these systems is very expensive and extends over a long period of time. Two-way and three-way arrays are built as the study progresses and one hopes to find some kind of latent variables that are useful for interpretation.

A typical study is that of Gemperline *et al.* [1992]. They measured a large number of elements by ICP-AES in tissues of crabs from three crab populations. The three-way array becomes individuals × elements × tissues (see Figure 4.11). The true model of how certain elements are biochemically related to different organs of diseased and healthy crabs is quite difficult to figure out, but Gemperline *et al.* managed to cluster groups of elements and to relate these to crab populations and affected organs just by making plots of the results of a Tucker3 model. The latent variables related to the diseased crabs contained trace elements showing the influence of mining on the water. It was also found that these elements are found mainly in the gill. This example is also discussed in Chapters 8 and 10.

PROPERTIES OF THE TUCKER3 MODEL

The Tucker3 model has rotational freedom. This can be seen by writing

$$
\begin{aligned}
\mathbf{X} &= \mathbf{AG}(\mathbf{C}' \otimes \mathbf{B}') + \mathbf{E} \\
&= \mathbf{A T_A T_A^{-1} G}(\mathbf{C}' \otimes \mathbf{B}') + \mathbf{E} \\
&= \tilde{\mathbf{A}}\tilde{\mathbf{G}}(\mathbf{C}' \otimes \mathbf{B}') + \mathbf{E}
\end{aligned}
\tag{4.13}
$$

where $\mathbf{T_A}$ is an arbitrary nonsingular matrix; $\tilde{\mathbf{A}} = \mathbf{A T_A}$ and $\tilde{\mathbf{G}} = \mathbf{T_A^{-1} G}$. Such a transformation of the loading matrix $\mathbf{A}$ can be defined similarly for $\mathbf{B}$ and $\mathbf{C}$, using $\mathbf{T_B}$ and $\mathbf{T_C}$, respectively. Hence, because there is rotational freedom, orthogonality of the component matrices can be obtained at no cost in fit by defining proper matrices $\mathbf{T_A}$, $\mathbf{T_B}$ and $\mathbf{T_C}$. The Tucker3 model, therefore, does not give unique component matrices because it has rotational freedom.

It is convenient to make the component matrices orthogonal, i.e. $\mathbf{A'A} = \mathbf{B'B} = \mathbf{C'C} = \mathbf{I}$. This allows for an easy interpretation of the elements of the core-array and of the loadings by the loading plots. The sum of the squared elements of the core-array associated with the combination of certain factors then represents the amount of variation explained by that combination of factors in the different modes. If $\underline{\mathbf{X}}$ ($I \times J \times K$) is the three-way array modeled by a (P,Q,R) Tucker3 model[2] and if $\hat{\underline{\mathbf{X}}}$ represents the fitted part of $\underline{\mathbf{X}}$, then the following holds [Kroonenberg 1984],

$$\|\underline{\mathbf{X}}\|^2 = \|\hat{\underline{\mathbf{X}}}\|^2 + \|\underline{\mathbf{E}}\|^2$$
$$\|\hat{\underline{\mathbf{X}}}\|^2 = \sum_{p=1}^{P} \sum_{q=1}^{Q} \sum_{r=1}^{R} g_{pqr}^2 \qquad (4.14)$$

where $\underline{\mathbf{E}}$ is the array of residuals. In words, Equation (4.14) means that the variation in $\underline{\mathbf{X}}$ is divided into unexplained variation and variation explained by the model. Furthermore, the fitted sum of squares can be divided in parts related to each combination of components in the different directions.

The rotational freedom of Tucker3 models can also be used to rotate the core-array to a simple structure as is also common in two-way analysis. This will be dealt with in Section 6.8. Imposing the restrictions $\mathbf{A'A} = \mathbf{B'B} = \mathbf{C'C} = \mathbf{I}$ is not sufficient for obtaining a unique solution. To obtain uniqe estimates of the parameters, not only the loading matrices should be orthogonal but $\mathbf{A}$ should also contain the eigenvectors of $\mathbf{X(CC' \otimes BB')X'}$ corresponding to decreasing eigenvalues of that same matrix; and similar restrictions should be put on $\mathbf{B}$ and $\mathbf{C}$ [De Lathauwer 1997, Kroonenberg et al. 1989].

There are relations between the Tucker3 model and the PARAFAC model. These will be dealt with in Chapter 5. Algorithms to calculate the parameters in a Tucker3 model will be discussed in Chapter 6.

ELABORATION **4.4**

Interpretation of Tucker models in terms of a new basis

There is a convenient interpretation of the Tucker model in terms of a new basis to express a three-way array. Consider the SVD of $\mathbf{X}$: $\mathbf{X} = \mathbf{USV'}$. The matrix $\mathbf{V}$ is an orthonormal basis for the row-space of $\mathbf{X}$. $\mathbf{X}$ can be expressed on this new basis by orthogonally projecting $\mathbf{X}$ on this basis; the new coordinates are found by regressing $\mathbf{X}$ on $\mathbf{V}$:

$$\mathbf{X} = \mathbf{TV'} \Rightarrow \mathbf{T} = \mathbf{XV(V'V)}^{-1} = \mathbf{XV} = \mathbf{USV'V} = \mathbf{US}$$

using the orthogonality of $\mathbf{V}$. The matrix $\mathbf{T}$ now contains the coordinates of $\mathbf{X}$ on the new basis $\mathbf{V}$. Likewise, the matrix $\mathbf{U}$ represents an orthonormal basis for the column-space of $\mathbf{X}$, and $\mathbf{X}$ can be expressed with respect to this basis as

$$\mathbf{X} = \mathbf{UP} \Rightarrow \mathbf{P} = \mathbf{U'X} = \mathbf{U'USV'} = \mathbf{SV'}$$

using the orthogonality of $\mathbf{U}$. The matrix $\mathbf{P}$ contains the coordinates on the new basis. It is also possible to express $\mathbf{X}$ on both bases $\mathbf{U}$ and $\mathbf{V}$ simultaneously. The new coordinates of $\mathbf{X}$ then become

---

[2] That is, a Tucker3 model with P components in the first mode, Q components in the second mode, and R in third mode.

$$\mathbf{X} = \mathbf{UDV'} \Rightarrow \mathbf{U'XV} = \mathbf{D} = \mathbf{S}$$

where $\mathbf{D}$ contains the new coordinates and equals $\mathbf{S}$. This shows an interpretation of the matrix $\mathbf{S}$ in the SVD of $\mathbf{X}$: it contains the coordinates of $\mathbf{X}$ expressed on the new bases $\mathbf{U}$ and $\mathbf{V}$. Due to the special structure of the bases $\mathbf{U}$ and $\mathbf{V}$, the new coordinates are very simple: a diagonal matrix. Note that all the variation present in $\mathbf{X}$ is 'transported' to $\mathbf{S}$

$$\|\mathbf{X}\|^2 = \|\mathbf{S}\|^2$$

which is a well-known property. The reason is that orthogonal transformations do no not change the scale (variation) of the data. Hence, variation is moved to the new coordinates ($\mathbf{S}$).

A similar interpretation is possible for Tucker3 models. If $\underline{\mathbf{X}}$ can be written exactly as

$$\mathbf{X} = \mathbf{AG}(\mathbf{C'} \otimes \mathbf{B'})$$

with orthogonal $\mathbf{A}$, $\mathbf{B}$ and $\mathbf{C}$ then

$$\mathbf{G} = \mathbf{A'X}(\mathbf{C} \otimes \mathbf{B})$$

as can be verified by matrix multiplication. The core-array $\mathbf{G}$ can be interpreted as new coordinates of $\mathbf{X}$ on the (possibly truncated) bases $\mathbf{A}$, $\mathbf{B}$ and $\mathbf{C}$. All the variation in $\mathbf{X}$ is transported to $\mathbf{G}$, $\|\mathbf{X}\|^2 = \|\mathbf{G}\|^2$.

TUCKER2 MODELS

In a (P,Q,R) Tucker3 model of a given array $\underline{\mathbf{X}}$ ($I \times J \times K$) all three modes are reduced, that is, usually $P < I$, $Q < J$ and $R < K$. There also exist models where only two of the three modes are reduced, which therefore are called *Tucker2* models. This gives rise to three special models, depending on which mode is not reduced. Suppose that a Tucker3 model is made for $\underline{\mathbf{X}}$ ($I \times J \times K$) but $\mathbf{C}$ is chosen to be the identity matrix $\mathbf{I}$, of size $K \times K$. Hence, there is no reduction sought in the third mode because the basis is not changed. Then this model, the *Tucker2 model*, can be written as

$$\mathbf{X}_{(K \times IJ)} = \mathbf{IG}_{(K \times PQ)}(\mathbf{B} \otimes \mathbf{A})' + \mathbf{E} = \mathbf{G}_{(K \times PQ)}(\mathbf{B} \otimes \mathbf{A})' + \mathbf{E} \quad (4.15)$$

where $\mathbf{G}_{(K \times PQ)}$ is now a ($K \times PQ$) matrix and this is the properly matricized version of the *extended* core-array $\underline{\mathbf{G}}$ ($P \times Q \times K$); $\mathbf{X}_{(K \times IJ)}$ is the properly matricized version of $\underline{\mathbf{X}}$ (see Section 2.3). In summation notation the Tucker2 model is

$$x_{ijk} = \sum_{p=1}^{P} \sum_{q=1}^{Q} a_{ip} b_{jq} g_{pqk} + e_{ijk} \quad (4.16)$$

where x_{ijk}, a_{ip}, b_{jq} are elements of $\underline{\mathbf{X}}$ ($I \times J \times K$), $\mathbf{A}$ ($I \times P$) and $\mathbf{B}$ ($J \times Q$), respectively. Moreover, g_{pqk} is an element of the three-way extended core-array $\underline{\mathbf{G}}$ ($P \times Q \times K$). Comparing Equation (4.16) with Equation (4.11) shows that one summation sign in Equation (4.16) is missing. This is a consequence of not reducing one of the modes. Other ways of writing the Tucker2 model are given in Appendix 4.B. Figure 4.12 visualizes the Tucker2 model. Equation (4.15) shows that the Tucker2 model also has rotational freedom because

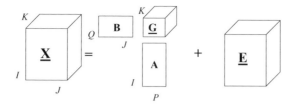

Figure 4.12. A pictorial presentation of the Tucker2 model.

$\mathbf{G}$ can be postmultiplied by $\mathbf{U} \otimes \mathbf{V}$ and $(\mathbf{B} \otimes \mathbf{A})'$ premultiplied by $(\mathbf{U} \otimes \mathbf{V})^{-1}$ resulting in $(\mathbf{B}(\mathbf{U}')^{-1} \otimes \mathbf{A}(\mathbf{V}')^{-1})'$ without changing the fit. Hence, the component matrices $\mathbf{A}$ and $\mathbf{B}$ can be made orthogonal without loss of fit. The parameters in a Tucker2 model can be estimated as explained in Chapter 6.

TUCKER1 MODELS

In the case of Tucker3 models, three modes are reduced. Tucker2 models reduce two modes. Taking this one step further, Tucker1 models can be defined that reduce only one of the modes. Three different Tucker1 models exist for a given array $\underline{\mathbf{X}}$ $(I \times J \times K)$, depending on which mode is being reduced. In deriving the Tucker2 model above, the third mode was not reduced by setting $\mathbf{C}$ to $\mathbf{I}$. If also $\mathbf{B}$ is set to $\mathbf{I}$, then both the second and the third mode are not reduced. The resulting Tucker1 model is then

$$\mathbf{X}_{(K \times IJ)} = \mathbf{IG}_{(K \times JP)}(\mathbf{I} \otimes \mathbf{A})' + \mathbf{E} \tag{4.17}$$

where $\mathbf{G}_{(K \times JP)}$ has now the size $(K \times JP)$ and only the third mode is reduced. Equation (4.26) becomes more readable if $\underline{\mathbf{X}}$ (and accordingly $\underline{\mathbf{G}}$) are matricized differently:

$$\mathbf{X}_{(I \times JK)} = \mathbf{AG}_{(P \times JK)} + \mathbf{E} \tag{4.18}$$

where $\mathbf{X}_{(I \times JK)}$ is the $(I \times JK)$ properly matricized $\underline{\mathbf{X}}$; $\mathbf{G}_{(P \times JK)}$ is the $(P \times JK)$ properly matricized $\underline{\mathbf{G}}$. If $\mathbf{X}_{(I \times JK)}$ is replaced by $\mathbf{X}$; $\mathbf{A}$ by $\mathbf{T}$ and $\mathbf{G}_{(P \times JK)}$ by $\mathbf{P}'$, then the usual PCA solution of $\mathbf{X}$ is obtained: $\mathbf{X} = \mathbf{TP}' + \mathbf{E}$. This is exactly what the Tucker1 model is: a (two-way) PCA on a properly matricized $\underline{\mathbf{X}}$. Hence, algorithms for finding $\mathbf{A}$ and $\mathbf{G}_{(P \times JK)}$ are widely available, as a singular value decomposition gives these matrices. This is one of the reasons for the popularity of Tucker1 models in chemometric applications.

EXAMPLE 4.5

Example of Tucker1 model from image analysis

Images are special three-way arrays. A simple flat greylevel image has two object modes, which are spatial coordinates. It is a bit like a black and white photographic picture. The image can be represented by a matrix of greyvalues, one for each coordinate pair $[x,y]$. This matrix has the size $(I \times J)$, where I and J can be very large. In the same way, three-dimensional image arrays with coordinates $[x,y,z]$ can be defined. These arrays are very rarely used for data analysis with the methods presented in this book.

A more interesting image array is the multivariate image. This image is a stack of congruent images measured for different variables, e.g. wavelengths. This is a three-way array with ways: x-coordinate, y-coordinate and variable. This is an array of size $(I \times J \times K)$. Most of the time, K is smaller than I and J. The structure in the image is complex, with many local phenomena. Therefore bilinear models are made on matricized images.

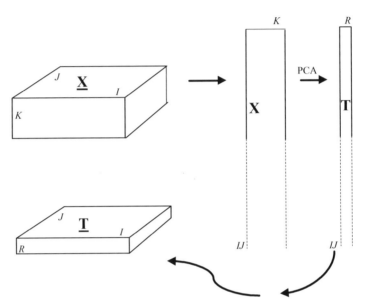

Figure 4.13. The $(I \times J \times K)$ image $\underline{\mathbf{X}}$ is matricized into a long $(IJ \times K)$ matrix $\mathbf{X}$. The variable mode K is reduced to R by PCA (SVD), forming a score matrix $\mathbf{T}$. Then the matrix is rearranged into an array $\underline{\mathbf{T}}$. In this array, all image detail is preserved. This is the same as making a Tucker1 model. $\underline{\mathbf{T}}$ is easier to interpret and the influence of noise and systematic errors from $\underline{\mathbf{X}}$ are reduced.

The array $(I \times J \times K)$ is matricized into $(IJ \times K)$ (Figure 4.13). This matrix has a low rank bilinear structure if the variables are properly chosen. Principal component analysis or Tucker1 is used to reduce K to $R < K$. I and J are not reduced, meaning that the resulting PCA scores still can be shown as images with full image quality. The reduction from K (wavelengths) images to R latent images makes the visual interpretation much easier and reduces noise and artifacts of the measuring equipment. This is useful in all forms of microscopy, satellite imaging and radiology [Geladi & Grahn 1996]

In summation notation this Tucker1 model for $\underline{\mathbf{X}}$ reads

$$x_{ijk} = \sum_{p=1}^{P} a_{ip} g_{pjk} + e_{ijk} \qquad (4.19)$$

where x_{ijk}, e_{ijk}, a_{ip}, g_{pjk} are elements of $\underline{\mathbf{X}}$, the error matrix $\underline{\mathbf{E}}$, the component matrix $\mathbf{A}$ $(I \times P)$ and the Tucker1 core-array $\underline{\mathbf{G}}$ $(P \times J \times K)$. Compared to Equation (4.11) two summation signs are missing. This is due to the fact that two of the modes are not reduced.

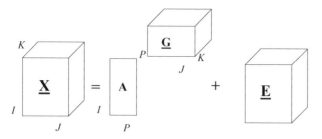

Figure 4.14. Pictorial presentation of a Tucker1 model with only the first mode reduced.

Equation (4.18) shows that the Tucker1 model also has rotational freedom. Hence, **A** can be chosen orthogonal without loss of fit. Other ways of describing a Tucker1 model are given in Appendix 4.B. A pictorial presentation of the Tucker1 model is given in Figure 4.14. As for Tucker2, there are several different Tucker1 models for a given data set depending on which modes are reduced. There is a clear difference between Tucker models: the number of estimated parameters and the risk of overfitting differs between the models. This subject will be treated in more detail in Chapter 5.

General framework for three-way component models

With all the presented three-way component models, it is convenient to have a general framework in which all these models can be expressed. Careful examination of the equations for the different models suggests the following [Kiers 1991, Smilde 1997]. Consider three-way component models for $\underline{\mathbf{X}}$ ($I \times J \times K$), and let **A** be the ($I \times P$) component matrix of the first (reduced) mode, then

$$
\begin{aligned}
&\mathbf{X}_{(I \times JK)} = \mathbf{AP}' + \mathbf{E} \\
&\text{PARAFAC: } \mathbf{P}' = \mathbf{G}_{(R \times RR)}(\mathbf{C} \otimes \mathbf{B})'; \mathbf{G}_{(R \times RR)} \text{ is matricized } \underline{\mathbf{I}} \\
&\text{Tucker3: } \mathbf{P}' = \mathbf{G}_{(P \times QR)}(\mathbf{C} \otimes \mathbf{B})' \\
&\text{Tucker2: } \mathbf{P}' = \mathbf{G}_{(P \times QK)}(\mathbf{I} \otimes \mathbf{B})' \\
&\text{Tucker1: } \mathbf{P}' = \mathbf{G}_{(P \times JK)}
\end{aligned}
\tag{4.20}
$$

where $\mathbf{X}_{(I \times JK)}$ ($I \times JK$) is the matricized $\underline{\mathbf{X}}$; **A**, **B**, **C** are component matrices; the **G** matrices represent the different matricized core-arrays and the array $\underline{\mathbf{I}}$ is a superdiagonal array of appropriate size with ones on the superdiagonal. The component matrices, core-arrays and residual error arrays are different for each model but this is not indicated for reasons of simplicity. Equation (4.20) also shows an alternative way of writing the PARAFAC model (similar to Equation (4.6)), which illustrates that the PARAFAC model is a special case of the Tucker3 model.

The similarity of the first line of Equation (4.20) with a two-way component model is striking. Hence, focusing on the first reduced mode, all three-way models can be seen as attempts to model $\underline{\mathbf{X}}$ as a multiplication of a score matrix **A** and a loading matrix **P'**. The structure of the loading matrix **P'** differs, however, depending on the three-way structure imposed. From Equation (4.20) it follows immediately that the Tucker2 model is a constrained Tucker1 model because in Tucker1, **P'** is unconstrained whereas in Tucker2,

$\mathbf{P}'$ has the structure $\mathbf{G}(\mathbf{I} \otimes \mathbf{B})'$. Thus, the Tucker1 model will always fit the data better than the Tucker2 model for equivalent numbers of components. However, if the structure of the Tucker2 model is appropriate, the Tucker2 model will provide a more accurate model with fewer parameters than the Tucker1 model. The Tucker3 model is a constrained version of the Tucker2 model and the PARAFAC model is a constrained Tucker3 model. Hence, there is a certain hierarchy between the different models as will be discussed in Section 5.1.

In Equation (4.20) the first mode is treated as the mode being reduced. Due to the symmetry of the PARAFAC and Tucker3 models equivalent descriptions exist for Equation (4.20) in which $\mathbf{B}$ or $\mathbf{C}$ acts as the object mode. This cannot be done for the Tucker2 and Tucker1 model without loss of generality as these models are not symmetric.

Extensions to multi-way arrays

Extensions of the three-way component models presented in the previous sections to higher than three-way arrays are straightforward [Bro 1997, Carroll & Chang 1970, Kapteyn *et al.* 1986, Lastovicka 1981]. For the Tucker models, the number of possible models, however, increases rapidly. Consider, e.g., a five-way array for which a Tucker1 model is sought. There exist five such types of models, depending on which mode is selected to be reduced and for each type, the actual number of componens to use has to be determined. Likewise, there exists ten different types of Tucker2 models for this five-way array; ten Tucker3 models and five Tucker4 models. There exists only one Tucker5 model, reducing all five modes simultaneously.

4.3   Models for Three-way Two-block Data: Three-way Regression Models

Some problems in chemistry can be cast as three-way regression problems. Consider a batch process in which a chemical product is made batchwise. For each batch run, J process variables are measured at K points in time. At the end of each batch run, the quality of the product (y) is measured. If I batch runs are available, then a three-way array $\underline{\mathbf{X}}$ ($I \times J \times K$) has the first mode in common with the quality measurements collected in $\mathbf{y}$ ($I \times 1$). It might be worthwhile to find a model predicting $\mathbf{y}$ from $\mathbf{X}$. This is a multi-way regression problem.

The three-way regression problem is concerned with finding a connection between a three-way array $\underline{\mathbf{X}}$ and a vector $\mathbf{y}$, a matrix $\mathbf{Y}$ or a three-way array $\underline{\mathbf{Y}}$. In this section, methods to deal with these situations will be described. The methods can be divided in two classes: sequential methods and simultaneous methods.

A sequential method calculates each component at a time, deflates the proper arrays and calculates the next component. In a simultaneous method all components are calculated simultaneously by minimizing a certain criterion, usually a function of sums of squared residuals.

In this section both simultaneous and sequential methods are discussed. In all the cases the example of predicting $\mathbf{y}$ ($I \times 1$) from $\underline{\mathbf{X}}$ ($I \times J \times K$) is used. It is indicated how this can be generalized to other situations such as multivariate $\mathbf{Y}$ or four-way $\underline{\mathbf{X}}$.

Simultaneous methods

Suppose that for each object i $(i=1,\ldots,I)$ a predictor matrix $\mathbf{X}_i$ $(J \times K)$ and a y_i value are available and y_i has to be predicted. An example is a batch process, where each $\mathbf{X}_i$ represents J process variables measured at K time points and y_i is the end-product quality (see Example 4.6). Collect all the $\mathbf{X}_i$ matrices in a three-way array $\underline{\mathbf{X}}$ $(I \times J \times K)$ and the y values in a vector $\mathbf{y}$ $(I \times 1)$. A natural way to find a model between $\underline{\mathbf{X}}$ and $\mathbf{y}$ is by decomposing $\underline{\mathbf{X}}$ first in, say, a $(P \times Q \times R)$ Tucker3 model:

$$\mathbf{X} = \mathbf{AG}(\mathbf{C} \otimes \mathbf{B})' + \mathbf{E_X} \qquad (4.21)$$

where the loading matrix $\mathbf{A}$ summarizes the information of the differences between the objects (i.e. batch runs). Subsequently, $\mathbf{y}$ can be regressed on $\mathbf{A}$ because the variation in $\mathbf{y}$ should be connected to the variation between objects:

$$\mathbf{y} = \mathbf{Aq} + \mathbf{e_y} \qquad (4.22)$$

where the regression coefficients $\mathbf{q}$ $(P \times 1)$ can be found easily by using the orthogonality of $\mathbf{A}$ $(\mathbf{A'A} = \mathbf{I})$:

$$\mathbf{q} = \mathbf{A'y} \qquad (4.23)$$

The residuals $\mathbf{e_y}$ and $\mathbf{E_X}$ can be used for diagnostic purposes as in, e.g, principal component regression and PLS.

Likewise, a PARAFAC model for $\mathbf{X}$ can be assumed and $\mathbf{y}$ can subsequently be regressed on the proper PARAFAC components [Bro 1997, Geladi *et al.* 1998]. Obviously, the number of PARAFAC components or the number of components in the Tucker3 model has to be found. This can be done, e.g., by cross-validation (see Chapter 7). The procedure of Equations (4.21)–(4.23) resembles principal component regression and can be generalized to higher-way $\underline{\mathbf{X}}$ and two-way $\mathbf{Y}$.

A drawback of the procedure above of regressing on the loadings obtained from a separate decomposition of the three-way array is that these components are not necessarily predictive for $\mathbf{y}$. To overcome this, a weighted criterion has been proposed [Smilde 1997, Smilde & Kiers 1999] that balances the selection of components between fitting $\mathbf{X}$ and predicting $\mathbf{y}$. This can be stated as selecting components $\mathbf{A} = \mathbf{XW}$ in the column-space of $\mathbf{X}$ $(I \times JK)$, that simultaneously fit $\underline{\mathbf{X}}$ and predict $\mathbf{y}$:

$$\min_{\mathbf{W}} \lfloor \alpha \|\mathbf{X} - \mathbf{AG}(\mathbf{C} \otimes \mathbf{B})'\|^2 + (1-\alpha)\|\mathbf{y} - \mathbf{Ap}\|^2 \rfloor =$$
$$\min_{\mathbf{W}} [\alpha \|\mathbf{X} - \mathbf{XWG}(\mathbf{C} \otimes \mathbf{B})'\|^2 + (1-\alpha)\|\mathbf{y} - \mathbf{XWp}\|] \qquad (4.24)$$

where $\mathbf{W}$ is the only free parameter to optimize, because given $\mathbf{W}$, the other parameters follow. The α weight is between 0 and 1 and weighs the importance of fitting $\mathbf{X}$ (α is high) or fitting $\mathbf{y}$ (α is low). The α value has to be chosen a priori or, alternatively, optimized in the same way as the number of components. A Tucker3 model for $\mathbf{X}$ is assumed, but also a PARAFAC or Tucker2 or Tucker1 model can be assumed. The method of Equation (4.24) is called multi-way covariates regression, because it is a direct extension of principal covariates regression [De Jong & Kiers 1992, Smilde 1997, Smilde & Kiers 1999]. By taking $\alpha = 1$, the multi-way covariates regression model equals the regression on loadings approach of Equations (4.21) and (4.22).

EXAMPLE 4.6

Predicting product quality with multi-way covariates regression

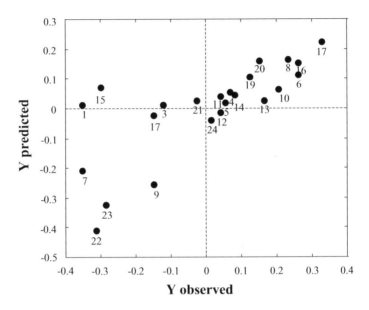

Figure 4.15. Cross-validated predicted versus observed **y** for the batch data.

An example is given of multi-way covariates regression using fat hardening data. Details can be found in [Smilde & Kiers 1999]. In a hydrogenation process, bean oil is hardened to obtain fat. The fat can then be used directly for consumption or as an intermediate product for further food processing. This fat-hardening process is a catalyzed batch process. The feed (bean oil) is pumped into the autoclave (the reactor) and during a certain period of time hydrogen is added to the autoclave until the reaction is finished. After completion of the reaction, the quality of the end product is measured. The quality parameters form a melting curve, expressing what percentage of the product is melted at a certain temperature. Quality limits are set for the amount of product melted at certain temperatures, hence, for points of the melting curve.

Data from 23 normal operating batch runs are available. The variables measured during the run of the batch process are *added amount of hydrogen*, *pressure* and *temperature*. The process data are collected at 101 equidistant points in time. The point of the melting curve related to a temperature of 35 °C is chosen as the quality variable. Hence, a problem results of relating a quality variable **y** (23 × 1) with a three-way array of process variables **X** (23 × 3 × 101). Prior to analysis, **X** and **y** are centered and scaled across the batch direction (see Chapter 9). Subsequently, both **X** and **y** are scaled to unit sum of squares.

Cross-validation of **X** shows that a (3,2,3) Tucker3 model is a reasonable model for **X** [Louwerse *et al.* 1999]. Hence, a multi-way covariates regression model relating **y** to **X**, and assuming a (3,2,3) Tucker3 structure for **X** was calculated. The optimal α was found by cross-validation to be 0.9. This indicates that stabilizing the predictions by extracting meaningful components from $\underline{\mathbf{X}}$ is important. The number of components of the Tucker3 model was varied, but this did not substantially improve the predictions of **y**, as evidenced by cross-validation. The (3,2,3) model explained 48 % of the variation in $\underline{\mathbf{X}}$ and 55 % of the variation in **y**. Moreover, the model was able to predict 45 % of the variation in **y** (estimated with cross-validation). A plot of the predicted versus observed values (Figure 4.15) shows large deviations, but considering that this is batch data the predictions are reasonable. Improvements of the predictions were reported later, when three outlying batches were removed and the scaling was done properly within the variable mode [Gurden *et al.* 2001]. This resulted in a (3,2,3) Tucker3 model that predicts 83 % of the variation in the **y** (estimated with cross-validation).

Multi-way covariates regression can be extended to multi-way arrays $\underline{\mathbf{X}}$ and $\underline{\mathbf{Y}}$ of arbitrary numbers of modes. For each array a specific multi-way structure can be assumed. For example, if both $\underline{\mathbf{X}}$ and $\underline{\mathbf{Y}}$ are three-way arrays and a Tucker3 structure is assumed for both arrays, then the multi-way covariates regression model is

$$\mathbf{X} = \mathbf{A}\mathbf{G}_X(\mathbf{C}_X \otimes \mathbf{B}_X)' + \mathbf{E}_X$$
$$\mathbf{Y} = \mathbf{A}\mathbf{G}_Y(\mathbf{C}_Y \otimes \mathbf{B}_Y)' + \mathbf{E}_Y$$
$$\min_{\mathbf{W}}[\alpha\|\mathbf{X} - \mathbf{X}\mathbf{W}\mathbf{G}_X(\mathbf{C}_X \otimes \mathbf{B}_X)'\|^2 + (1 - \alpha)\|\mathbf{Y} - \mathbf{X}\mathbf{W}\mathbf{G}_Y(\mathbf{C}_Y \otimes \mathbf{B}_Y)'\|^2] \quad (4.25)$$
$$\mathbf{A} = \mathbf{X}\mathbf{W}$$

Sequential methods

The similarity between a Tucker1 model of a three-way array $\underline{\mathbf{X}}$ and a PCA model of a two-way array **X** has already been mentioned in Section 4.1. This similarity is the basis for the most popular three-way regression model, the Tucker1-PLS model. In the literature this regression method has also been called three-way partial least squares regression [Wold *et al.* 1987]. This name should be avoided for the Tucker1 version specifically, because it gives rise to much confusion as to what is multi-way methodology. Even though the Tucker1 version of three-way partial least squares regression is indeed one way to implement a three-way method, it is actually the model that assumes the least three-way structure in the data.

Consider the problem of regressing **y** ($I \times 1$) on $\underline{\mathbf{X}}$ ($I \times J \times K$) and assume that the first mode contains measurements on the same objects in both $\underline{\mathbf{X}}$ and **y**. Then the Tucker1-PLS solution is found by properly matricizing $\underline{\mathbf{X}}$ to **X** ($I \times JK$) and performing PLS1 on **X** and **y**.

The Tucker1-PLS model can easily be extended to the case of regressing a matrix **Y** on $\underline{\mathbf{X}}$, or regressing a three-way array $\underline{\mathbf{Y}}$ ($I \times M \times N$) on $\underline{\mathbf{X}}$ ($I \times J \times K$) by properly matricizing both $\underline{\mathbf{X}}$ and $\underline{\mathbf{Y}}$ to **X** ($I \times JK$) and **Y** ($I \times MN$), respectively, and using PLS2 to model the relation between **Y** and **X**. All the properties for the ordinary PLS method hold for the matricized arrays, e.g., successive components are orthogonal to each other.

A sequential method that predicts **y** and decomposes **X** in a PARAFAC-like model is called multi-linear PLS or simply N-PLS regression. This method is developed by Bro [Bro 1996a] and commented on by Smilde [Smilde 1997] and De Jong [De Jong 1998]. The N-PLS method is a generalization of the nonorthogonal-t PLS1 method (see Section 3.2). There are two versions of N-PLS: the orginal version [Bro 1996b] and an improved version [Bro *et al.* 2001]. The original version is discussed first.

Before explaining N-PLS the following observation is important. Consider a general matrix $\mathbf{X}$ ($I \times J$). Define a component vector $\mathbf{t}$ as $\mathbf{t} = \mathbf{Xw}$ for a given $\mathbf{w}$ of length one. Then $\mathbf{t}$ solves the problem

$$\min_{\mathbf{s}} \|\mathbf{X} - \mathbf{sw}'\|^2 \tag{4.26}$$

where $\mathbf{s}$ is an arbitrary ($I \times 1$) vector and $\mathbf{w}$ is given and $\|\mathbf{w}\| = 1$. This is easily seen by solving problem (4.26) by ordinary least squares: $\mathbf{t}' = (\mathbf{w}'\mathbf{w})^{-1}\mathbf{w}'\mathbf{X}'$, hence, $\mathbf{t} = \mathbf{Xw}$. This is an important and simple result meaning that a linear combination of the columns of $\mathbf{X}$ defined using a weighing vector $\mathbf{w}$ of length one has least squares properties in terms of approximating $\mathbf{X}$. This can be generalized to multiple components. Suppose that $\mathbf{t}_r = \mathbf{Xw}_r$; $\mathbf{T} = [\mathbf{t}_1, \ldots, \mathbf{t}_R]$, $\mathbf{W} = [\mathbf{w}_1, \ldots, \mathbf{w}_R]$ with $\mathbf{W}'\mathbf{W} = \mathbf{I}$, then $\mathbf{T} = \mathbf{XW}$ minimizes $\|\mathbf{X} - \mathbf{TW}'\|^2$ for a given $\mathbf{W}$, which follows directly from the column-orthogonality of $\mathbf{W}$. The beauty of this result is that if a $\mathbf{w}$ with length one is found according to some criterion (e.g. maximizing covariation of $\mathbf{t}$ with a $\mathbf{y}$) then automatically $\mathbf{tw}'$ with $\mathbf{t} = \mathbf{Xw}$ gives the best fit of $\mathbf{X}$ for that $\mathbf{w}$.

It is possible to generalize the notion above to three-way arrays. Consider an array $\underline{\mathbf{X}}$ ($I \times J \times K$) and weighing vectors $\mathbf{w}^J$ ($J \times 1$), $\mathbf{w}^K$ ($K \times 1$), where the superscripts of the ws refer to the mode for which these vectors are defined. Let $\|\mathbf{w}^J\| = \|\mathbf{w}^K\| = 1$, then $\mathbf{t} = \mathbf{X}(\mathbf{w}^K \otimes \mathbf{w}^J)$ solves the problem

$$\min_{\mathbf{s}} \|\mathbf{X} - \mathbf{s}(\mathbf{w}^K \otimes \mathbf{w}^J)'\|^2 \tag{4.27}$$

where $\mathbf{s}$ is an arbitrary ($I \times 1$) vector and $\mathbf{X}$ is the properly rearranged array $\underline{\mathbf{X}}$. The proof of this result is straightforward [Smilde 1997].

N-PLS uses this result and solves the problem

$$\max_{\mathbf{w}^J, \mathbf{w}^K} [\text{cov}(\mathbf{t}, \mathbf{y}) | \mathbf{t} = \mathbf{X}(\mathbf{w}^K \otimes \mathbf{w}^J); \|\mathbf{w}^J\| = \|\mathbf{w}^K\| = 1] \tag{4.28}$$

and the resulting $\mathbf{t}$, $\mathbf{w}^J$ and $\mathbf{w}^K$ have the least squares property as shown in Equation (4.27). In words, Equation (4.28) means that a specific vector $\mathbf{t}$ in the column space of $\mathbf{X}$ is sought, that covaries maximally with $\mathbf{y}^3$, and is constructed such that it also approximates $\mathbf{X}$ well. This results in the following model

$$\begin{aligned} \mathbf{X} &= \mathbf{t}(\mathbf{w}^K \otimes \mathbf{w}^J)' + \mathbf{E}_\mathbf{X} \\ \mathbf{y} &= \mathbf{t}b + \mathbf{e}_\mathbf{y} \end{aligned} \tag{4.29}$$

where b is the regression coefficient obtained from regressing $\mathbf{y}$ on $\mathbf{t}$ and $\mathbf{E}_\mathbf{X}$ and $\mathbf{e}_\mathbf{y}$ contain the residuals of the respective models. Equation (4.29) is a generalization of PLS where $\underline{\mathbf{X}}$ is modeled with a one component trilinear (PARAFAC) model.

The next step in N-PLS is a deflation step. Both $\mathbf{X}$ and $\mathbf{y}$ are deflated using $\mathbf{t}$, $\mathbf{w}^K$, $\mathbf{w}^J$ and $\mathbf{t}$, b. The algorithm continues finding weights $\mathbf{w}_2^K$, $\mathbf{w}_2^J$ such that $\mathbf{t}_2 = \mathbf{E}_\mathbf{X}(\mathbf{w}_2^K \otimes \mathbf{w}_2^J)$ maximizes the covariance of $\mathbf{t}_2$ and $\mathbf{e}_\mathbf{y}$. This process is continued until enough components are calculated. As a criterion for the number of components to use in the final model, cross-validation can be used.

The total N-PLS model in the original version is now

---

[3] Only: $\mathbf{f}_\mathbf{y}$ and $\mathbf{t}$ are mean centered.

$$\mathbf{T} = \mathbf{XV}$$
$$\mathbf{X} = \mathbf{TW'} + \mathbf{E_X}; \mathbf{W} = \left[\mathbf{w}_1^K \otimes \mathbf{w}_1^J \mid \cdots \mid \mathbf{w}_R^K \otimes \mathbf{w}_R^J\right]$$
$$\mathbf{y} = \mathbf{T}\mathbf{b}_R + \mathbf{e_y} \tag{4.30}$$
$$\max_{\mathbf{w}_r^J, \mathbf{w}_r^K} \mathrm{cov}(\mathbf{t}_r, \mathbf{y}^{(r-1)}); r = 1, \ldots, R$$

where $\mathbf{V}$ is a matrix of weighing coefficients which can be written in terms of $\mathbf{w}_1 = (\mathbf{w}_1^K \otimes \mathbf{w}_1^J)$ to $\mathbf{w}_R = (\mathbf{w}_R^K \otimes \mathbf{w}_R^J)$:

$$\mathbf{V} = [\mathbf{w}_1 \, (\mathbf{I} - \mathbf{w}_1 \mathbf{w}_1')\mathbf{w}_2 \cdots (\mathbf{I} - \mathbf{w}_1\mathbf{w}_1')(\mathbf{I} - \mathbf{w}_2\mathbf{w}_2') \cdots (\mathbf{I} - \mathbf{w}_{R-1}\mathbf{w}_{R-1}')\mathbf{w}_R] \tag{4.31}$$

This is also discussed elsewhere [Smilde 1997]. N-PLS can be generalized for higher-order arrays $\underline{\mathbf{X}}$ and $\underline{\mathbf{Y}}$ and in all cases a PARAFAC-like multilinear model is assumed for all the multi-way arrays involved [Bro 1996a].

The matrix $\mathbf{W}$ has no orthogonality properties. The columns of $\mathbf{W}$ have length one, but are not orthogonal. Hence, $\mathbf{TW'}$ is *not* the least squares fit for $\mathbf{X}$, given $\mathbf{W}$. In that respect, N-PLS does not generalize nonorthogonal-t PLS (see Section 3.2). The least squares property does hold, however, for each separate dimension in the N-PLS model. The matrix $\mathbf{T}$ has no orthogonality properties either. Hence, the regression coefficient $\mathbf{b}_r$ has to be calculated with the current and all previous $\mathbf{t}$-vectors.

It is possible to write Equation (4.31) in the closed form $\hat{\mathbf{y}} = \mathbf{X}\mathbf{b}_{\text{N-PLS}}$ [Smilde 1997]. Moreover, deflating $\mathbf{X}$ *and* $\mathbf{y}$ is not necessary [De Jong 1998]. Only deflating $\mathbf{y}$ is sufficient to calculate $\mathbf{b}_{\text{N-PLS}}$. In that case, however, the residuals in the $\mathbf{X}$ block are not available.

The original N-PLS model has disadvantages which can be circumvented by building in an extra step after the components are found [Bro *et al.* 2001]. This extra step is introducing a core-array in the X-model part of N-PLS *given* the estimated components. Hence, the problem to solve is

$$\min_{\mathbf{G}} \|\mathbf{X} - \mathbf{TG}(\mathbf{W}^K \otimes \mathbf{W}^J)'\|^2 \tag{4.32}$$

where the minimization is *only* over $\mathbf{G}$. The solution of this problem is

$$\mathbf{G} = \mathbf{T}^+\mathbf{X}((\mathbf{W}^K)^+ \otimes (\mathbf{W}^J)^+)' \tag{4.33}$$

and this results in a Tucker3 model of $\mathbf{X}$:

$$\mathbf{T} = \mathbf{XV}$$
$$\mathbf{X} = \mathbf{TG}(\mathbf{W}^K \otimes \mathbf{W}^J)' + \tilde{\mathbf{E}}_X$$
$$\mathbf{y} = \mathbf{T}\mathbf{b}_R + \mathbf{e_y} \tag{4.34}$$
$$\max_{\mathbf{w}_r^J, \mathbf{w}_r^K} \mathrm{cov}(\mathbf{t}_r, \mathbf{y}^{(r-1)}); r = 1, \ldots, R$$

where $\mathbf{T}$, $\mathbf{W}^K$ and $\mathbf{W}^J$ are exactly the same as in Equation (4.30). Due to the extra fitting, the residuals $\tilde{\mathbf{E}}_X$ have changed and have a lower (or at least not higher) sum of squares than their counterparts in Equation (4.30). This was actually one of the incentives for improving N-PLS. The predicted $\mathbf{y}$ values do not change in this operation, hence, in terms of predictive performance there is no difference between the original and improved version of N-PLS.

The essential difference between the original and the improved *N*-PLS is that the orginal version gives a *low-rank* approximation of **X** in terms of the triads of the **t**s and corresponding **w**s. The improved version gives a *subspace* approximation of **X** using a proper multiplication of the component matrices. For the model of **X**, this is an essential difference. An algorithm for *N*-PLS is described in Chapter 6.

Ståhle developed a linear three-way decomposition for the problem of regressing **y** ($I \times 1$) on $\underline{\mathbf{X}}$ ($I \times J \times K$), where a bilinear (Tucker1) model for $\underline{\mathbf{X}}$ was imposed, but the weight vector defining the components has the same form as in *N*-PLS [Ståhle 1989]. This method also works sequentially and is a hybrid form between Tucker1-PLS and *N*-PLS. It can be shown that linear three-way decomposition and *N*-PLS give exactly the same results in terms of predicted **y**s and the component vectors obtained from linear three-way decomposition and *N*-PLS span the same space [De Jong 1998]. Linear three-way decomposition produces orthogonal **t** scores and can be seen as a generalization of the orthogonal-*t* PLS method (see Section 3.2).

EXAMPLE 4.7

Example of *N*-PLS

Table 4.1. Leave-one-out cross-validation results from Tucker1- and *N*-PLS on sensory data for one to four components (#LV) for prediction of salt content. The percentage of variation explained (sum-squared residuals versus sum-squared centered data) is shown for fitted modes (Fit) and for cross-validated models (Xval) for both $\underline{\mathbf{X}}$ (sensory data) and **Y** (salt). The root mean squared error of cross-validation (RMSECV) of salt (weight %) is also provided.

	#LV	Fit **X**	Xval **X**	Fit **Y**	Xval **Y**	RMSECV
Tucker1-PLS	1	39.7	5.6	79.7	61.7	0.291
	2	58.2	17.4	95.3	75.8	0.231
	3	73.5	36.5	99.5	83.8	0.190
	4	78.1	37.9	100.0	84.0	0.188
N-PLS	1	31.2	4.4	75.4	59.6	0.299
	2	45.7	20.5	93.1	82.4	0.198
	3	55.2	33.4	97.6	90.5	0.145
	4	59.2	34.3	99.6	90.7	0.144

Five different breads were baked in duplicate giving a total of ten samples. Eight different judges assessed the breads with respect to eleven different attributes. The data can be regarded as a three-way array ($10 \times 11 \times 8$) or alternatively as an ordinary two-way matrix (10×88). The data are quite noisy as opposed to, e.g., spectral data. It may be speculated, that for matricizing to be appropriate it must hold that the judges use individually defined latent variables. Using a trilinear model, the approximation is that the judges use the same latent variables, only in different proportions.

For every bread the salt content (in weight %) is known. The minimum, maximum and average salt content is 0.6, 1.9 and 1.22 wt %, respectively. To predict the salt from the sensory data two different calibration models are calculated: *N*-PLS regression on the 10 × 11 × 8 data array and Tucker1-PLS on the 10 × 88 matrix obtained by matricizing the three-way array. Several conclusions can be drawn from the results in Table 4.1. The most important conclusion is that Tucker1-PLS is clearly the poorest model. It has the lowest predictability (using the solution of three or four latent variables) and it also amply overfits the models of both **X** and **Y**. For example, the three-component model fits 73.5 % of the variation of **X** but explains only 36.5 % in terms of cross-validation. This is a common phenomenon in multi-way analysis using matricizing methods. It also points to the important fact that interpretation of matricized models should be performed with care. Overfitting can sometimes lead to noisy model parameters. *N*-PLS behaves more realistically in this respect. The fitted and cross-validated explained sum of squares of **X** and **Y** are more similar than for Tucker1-PLS.

Differences and similarities between the regression models

The most significant difference between Tucker1-PLS and *N*-PLS on the one hand and multi-way covariates regression models on the other hand is that Tucker1-PLS and *N*-PLS models are calculated sequential and the multi-way covariates regression model is derived in a simultaneous fashion. This has several consequences:

1. The solution of a Tucker1-PLS, *N*-PLS and a multi-way covariates regression model for a given data set will be different.
2. The solutions of Tucker1-PLS models with an increasing number of components are nested; this also holds for *N*-PLS models but not for multi-way covariates regression models.
3. The algorithms of Tucker-1-PLS, *N*-PLS and multi-way covariates regression are different and this will result in differences in speed and properties of the estimated parameters.

It is hard to make a general statement about which models are best for a given situation partly because the experience so far is limited. One study compares Tucker1-PLS, *N*-PLS (old version) and multi-way covariates regression on three very different data sets [Gurden *et al.* 2001]. Apart from comparing methods, the study also compares different kinds of scaling. The conclusion is that the predictive ability of the models depends much more on the quality of the data and a proper scaling, than on the prediction method used. For interpreting the model, the multilinear prediction methods (especially multi-way covariates regression) were found more useful than Tucker1-PLS. Knowledge of the data and testing the quality of alternative models is a way to identify which model to use.

4.4 Summary

Important multi-way component and regression models have been described in this chapter. PARAFAC and Tucker3 are the best-known methods which can both be viewed as extensions of ordinary two-way PCA. PARAFAC is an extension in the sense that it provides the best-fitting rank *R* component model of a three-way data set, and Tucker3 is an extension of PCA in the sense that it provides the best-fitting model of the three-way data set within a finite-dimensional subspace in each way. Additionally, Tucker2 and Tucker1 have been mentioned, both being special cases of Tucker3, and Tucker1 being identical to what is known as unfold-PCA, i.e., two-way principal component analysis on matricized three-way data.

Further, different regression models have been described and a distinction between sequential and simultaneous models was made. Simultaneous methods include using scores from decomposition models such as PARAFAC or Tucker, or building the regression model directly using multi-way covariates regression. For component-wise regression models, bilinear and trilinear partial least squares regression was discussed.

Appendix 4.A: Alternative Notation for the PARAFAC Model

PARAFAC in Kronecker products notation

It is possible to write Equation (4.7) in terms of Kronecker products (see Chapter 2) by using $\mathbf{X} = [\mathbf{X}_1 \, \mathbf{X}_2 \ldots \mathbf{X}_K]$; $\mathbf{E} = [\mathbf{E}_1 \, \mathbf{E}_2 \ldots \mathbf{E}_K]$ and $\mathbf{D} = [\mathbf{D}_1 \, \mathbf{D}_2 \ldots \mathbf{D}_K]$:

$$\mathbf{X} = \mathbf{AD}(\mathbf{I} \otimes \mathbf{B}') + \mathbf{E} \tag{4.35}$$

or, by using the unit superdiagonal array $\underline{\mathbf{I}}$ $(R \times R \times R)$ properly matricized to $\mathbf{H}$ $(R \times R^2)$

$$\mathbf{X} = \mathbf{AH}(\mathbf{C}' \otimes \mathbf{B}') + \mathbf{E} \tag{4.36}$$

where for both Equations (4.35) and (4.36) there are two equivalent alternatives due to the symmetric structure of the PARAFAC model.

PARAFAC model in tensor products notation

Using the tensor products notation (see Chapter 2) an alternative expression of the PARAFAC model is:

$$\underline{\mathbf{X}} = \sum_{r=1}^{R} \mathbf{a}_r \Delta \mathbf{b}_r \Delta \mathbf{c}_r + \underline{\mathbf{E}} \tag{4.37}$$

where $\mathbf{a}_r$, $\mathbf{b}_r$ and $\mathbf{c}_r$ are the rth columns of $\mathbf{A}$, $\mathbf{B}$ and $\mathbf{C}$, respectively.

PARAFAC model in vec and Kronecker products notation

The PARAFAC model can also be written by using the vec-operator and the Kronecker product. This is given by

$$\text{vec} \, \underline{\mathbf{X}} = \sum_{r=1}^{R} \mathbf{a}_r \otimes \mathbf{b}_r \otimes \mathbf{c}_r + \text{vec} \, \underline{\mathbf{E}} \tag{4.38}$$

where $\underline{\mathbf{X}}$ is the $(I \times J \times K)$ three-way array. Vec $\underline{\mathbf{X}} = \text{vec} \, \mathbf{X}$, where $\mathbf{X}$ is the properly matricized $\underline{\mathbf{X}}$.

PARAFAC model in Khatri–Rao products notation

A final way of writing the PARAFAC model is given by the Khatri–Rao product.

$$\mathbf{X} = \mathbf{A}(\mathbf{C} \odot \mathbf{B})' + \mathbf{E} \tag{4.39}$$

where $\mathbf{X}$ and $\mathbf{E}$ are as in Equation (4.35) (see also Chapter 2).

It is instructive to have an understanding of why the Khatri–Rao product is useful for multi-linear models. Starting with a bilinear two-way model

$$\mathbf{X} = \mathbf{AB}' \tag{4.40}$$

this model may be vectorized as

$$\text{vec}\,\mathbf{X} = \text{vec}(\mathbf{AB}') \tag{4.41}$$

which means that the matrices are simply rearranged into vectors. For example, vec $\mathbf{X}$ is the IJ vector obtained as

$$\begin{bmatrix} \mathbf{x}_1 \\ \mathbf{x}_2 \\ \vdots \\ \mathbf{x}_J \end{bmatrix} \tag{4.42}$$

where $\mathbf{x}_j$ is the jth column of $\mathbf{X}$. The bilinear part of this model can be written correspondingly as

$$\begin{aligned} \text{vec}(\mathbf{AB}') = \\ \text{vec}(\mathbf{a}_1\mathbf{b}_1') + \text{vec}(\mathbf{a}_2\mathbf{b}_2') + \cdots \text{vec}(\mathbf{a}_R\mathbf{b}_R') = \\ \begin{bmatrix} b_{11}\mathbf{a}_1 \\ b_{21}\mathbf{a}_1 \\ \vdots \\ b_{J1}\mathbf{a}_1 \end{bmatrix} + \begin{bmatrix} b_{12}\mathbf{a}_2 \\ b_{22}\mathbf{a}_2 \\ \vdots \\ b_{J2}\mathbf{a}_2 \end{bmatrix} + \cdots + \begin{bmatrix} b_{1R}\mathbf{a}_R \\ b_{2R}\mathbf{a}_R \\ \vdots \\ b_{JR}\mathbf{a}_R \end{bmatrix} \end{aligned} \tag{4.43}$$

Hence, every bilinear component $\text{vec}(\mathbf{a}_r\mathbf{b}_r')$ can be written as a well-defined vector that is a function of $\mathbf{a}_r$ and $\mathbf{b}_r$ only. This vector is the Kronecker product of $\mathbf{b}_r$ and $\mathbf{a}_r$. Define the Khatri–Rao product of two matrices, $\mathbf{A}$ and $\mathbf{B}$, with the same number of columns as

$$\begin{aligned} \mathbf{B} \odot \mathbf{A} \\ = \begin{bmatrix} b_{11}\mathbf{a}_1 & \dots & b_{1R}\mathbf{a}_R \\ b_{21}\mathbf{a}_1 & \dots & b_{2R}\mathbf{a}_R \\ \cdot & \cdots & \cdot \\ \cdot & \cdots & \cdot \\ \cdot & \cdots & \cdot \\ b_{J1}\mathbf{a}_1 & \dots & b_{JR}\mathbf{a}_R \end{bmatrix} = [\,\mathbf{b}_1 \otimes \mathbf{a}_1 \,..\, \mathbf{b}_R \otimes \mathbf{a}_R\,] \end{aligned} \tag{4.44}$$

Then the model of Equation (4.41) may be written as

$$\text{vec}\,\mathbf{X} = (\mathbf{B} \odot \mathbf{A})\mathbf{1}, \tag{4.45}$$

where $\mathbf{1}$ is an R-vector of ones. This is the principle used in Equation (4.39) expressing the second and third modes simultaneously in the matricized array.

Appendix 4.B: Alternative Notations for the Tucker3 Model

Tucker3 model in simultaneous components notation

In simultaneous components notation the Tucker3 model reads:

$$\mathbf{X}_k = \mathbf{A} \left[\sum_{r=1}^{R} c_{kr}\mathbf{G}_r \right] \mathbf{B}' + \mathbf{E}_k; \; k = 1, \dots, K \tag{4.46}$$

where $\mathbf{X}_k$ and $\mathbf{E}_k$ are defined in the same way as in Equation (4.12); $\mathbf{G}_r$ is the rth slice of the third mode of $\underline{\mathbf{G}}$ with size $(P \times Q)$ and c_{kr} is the krth element of $\mathbf{C}$.

Using this notation it is clear that all slabs $\mathbf{X}_k$ are modeled with the same component matrices $\mathbf{A}$ and $\mathbf{B}$, but different core-matrices. There are further two fully equivalent ways of writing the Tucker3 model in simultaneous components notation form.

Tucker3 model in tensor product notation

In tensor products the Tucker3 model can be written as:

$$\underline{\mathbf{X}} = \sum_{p=1}^{P} \sum_{p=1}^{Q} \sum_{r=1}^{R} g_{pqr}\mathbf{a}_p \Delta \mathbf{b}_q \Delta \mathbf{c}_r + \underline{\mathbf{E}} \tag{4.47}$$

where $\mathbf{a}_p$ is the pth column of $\mathbf{A}$; $\mathbf{b}_q$ the qth column of $\mathbf{B}$; $\mathbf{c}_r$ the rth column of $\mathbf{C}$ and g_{pqr} the pqrth typical element of $\underline{\mathbf{G}}$.

Using the tensor product notation it becomes clear that the Tucker3 model generalizes the PARAFAC model. Not only tensor products of the same components in all three modes are taken, but also their cross-products. Moreover, the Tucker3 model allows for a different number of components in all three modes.

Tucker2 model in simultaneous components notation

In simultaneous components notation the Tucker2 model for the $(I \times J \times K)$ array $\underline{\mathbf{X}}$ reads

$$\mathbf{X}_k = \mathbf{A}\mathbf{G}_k\mathbf{B}' + \mathbf{E}_k; k = 1, \dots, K$$
$$\mathbf{G} = [\mathbf{G}_1 \; \cdot \; \cdot \; \mathbf{G}_K] \tag{4.48}$$

where $\mathbf{X}_k$ is the kth slice of $\underline{\mathbf{X}}$ in the third mode (K-mode); $\mathbf{A}$ ($I \times P$) and $\mathbf{B}$ ($J \times Q$) are the component matrices of the (reduced) first and second mode, respectively; $\mathbf{E}_k$ is the kth slice of the error array $\underline{\mathbf{E}}$ ($I \times J \times K$) in the third mode; $\mathbf{G}_k$ ($P \times Q$) is the kth slice of the third mode of the extended core-array $\underline{\mathbf{G}}$; and $\mathbf{G}$ ($P \times QK$) is a matricized version of $\underline{\mathbf{G}}$.

Tucker2 model in tensor product notation

The tensor product notation of the Tucker2 model is

$$\underline{\mathbf{X}} = \sum_{p=1}^{P} \sum_{q=1}^{Q} \mathbf{a}_p \Delta \mathbf{b}_q \Delta \mathbf{g}_{pq} + \underline{\mathbf{E}} \tag{4.49}$$

where $\underline{\mathbf{X}}$, $\underline{\mathbf{E}}$ are the same as before; $\mathbf{a}_p$, $\mathbf{b}_q$ are the pth and qth columns of $\mathbf{A}$ and $\mathbf{B}$, respectively; and $\mathbf{g}_{pq} = (g_{pq1}, \ldots, g_{pqK})'$, the pqth tube in $\underline{\mathbf{G}}$.

Tucker1 model in simultaneous components notation

In simultaneous notation the Tucker1 model for $\underline{\mathbf{X}}$ ($I \times J \times K$) is

$$\begin{aligned} \mathbf{X}_k &= \mathbf{A}\mathbf{G}_k + \mathbf{E}_k; k = 1, \ldots, K \\ \mathbf{G} &= [\mathbf{G}_1 \cdot \cdot \mathbf{G}_K] \end{aligned} \tag{4.50}$$

where $\mathbf{X}_k$ and $\mathbf{E}_k$ are the kth slices of the third mode in $\underline{\mathbf{X}}$ and $\underline{\mathbf{E}}$, respectively. In other words Equation (4.50) means that each slice $\mathbf{X}_k$ is modeled as a product of a common component matrix ($\mathbf{A}$) and its own 'loading' matrix $\mathbf{G}_k$.

Tucker1 model in Kronecker product notation

The Tucker1 model in Kronecker products notation reads:

$$\mathbf{X} = \mathbf{A}\mathbf{G}(\mathbf{I} \otimes \mathbf{I}) + \mathbf{E} \tag{4.51}$$

where $\mathbf{A}$ and $\mathbf{G}$ the same as in Equation (4.50) and $\mathbf{X}$, $\mathbf{E}$ are the properly matricized versions of $\underline{\mathbf{X}}$ and $\underline{\mathbf{E}}$, respectively.

Studying Equations (4.12) (Tucker3), (4.15) (Tucker2) and (4.51) (Tucker1) it is evident that going from Tucker1 to Tucker3 models, more structure is imposed on $\underline{\mathbf{X}}$ by reducing more of its modes.

5

SOME PROPERTIES OF THREE-WAY COMPONENT MODELS

5.1 Relationships Between Three-way Component Models

Three-way models and their properties were introduced in the previous chapters. It may be difficult for the newcomer to choose between them, just as it is difficult to choose the proper classification or regression method for two-way data. There are many three-way component and regression models to choose from. In order to decide which model to use in which situation, it is important to have a good understanding of the differences between and similarities of the models. The purpose of this chapter is to provide such an understanding.

An important question is how the PARAFAC and Tucker3 models are related. PARAFAC models provide unique axes, while Tucker3 models do not. A Tucker model may be transformed (rotated) and simplified to look more like PARAFAC models. This can sometimes be done with little or no loss of fit. There is a hierarchy e.g. within the family of Tucker models, Tucker3, Tucker2 and Tucker1, which is worth studying in more detail. PARAFAC models may be difficult or impossible to fit due to so-called degeneracies (Section 5.4), in which case a Tucker3 model is usually a better a choice. Further, the statistical properties of the data – noise and systematic errors – also play an important role in the choice of model.

There is not always a clear answer or a strictly defined proper choice of a three-way model for a given application. Sometimes more than one three-way model has to be tried in order to find the proper one. In many cases, the models appear equally good and the selection may be based on practical criteria such as algorithm precision or speed. Knowing the palette of three-way models along with their properties and relationships serves as a basis for making objective choices.

Multi-way Analysis With Applications in the Chemical Sciences.   A. Smilde, R. Bro and P. Geladi
© 2004 John Wiley & Sons, Ltd   ISBN: 0-471-98691-7

Nestedness and hierarchy

In the case of linear polynomial models that are used for example in response surface methodology [Box & Draper 1987], there is a clear hierarchy between different models. Suppose that a yield y of a chemical reaction has to be modeled in terms of pressure (p) and temperature (T), then a first order response surface model for y is

$$y = \beta_0 + \beta_1 p + \beta_2 T + \varepsilon \tag{5.1}$$

where β_0 is an intercept, β_1 and β_2 are regression coefficients and ε is an error term. The second-order response surface model is

$$y = \beta_0 + \beta_1 p + \beta_2 T + \beta_{11} p^2 + \beta_{22} T^2 + \beta_{12} pT + \varepsilon \tag{5.2}$$

in which β_{11}, β_{22} and β_{12} are regression coefficients describing the influence of the quadratic and interaction terms, respectively. Similarly, third and higher order models can be defined. There is a clear relationship between the models in Equations (5.1) and (5.2): if the parameters β_{11}, β_{22} and β_{12} are assumed to be zero, then the two equations become identical. Stated otherwise, model (5.1) is a constrained version of model (5.2).

Unfortunately, there is a discrepancy between the concept of nestedness in chemometrics and statistics. In statistics, nestedness means that one model can be derived from the other by imposing certain restrictions on the *true* parameters [Judge *et al.* 1985]. In the case of Equations (5.1) and (5.2), the restriction that β_{11}, β_{22} and β_{12} are zero makes model (5.2) equal to model (5.1). Hence, the models are nested. In chemometrics, the term nested is often used in a more strict sense: two models are nested if one can be derived from the other by setting certain *estimated* parameters to zero. To explain this, consider again models (5.1) and (5.2). In general, the estimated parameters b_0, b_1 and b_2 (estimates for β_0, β_1 and β_2, respectively) differ between models (5.1) and (5.2). This makes the models nonnested in the chemometric sense, but still nested in the statistical sense.

The term hierarchy will be used here for the statistical notion of nestedness, and the term nestedness will be used in the strict (chemometric) sense. In this terminology PCA models are nested in the number of components, because the first R components remain exactly the same upon going to an $R + 1$ component model. PARAFAC models are hierarchical because an R-component model can be derived from an $R + 1$ component model in the sense of the true parameters by setting the true $(R + 1)$th component parameters to zero. PARAFAC models are not nested in general: the estimates of the R components change when going to an $(R + 1)$ component model.

All component-wise calculated families of models (PCA, PLS, etc) are by definition nested. Nestedness is computationally convenient, but not by definition a desirable property. Hierarchical relationships between models are convenient because they allow for a general framework. It is then possible to think of a continuum of models, with increasing complexity, where complexity is defined as the number of (free) parameters which have to be estimated. For example, model (5.1) is less complex than model (5.2) and if model (5.1) can describe the variation well, there is no need for the added complexity of model (5.2). Given a particular data set, it holds in general that adding complexity to the model increases the fit to the data but also increases the variance of the estimated parameters. Hence, there is an optimal model complexity balancing both properties. This is the basic rationale in many statistical tests of model complexity [Fujikoshi & Satoh 1997, Mallows 1973]. Hierarchy is a desirable property from a statistical point of view, because it makes comparisons between the performances of models easier. The concept is formalized in statistical decision theory [Judge *et al.* 1985].

In the following sections, relationships between three-way methods are outlined, facilitating comparisons between different models used on the same data set. An overview of hierarchical relationships between the major three-way models is given by Kiers [1991a]. While a mathematical viewpoint is taken in this chapter, a more practical viewpoint for choosing between competing models is taken in Chapter 7.

Hierarchy of PARAFAC and Tucker3 models

The general R-component PARAFAC model of a three-way array $\underline{\mathbf{X}}$ ($I \times J \times K$) is given by

$$x_{ijk} = \sum_{r=1}^{R} a_{ir}b_{jr}c_{kr} + e_{ijk} \tag{5.3}$$

where x_{ijk}, a_{ir}, b_{jr}, c_{kr} ($i = 1, \ldots I; j = 1, \ldots J; k = 1, \ldots K$) and e_{ijk} are the typical elements of $\underline{\mathbf{X}}$, $\mathbf{A}$ ($I \times R$), $\mathbf{B}$ ($J \times R$), $\mathbf{C}$ ($K \times R$) and $\underline{\mathbf{E}}$ ($I \times J \times K$), respectively. Here and in the following a distinction is not made between true and estimated values. The difference should be evident in context. Introducing the term g_{pqr} in the summation sign with $g_{pqr} = 1$ if and only if $p = q = r$ and 0 otherwise, Equation (5.3) can be rewritten as

$$x_{ijk} = \sum_{p=1}^{P}\sum_{q=1}^{Q}\sum_{r=1}^{R} a_{ip}b_{jq}c_{kr}g_{pqr} + e_{ijk} \tag{5.4}$$

with $P = Q = R$ and where all typical elements are defined as before. Equation (5.4) is similar to the Tucker3 model of $\underline{\mathbf{X}}$ (see Chapter 4). Hence, the PARAFAC model can be understood as a constrained Tucker3 model (Figure 5.1). Therefore, there exists a hierarchical relationship between the PARAFAC and the Tucker3 model. This has repercussions for the fit of PARAFAC and Tucker3 models, as was explained already for the polynomial models of Equations (5.1) and (5.2). A PARAFAC model with R components always fits worse than or equal to a Tucker3 (R,R,R) model. Yet, this does not necessarily mean that the Tucker3 (R,R,R) model is preferred. There is more to consider than fit alone (see Chapter 7).

It is less intuitive but nevertheless instructive to see that a Tucker3 model can also be represented as a constrained PARAFAC model albeit using many more rank one components. Due to the way these components are constrained, this representation does not in itself imply any hierarchical relation from PARAFAC to Tucker3. Suppose that a (2,2,3)

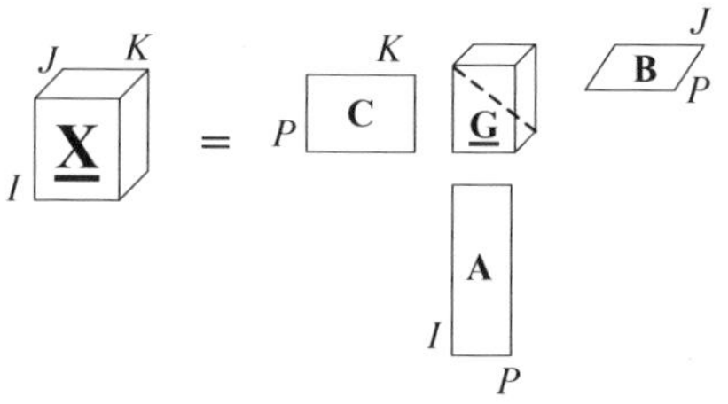

Figure 5.1. PARAFAC model written as a Tucker3 model. The superdiagonal part of $\underline{\mathbf{G}}$ consists of ones and the remaining elements are zero. P = Q = R.

Tucker3 model is fitted to a three-way array $\underline{\mathbf{X}}$. That is,

$$x_{ijk} = \sum_{p=1}^{2}\sum_{q=1}^{2}\sum_{r=1}^{3} a_{ip}b_{jq}c_{kr}g_{pqr} + e_{ijk} \tag{5.5}$$

where x_{ijk}, a_{ip}, b_{jq}, c_{kr}, g_{pqr} and e_{ijk} are typical elements of $\underline{\mathbf{X}}$, $\mathbf{A}$ ($I \times 2$), $\mathbf{B}$ ($J \times 2$), $\mathbf{C}$ ($K \times 3$), $\underline{\mathbf{G}}$ ($2 \times 2 \times 3$) and $\underline{\mathbf{E}}$, respectively. Suppose that all elements g_{pqr} are arranged on a superdiagonal in an $S \times S \times S$ core-array $\underline{\tilde{\mathbf{G}}}$, with $S = 2 \times 2 \times 3 = 12$, in the order g_{111}, g_{211}, g_{121}, g_{221}, g_{112}, g_{212}, g_{122}, g_{222}, g_{113}, g_{213}, g_{123}, g_{223}; and new loading matrices $\tilde{\mathbf{A}}$ ($I \times 12$), $\tilde{\mathbf{B}}$ ($J \times 12$) and $\tilde{\mathbf{C}}$ ($K \times 12$) are formed, with

$$\tilde{\mathbf{A}} = [\mathbf{a}_1 \quad \mathbf{a}_2 \quad \mathbf{a}_1 \quad \mathbf{a}_2 \quad \mathbf{a}_1 \quad \mathbf{a}_2 \quad \mathbf{a}_1 \quad \mathbf{a}_2 \quad \mathbf{a}_1 \quad \mathbf{a}_2 \quad \mathbf{a}_1 \quad \mathbf{a}_2]$$

$$\tilde{\mathbf{B}} = [\mathbf{b}_1 \quad \mathbf{b}_1 \quad \mathbf{b}_2 \quad \mathbf{b}_2 \quad \mathbf{b}_1 \quad \mathbf{b}_1 \quad \mathbf{b}_2 \quad \mathbf{b}_2 \quad \mathbf{b}_1 \quad \mathbf{b}_1 \quad \mathbf{b}_2 \quad \mathbf{b}_2] \tag{5.6}$$

$$\tilde{\mathbf{C}} = [\mathbf{c}_1 \quad \mathbf{c}_1 \quad \mathbf{c}_1 \quad \mathbf{c}_1 \quad \mathbf{c}_2 \quad \mathbf{c}_2 \quad \mathbf{c}_2 \quad \mathbf{c}_2 \quad \mathbf{c}_3 \quad \mathbf{c}_3 \quad \mathbf{c}_3 \quad \mathbf{c}_3]$$

where $\mathbf{a}_i$, $\mathbf{b}_j$ and $\mathbf{c}_k$ are the ith, jth and kth column of $\mathbf{A}$, $\mathbf{B}$ and $\mathbf{C}$, respectively, then Equation (5.5) can be rewritten as

$$x_{ijk} = \sum_{s=1}^{S} \tilde{a}_{is}\tilde{b}_{js}\tilde{c}_{ks}\tilde{g}_{sss} e_{ijk} \tag{5.7}$$

where the index s indicates the elements on the superdiagonal of $\underline{\tilde{\mathbf{G}}}$ and runs from $p = q = r = 1$ to $p = 2$, $q = 2$ and $r = 3$. If the elements on the superdiagonal of $\underline{\tilde{\mathbf{G}}}$ are absorbed in $\tilde{\mathbf{A}}$, then the equation for a 12-component constrained PARAFAC model is obtained. The process as described above can be done for every Tucker3 model. Hence, every Tucker3 model can be described as a very special constrained PARAFAC model with many components. The resulting PARAFAC model is constrained, because some columns in $\mathbf{A}$, $\mathbf{B}$ and $\mathbf{C}$ are forced to be equal as described indirectly in Equation (5.7). Due to the constraints, this particular PARAFAC model does not possess the uniqueness properties that PARAFAC usually does.

EXAMPLE 5.1

PARAFAC and Tucker3 representations of the FIA example

In Chapter 2 the FIA example was used to explain rank deficiency. The same example is used here to show that a constrained PARAFAC model can also be represented by a constrained Tucker3 model. For convenience, some of the equations of Chapter 2 are repeated. Suppose that the response of the analyte 2-hydroxybenzaldehyde ($\mathbf{X}_{2\text{-HBA}}$) can be modeled by

$$\mathbf{X}_{2\text{-HBA}} = \mathbf{c}_{a,2}\mathbf{s}'_{a,2} + \mathbf{c}_{b,2}\mathbf{s}'_{b,2} + \mathbf{E}_{2\text{-HBA}}$$

where the symbols are the same as in Chapter 2 ($\mathbf{c}_{a,2}$ is the concentration profile of 2-HBA in acidic form and $\mathbf{s}_{a,2}$ is the corresponding spectrum. Subscript b refers to the analyte in basic form). The response of a new mixture $\mathbf{X}_{\text{new}}$ containing 2-hydroxybenzaldehyde and an unknown interferent can be modeled as

$$\mathbf{X}_{\text{new}} = \gamma\mathbf{c}_{a,2}\mathbf{s}'_{a,2} + \gamma\mathbf{c}_{b,2}\mathbf{s}'_{b,2} + \mathbf{c}_{a,i}\mathbf{s}'_{a,i} + \mathbf{c}_{b,i}\mathbf{s}'_{b,i} + \mathbf{E}_{\text{new}}$$

where $\mathbf{c}_{a,i}$ and $\mathbf{c}_{b,i}$ are the concentration profiles of the acidic and basic form of the interferent, respectively, and $\mathbf{s}_{a,i}$ and $\mathbf{s}_{b,i}$ are the spectra of the acidic and basic form of the interferent, respectively. There is a restriction on the total concentration profiles, that the sum of the basic and acidic profile must be the same shape for every analyte. This can be expressed as $\mathbf{c}_{a,2} + \mathbf{c}_{b,2} = \alpha(\mathbf{c}_{a,i} + \mathbf{c}_{b,i})$ (see Chapter 2). Define now the matrices

$$\mathbf{S} = [\, \mathbf{s}_{a,2} \quad \mathbf{s}_{b,2} \quad \mathbf{s}_{a,i} \quad \mathbf{s}_{b,i} \,], \quad \mathbf{C} = [\, \mathbf{c}_{a,2} \quad \mathbf{c}_{b,2} \quad \mathbf{c}_{a,i} \quad \mathbf{c}_{b,i} \,], \quad \mathbf{D}_1 = \mathrm{diag}(1,1,0,0),$$
$$\mathbf{D}_2 = \mathrm{diag}(\gamma,\gamma,1,1),$$

then

$$\mathbf{X}_{2\text{-HBA}} = \mathbf{CD}_1\mathbf{S}' + \mathbf{E}_{2\text{-HBA}}$$
$$\mathbf{X}_{\text{new}} = \mathbf{CD}_2\mathbf{S}' + \mathbf{E}_{\text{new}}$$

which shows that the three-way array $\underline{\mathbf{X}}$, resulting from stacking $\mathbf{X}_{2\text{-HBA}}$ and $\mathbf{X}_{\text{new}}$ on top of each other, can be modeled with a PARAFAC model. The restriction on the total concentration profiles translates to a restriction on the columns of $\mathbf{C}$.

The restriction on the columns of $\mathbf{C}$ can also be explicitly accounted for by rewriting $\mathbf{X}_{2\text{-HBA}}$ and $\mathbf{X}_{\text{new}}$ using $\mathbf{c}_{a,2} = \alpha(\mathbf{c}_{a,i} + \mathbf{c}_{b,i}) - \mathbf{c}_{b,2}$:

$$\mathbf{X}_{2\text{-HBA}} = \alpha\mathbf{c}_{a,i}\mathbf{s}'_{a,2} + \alpha\mathbf{c}_{b,i}\mathbf{s}'_{a,2} - \mathbf{c}_{b,2}\mathbf{s}'_{a,2} + \mathbf{c}_{b,2}\mathbf{s}'_{b,2} + \mathbf{E}_{2\text{-HBA}}$$
$$\mathbf{X}_{\text{new}} = \gamma\alpha\mathbf{c}_{a,i}\mathbf{s}'_{a,2} + \gamma\alpha\mathbf{c}_{b,i}\mathbf{s}'_{a,2} - \gamma\mathbf{c}_{b,2}\mathbf{s}'_{a,2} + \gamma\mathbf{c}_{b,2}\mathbf{s}'_{b,2} + \mathbf{c}_{a,i}\mathbf{s}'_{a,i} + \mathbf{c}_{b,i}\mathbf{s}'_{b,i} + \mathbf{E}_{\text{new}}$$

By collecting $\mathbf{c}_{b,2}$, $\mathbf{c}_{a,i}$, $\mathbf{c}_{b,i}$ in $\mathbf{C}^*$ and $\mathbf{s}_{a,2}$, $\mathbf{s}_{b,2}$, $\mathbf{s}_{a,i}$, $\mathbf{s}_{b,i}$ in $\mathbf{S}$; defining

$$\mathbf{D} = \begin{bmatrix} 1 & 0 \\ \gamma & 1 \end{bmatrix}$$

and a proper core-array $\underline{\mathbf{G}}$, the three-way array $\underline{\mathbf{X}}$ can be written as a Tucker3 model. Actually, due to restrictions on the elements of the core-array $\underline{\mathbf{G}}$, the result is a constrained Tucker3 model.

This example shows that there are alternative ways to represent a three-way array, and Tucker3 and PARAFAC models are in some way exchangeable (see also Example 5.2).

Suppose that a three-way array $\underline{\mathbf{X}}$ can be fitted exactly with a (2,2,3) Tucker3 model. Then it can also be fitted with a 12-component PARAFAC model. Hence, the rank of the three-way array has a maximum of 12. This holds in general: an upper bound for rank $(\underline{\mathbf{X}})$, if $\underline{\mathbf{X}}$ can be fit exactly with an $P \times Q \times R$ Tucker3 model, is PQR. This upper bound is higher than the one given in Chapter 2 and therefore of little use in practical data analysis.

Hierarchy of Tucker3, Tucker2 and Tucker1 models

Consider a three-way array $\underline{\mathbf{X}}$ $(I \times J \times K)$ and different Tucker models for this array. A Tucker3 model for this array is

$$x_{ijk} = \sum_{p=1}^{P}\sum_{q=1}^{Q}\sum_{r=1}^{R} a_{ip}b_{jq}c_{kr}g_{pqr} + e_{ijk} \tag{5.8}$$

where x_{ijk}, a_{ip}, b_{jq}, c_{kr}, g_{pqr} and e_{ijk} are the typical elements of $\underline{\mathbf{X}}$, $\mathbf{A}$ $(I \times P)$, $\mathbf{B}$ $(J \times Q)$, $\mathbf{C}$ $(K \times R)$, $\underline{\mathbf{G}}$ $(P \times Q \times R)$ and $\underline{\mathbf{E}}$ $(I \times J \times K)$, respectively. The matrices $\mathbf{A}$, $\mathbf{B}$ and $\mathbf{C}$ are loading matrices, $\underline{\mathbf{G}}$ is the core-array and $\underline{\mathbf{E}}$ is the matrix of residuals.

A Tucker2 model can be obtained from a Tucker3 model by setting the number of components in one mode to the dimension of that mode. For example, by setting the number of components in the third mode to K the Tucker3 model reads

$$x_{ijk} = \sum_{p=1}^{P} \sum_{q=1}^{Q} \sum_{r=1}^{K} a_{ip} b_{jq} c_{kr} g_{pqr} + e_{ijk} \tag{5.9}$$

Due to the rotational freedom of the Tucker3 model, the loading matrix $\mathbf{C}$ $(K \times K)$ can be rotated to the identity matrix $\mathbf{I}$ $(K \times K)$. Then c_{kr} is one for $k = r$ and zero elsewhere. Rewriting Equation (5.9) (using the same symbols for the nonrotated and rotated forms for simplicity) gives

$$x_{ijk} = \sum_{p=1}^{P} \sum_{q=1}^{Q} a_{ip} b_{jq} c_{kk} g_{pqk} + e_{ijk} = \sum_{p=1}^{P} \sum_{q=1}^{Q} a_{ip} b_{jq} f_{pqk} + e_{ijk} \tag{5.10}$$

where the term f_{pqk} is defined implicitly. In matrix notation, this can be written as

$$\begin{aligned} \mathbf{X}_{(K \times IJ)} &= \mathbf{CG}_{(K \times PQ)}(\mathbf{B} \otimes \mathbf{A})' + \mathbf{E} \\ &= \mathbf{F}_{(K \times PQ)}(\mathbf{B} \otimes \mathbf{A})' + \mathbf{E} \end{aligned} \tag{5.11}$$

where $\mathbf{F}_{(K \times PQ)} = \mathbf{CG}_{(K \times PQ)} = \mathbf{IG}_{(K \times PQ)}$ where both $\underline{\mathbf{F}}$ and $\underline{\mathbf{G}}$ are of size $P \times Q \times K$. There are no third mode loadings as the third mode is not compressed. Again, $\mathbf{A}$ and $\mathbf{B}$ are loading matrices and $\underline{\mathbf{E}}$ is the three-way array of residuals, but these are different from the corresponding ones of the Tucker3 model. The three-way array $\underline{\mathbf{F}}$ is called the *extended* core-array in the Tucker2 model. There are two additional types of Tucker2 models: one in which the first mode is not compressed, and one in which the second mode is not compressed.

To explain the hierarchy of Tucker3 and Tucker2 models, assume a Tucker3 model $\mathbf{X} = \mathbf{AG}(\mathbf{C} \otimes \mathbf{B})' + \mathbf{E}$ and then a similar Tucker2 model where the first mode is not reduced $\mathbf{X} = \mathbf{F}(\mathbf{C} \otimes \mathbf{B})' + \mathbf{E}$ where $\mathbf{F}$ is the Tucker2 core array. The parameters in $\mathbf{B}$ and $\mathbf{C}$ will differ between the two models, but the important thing to note here is that the Tucker2 model will be more flexible than the Tucker3 model because in Tucker3 the part corresponding to $\mathbf{F}$ is specifically parameterized as $\mathbf{AG}$. Thus, the Tucker2 model will fit better than the Tucker3 model. Hence, a Tucker3 model can be seen as a constrained Tucker2 model, in much the same way as the linear models in Section 5.1.

The Tucker2 model can even be made less restrictive, e.g., by not compressing the second mode. This can be done by formulating the Tucker1 model

$$x_{ijk} = \sum_{p=1}^{P} a_{ip} h_{pjk} + e_{ijk} \tag{5.12}$$

where x_{ijk}, a_{ip} and e_{ijk} are the elements of $\underline{\mathbf{X}}$, $\mathbf{A}(I \times P)$ and $\underline{\mathbf{E}}$ $(I \times J \times K)$, respectively. The values h_{pjk} are the elements of the array $\underline{\mathbf{H}}$ $(P \times J \times K)$. The second and third mode of $\underline{\mathbf{X}}$ are not compressed, because loadings are not estimated in these modes. This model is even less constrained than the Tucker2 model and will therefore fit better.

Summarizing, a Tucker3 model compresses all three modes, a Tucker2 model compresses only two of the three modes, and a Tucker1 model compresses only one of the three modes. Hence the names Tucker1, Tucker2 and Tucker3 (and, for example, a four-way Tucker model with compression in all modes would therefore be a Tucker4 model [Kiers 2000]).

The model complexity increases in the order Tucker3, Tucker2, Tucker1, for a comparable set of models.

Hybrid forms: constrained Tucker3 models

In the previous section it was shown that a PARAFAC model can be regarded as a constrained Tucker3 model. Taking, e.g., an R-component PARAFAC model and a (P,Q,R) Tucker3 model of $\underline{\mathbf{X}}$ $(I \times J \times K)$ with $P = Q = R$, the connection between the two models is that the Tucker3 model reduces to the PARAFAC model by taking $g_{pqr} = 1$ if and only if $p = q = r$, and $g_{pqr} = 0$ elsewhere.

There is a whole range of models in between the R-component PARAFAC model and the (R, R, R) Tucker3 model above. If not only the superdiagonal elements of the $(R \times R \times R)$ core-array $\underline{\mathbf{G}}$ of the Tucker3 model are allowed to be nonzero, but also other specified elements, then a hybrid model is obtained. For example, if $R = 3$ and only the core-array elements $g_{111}, g_{222}, g_{333}$ and g_{123} are allowed to be nonzero then a hybrid form between a three-component PARAFAC model and a (3,3,3) Tucker3 model is obtained. By selecting other core elements of the Tucker3 model to be nonzero, a range of constrained Tucker3 models is obtained.

The property of hierarchy does not necessarily hold for the whole range of constrained Tucker3 models. Compare, e.g., two (3,3,3) constrained Tucker3 models. The first model has nonzero core elements $g_{111}, g_{222}, g_{333}$ and g_{123}. The second model has nonzero core elements $g_{111}, g_{222}, g_{333}$ and g_{223}. These models are not hierarchical because the first one cannot be derived by setting parameters to zero of the second one and vice versa. A third model with nonzero core elements $g_{111}, g_{222}, g_{333}, g_{123}$ and g_{223} has the first and second model as a special case. Hence, there exists a hierarchical structure connecting the first and third model, and between the second and third model.

A constrained Tucker3 model can also be regarded as a constrained PARAFAC model. In the previous section it was shown that a general Tucker3 model can be regarded as a constrained PARAFAC model. Hence, this also holds for constrained Tucker3 models. Consider, e.g., the (2,2,2) constrained Tucker3 model in which g_{111}, g_{222} and g_{112} are nonzero and with loading matrices $\mathbf{A} = [\mathbf{a}_1\ \mathbf{a}_2]$, $\mathbf{B} = [\mathbf{b}_1\ \mathbf{b}_2]$ and $\mathbf{C} = [\mathbf{c}_1\ \mathbf{c}_2]$. Then a new $(3 \times 3 \times 3)$ core-array $\underline{\tilde{\mathbf{G}}}$ can be defined with g_{111}, g_{222} and g_{112} on its superdiagonal and zero elsewhere, and new loading matrices $\tilde{\mathbf{A}} = [\mathbf{a}_1\ \mathbf{a}_1\ \mathbf{a}_2]$, $\tilde{\mathbf{B}} = [\mathbf{b}_1\ \mathbf{b}_1\ \mathbf{b}_2]$, $\tilde{\mathbf{C}} = [\mathbf{c}_1\ \mathbf{c}_2\ \mathbf{c}_2]$. The original constrained Tucker3 model is the same as a PARAFAC model with superdiagonal core-array $\underline{\tilde{\mathbf{G}}}$ and loading matrices $\tilde{\mathbf{A}}, \tilde{\mathbf{B}}, \tilde{\mathbf{C}}$.

Constrained or restricted Tucker models have found use in analytical chemistry, specifically, in second-order calibration [Kiers & Smilde 1998, Smilde *et al.* 1994a, Smilde *et al.* 1994b, Tauler *et al.* 1994] as well as in batch process modelling [Gurden *et al.* 2001, Gurden *et al.* 2002].

EXAMPLE 5.2

Constrained Tucker3 model for the FIA example

The FIA example, as used in earlier examples, can be used to illustrate a constrained Tucker3 model. The equations for $\mathbf{X}_{2\text{-HBA}}$ and $\mathbf{X}_{\text{new}}$ are repeated for convenience

$$\mathbf{X}_{2\text{-HBA}} = \alpha \mathbf{c}_{a,i} s'_{a,2} + \alpha \mathbf{c}_{b,i} s'_{a,2} - \mathbf{c}_{b,2} s'_{a,2} + \mathbf{c}_{b,2} s'_{b,2} + \mathbf{E}_{2\text{-HBA}}$$

$$\mathbf{X}_{\text{new}} = \gamma \alpha \mathbf{c}_{a,i} s'_{a,2} + \gamma \alpha \mathbf{c}_{b,i} s'_{a,2} - \gamma \mathbf{c}_{b,2} s'_{a,2} + \gamma \mathbf{c}_{b,2} s'_{b,2} + \mathbf{c}_{a,i} s'_{a,i} + \mathbf{c}_{b,i} s'_{b,i} + \mathbf{E}_{\text{new}}$$

Collecting $\mathbf{c}_{b,2}$, $\mathbf{c}_{a,i}$, $\mathbf{c}_{b,i}$ in $\mathbf{C}^*$ and $s_{a,2}$, $s_{b,2}$, $s_{a,i}$, $s_{b,i}$ in $\mathbf{S}$ and defining

$$\mathbf{D} = \begin{bmatrix} 1 & 0 \\ \gamma & 1 \end{bmatrix}$$

gives the possibility to express $\mathbf{X} = [\mathbf{X}_{2\text{-HBA}} \ \mathbf{X}_{\text{new}}]$, which is the matricized three-way array $\underline{\mathbf{X}}$, in terms of a constrained Tucker3 model. The only question is which core-array elements of that model are zero and nonzero. This can be deduced by carefully considering the equations for $\mathbf{X}_{2\text{-HBA}}$ and $\mathbf{X}_{\text{new}}$. The term $\mathbf{c}_{a,i} \ s'_{a,2}$ is present in both $\mathbf{X}_{2\text{-HBA}}$ and $\mathbf{X}_{\text{new}}$ in the ratio of $1/\gamma$. Hence, the second column of $\mathbf{C}^*$, the first column in $\mathbf{S}$ and the first column in $\mathbf{D}$ have to be connected with a nonzero core-array element g_{211}. In the same way, all other nonzero elements of the core-array $\underline{\mathbf{G}}$ can be found; these are g_{311}, g_{121}, g_{121}, g_{232}, g_{342}.

Fitting errors, degrees of freedom, number of parameters and model complexity

The hierarchical relationships between the three-way models imply relationships between the fitting errors of models for a particular dataset because some of the three-way models can be regarded as constrained versions of one another [Kiers 1991a]. To explain these consequences the polynomial example in the beginning of Section 5.1 is considered again.

Suppose a given data set consisting of 20 measurements of the yield y, at different pressures (p) and temperatures (T) is available. Fit the model (5.1) to the data; this results in a fit error SSE_1, where SSE stands for sum of squared residuals (e). If the same data set is fitted to model (5.2), then this results in a fit error SSE_2. Then $\text{SSE}_2 \leq \text{SSE}_1$, because model (5.2) will always contain model (5.1). Model (5.1) is a constrained version of model (5.2), hence, there is a relationship between the fit errors of both models. This does not mean, however, that model (5.2) is better than model (5.1). If model (5.1) can describe the data well, the increased complexity of model (5.2) will only lead to overfitting, i.e., fitting of noise.

Comparing models (5.1) and (5.2), it is also clear that model (5.2) has more parameters (six) to estimate than model (5.1), which has only three parameters to estimate. Given the same amount of data, the parameters in model (5.1) are estimated with more degrees of freedom. Hence, the variances of the estimated parameters in model (5.1) are lower than the variances of the analogous parameters in model (5.2). This points to a trade-off between fit and variance, and this is indeed one of the leading principles of model building. If model complexity is defined in terms of the number of parameters which have to be estimated, then model (5.2) is more complex than model (5.1).

A similar reasoning as above can be applied to three-way models. Consider a data set $\underline{\mathbf{X}}$ ($50 \times 10 \times 20$) of 50 objects where 10 variables are measured in the second mode and 20 in the third mode. Several three-way models can be built for the systematic part of $\underline{\mathbf{X}}$. For illustrative reasons, consider possible three-way models as summarized in Table 5.1. This table also gives an idea of the model complexity. In the constrained (2,2,2) Tucker3 model it is assumed that only the elements g_{111}, g_{222} and g_{112} are nonzero. The (2,2,20) Tucker2 model

Table 5.1. Complexity of different three-way models of a data array with 10,000 elements

Model	Number of parameters				Total number of parameters
	<u>G</u>	A	B	C	
Two-component PARAFAC	0	100	20	40	160
Constrained-(2,2,2) Tucker3	3	100	20	40	163
(2,2,2) Tucker3	8	100	20	40	168
(2,2,20) Tucker2	80	100	20	0	200
(2,10,20) Tucker1	400	100	0	0	500

assumes that the third mode is not compressed, hence, this model has an extended core-array of size $2 \times 2 \times 20$ and no loading matrix **C**. In the results reported for the (2,10,20) Tucker1 model it is assumed that the second and third mode are not compressed, hence, this model has an extended core-array of size $2 \times 10 \times 20$ and no loading matrices **B** and **C**.

In Table 5.1, the number of parameters is not corrected for rotational freedom, scaling indeterminacies and possibilities to fix core-elements without losing fit [Louwerse *et al.* 1999]. Hence, the number of parameters of the models reported are only approximately equal to the number of free parameters but the corrections are small.

Going from the two-component PARAFAC model to the (2,10,20) Tucker1 model the number of parameters to estimate increases and hence also the model complexity increases. Obviously, the two-component PARAFAC model is the least complex of the models considered. If one of the modes (or two of the modes) is not compressed in Tucker models, this increases the number of parameters to be estimated considerably.

Because the five models are hierarchically ordered, the fit will improve from PARAFAC (2) to Tucker1 (2,10,20). The number of degrees of freedom associated with the residual sum of squares is not strictly defined in such nonlinear models, but it can be approximated. If three-way models are not hierarchically ordered, then comparing fit values of those alternative models is problematic [Barbieri *et al.* 1999, Louwerse *et al.* 1999].

ELABORATION **5.1**

Degrees of freedom of a three-way array

Consider a $2 \times 2 \times 2$ array. Such an array will contain eight elements, hence eight degrees of freedom. Each PARAFAC component of this array will consist of six parameters (two in each mode). Even though a two-component model will comprise 12 parameters, this is not always enough to describe all the variation in the data, as some $2 \times 2 \times 2$ arrays have rank three. Thus, simple counting of parameters does not lead to any good approximation of degrees of freedom. In fact, the problem of finding explicit rules for the maximal rank of arrays directly points to the difficulty of defining degrees of freedom. Also, the fact that different randomly made arrays can have different ranks indicate that degrees of freedom do not exist a priori, but have to be determined from the specific noise structure. This is very different from standard linear models, where degrees of freedom have a general and more easily understood definition.

5.2 Rotational Freedom and Uniqueness in Three-way Component Models

Uniqueness of models is an important issue. A unique model is a model where all parameters are identified under the said premises, e.g. structure and additional constraints such as orthogonality. Requiring a model of a matrix to be bilinear is not sufficient for identifying the parameters of the model uniquely. However, if additional constraints are used, the model can become unique. In principal component analysis it is additionally required that the components are orthogonal and extracted such that the first component explains as much variation as possible; the second explains as much variation of the yet unexplained part, etc. Applying these additional constraints makes the bilinear model unique (apart from sign indeterminacies).

Tucker models

In order to discuss uniqueness and rotational freedom, a clear definition of uniqueness is necessary. There are different levels of uniqueness. Consider, e.g., a PCA model of $\mathbf{X}\,(I \times J)$

$$\begin{aligned} &\mathbf{X} = \mathbf{TP}' + \mathbf{E} \text{ or} \\ &\mathbf{X} = \mathbf{t}_1\mathbf{p}_1' + \cdots + \mathbf{t}_R\mathbf{p}_R' + \mathbf{E} \end{aligned} \qquad (5.13)$$

where $\mathbf{T}$ is an $(I \times R)$ matrix of scores with elements columns $\mathbf{t}_r$ $(r = 1, \ldots, R)$, $\mathbf{P}$ is a $(J \times R)$ matrix of loadings with columns $\mathbf{p}_r$ $(r = 1, \ldots, R)$ and $\mathbf{E}$ is an $(I \times J)$ matrix of residuals (see a more thorough description of PCA in Chapter 3). Obviously, multiplying $\mathbf{t}_r$ by $c \neq 0$ and dividing $\mathbf{p}_r$ by the same scalar c gives the same result for the fit part of $\mathbf{X}$. Likewise, exchanging $\mathbf{t}_r$ with $\mathbf{t}_k$, and simultaneously $\mathbf{p}_r$ with $\mathbf{p}_k$, also gives the same fit values of $\mathbf{X}$. The first type of nonuniqueness is called the scaling nonuniqueness and the second type is called the permutational nonuniqueness (which is usually eliminated in PCA by requiring components to come in order of variance explained). It is customary to refer to uniqueness without considering the above (trivial) forms of nonuniqueness. Often this is stated as: a model is unique up to permutation and scaling differences. From now on, uniqueness will be meant in this way: the trivial scaling and permutation nonuniqueness are not considered relevant.

For convenience the equation of a (P, Q, R) Tucker 3 model of the three-way array $\underline{\mathbf{X}}$ $(I \times J \times K)$ is repeated:

$$\mathbf{X} = \mathbf{AG}(\mathbf{C}' \otimes \mathbf{B}') + \mathbf{E} \qquad (5.14)$$

where $\mathbf{X}\,(I \times JK)$, $\mathbf{G}\,(P \times QR)$ and $\mathbf{E}\,(I \times JK)$ are the properly matricized versions of $\underline{\mathbf{X}}$, $\underline{\mathbf{G}}$ and $\underline{\mathbf{E}}$, respectively, and $\mathbf{A}\,(I \times P)$, $\mathbf{B}\,(J \times Q)$, $\mathbf{C}\,(K \times R)$ are loading matrices. If a nonsingular matrix $\mathbf{S}\,(P \times P)$ is selected, then the following holds

$$\mathbf{X} = \mathbf{ASS}^{-1}\mathbf{G}(\mathbf{C}' \otimes \mathbf{B}') + \mathbf{E} = \tilde{\mathbf{A}}\tilde{\mathbf{G}}(\mathbf{C}' \otimes \mathbf{B}') + \mathbf{E} \qquad (5.15)$$

where $\tilde{\mathbf{A}}, \tilde{\mathbf{G}}$ are implicitly defined. Hence, the transformation with $\mathbf{S}$ does not change the fit and gives exactly the same residuals. The model with parameters $\mathbf{A}, \mathbf{B}, \mathbf{C}$ and $\mathbf{G}$ is not distinguishable from the model with parameters $\tilde{\mathbf{A}}, \mathbf{B}, \mathbf{C}$ and $\tilde{\mathbf{G}}$: both models are equivalent. The Tucker3 model can be written in two other, but completely equivalent, ways in terms of fit and parameters (see Section 4.1). In these two alternative ways of writing the same model, the row-mode of the matricized $\mathbf{X}$ is the second and the third mode of the three-way array respectively. Therefore, transformations of **B** and **C** can be defined in the same way as they are defined for **A** above.

If the matrix **S** is an orthogonal matrix ($\mathbf{S'S} = \mathbf{SS'} = \mathbf{I}$), then the transformation is called an orthogonal transformation. An orthogonal transformation is a reflection and/or a rotation. The property, as exemplified in and below Equation (5.15), is called rotational freedom and although this term is rather sloppy, it is commonly used.

The rotational freedom in Tucker3 models gives room for a convenient way to express the model. A given Tucker3 model can always be rotated to a model with columnwise orthonormal **A**, **B** and **C**. This has advantages in terms of interpreting the core-array elements because then the squared value of the core-array element g_{pqr} is the amount of variation in $\underline{\mathbf{X}}$ explained by the triad consisting of $\mathbf{a}_p$, $\mathbf{b}_q$, $\mathbf{c}_r$ [Kroonenberg 1983]. This is understandable due to the property of column-orthogonal matrices:

$$\|\mathbf{X}\| = \|\mathbf{AX}\| \qquad (5.16)$$

where **X** ($I \times J$) is an arbitrary matrix and **A** ($K \times I$) is a column-orthogonal matrix ($\mathbf{A'A} = \mathbf{I}$). In other words, Equation (5.16) says that variation in **X** is not changed if **X** is multiplied by a columnwise orthonormal matrix. Hence, writing the Tucker3 model with orthogonal loading matrices, ensures that

$$\|\hat{\mathbf{X}}\| = \|\mathbf{AG}(\mathbf{C'} \otimes \mathbf{B'})\| = \|\mathbf{G}\| \qquad (5.17)$$

All the variation in $\hat{\mathbf{X}}$ is contained in $\underline{\mathbf{G}}$ as explained in Elaboration 4.4.

The constraint of having orthogonal loading matrices **A**, **B** and **C** does not give a unique way of writing the Tucker3 model: even then it is possible to find (an infinite number of) orthogonal transformations that produce exactly the same fit of $\underline{\mathbf{X}}$. As already discussed in Chapter 4, the additional constraint on **A** to contain the eigenvectors of $\mathbf{X}(\mathbf{CC'} \otimes \mathbf{BB'})$ in order of decreasing eigenvalues, and similar constraints for **B** and **C** makes the Tucker3 decomposition unique.

Regardless of whether the parameters of the Tucker3 model are uniquely estimated, it can be shown that, like PCA, the Tucker3 model provides unique subspaces [Kroonenberg *et al.* 1989, Lathauwer 1997].

The same type of nonuniqueness as is present in Tucker3 models is present in Tucker2 and Tucker1 models.

Simplifying the core-array of Tucker3 models

A four-component PARAFAC model has four loading vectors in each mode and no inter-action between them. Therefore, interpretation is straightforward in the sense that loading vectors with the same component number vary independently of other factors and can be interpreted separately from these. A (4,4,4) Tucker3 model, on the other hand, has 64 ($4 \times 4 \times 4$) core-array elements (with 60 interactions!) and this makes it more difficult to interpret such a model.

If all core elements of a Tucker3 model are of a significant size, then all combinations of factor loadings must be interpreted. However, it may be possible to rotate the core (with appropriate counter-rotation of the loading matrices), in such a way that only a few core entries are significant. If this is the case, then interpretation can be simplified, because fewer combinations of factors have to be understood. There are three different issues playing a role in simplifying core-arrays (Figure 5.2). These are

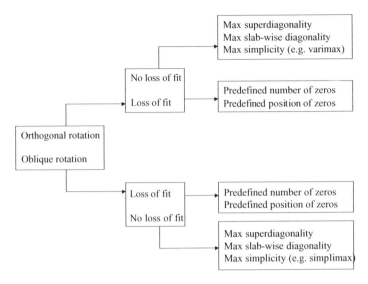

Figure 5.2. Different criteria can be used for selecting the appropriate type of rotation of a Tucker3 model.

- type of rotation (oblique or orthogonal);
- type of criterion (e.g. predefined zero elements, superdiagonality or just simplicity without preassumed structure);
- degree of loss of fit (no loss of fit or loss of fit).

The first choice which has to be made is whether the transformation is orthogonal or oblique. Columnwise orthonormal transformation matrices have the advantage that the component matrices remain orthogonal after the transformation, facilitating the interpretation of the transformed model. Additionally, columnwise orthonormal transformations effectively avoid problems with degenerate and meaningless solutions, and are also easier to implement and faster to compute [Kiers 1992]. Finally, rotating a Tucker3 model successfully towards superdiagonality using orthogonal transformations has advantages from an interpretational point of view. This is so because having orthogonal component matrices and only significant core elements on the superdiagonal, makes visualization of the model more straightforward (see Chapter 8). This has been discussed e.g. by Henrion [1993].

For the above reasons, orthogonal transformations are generally used for rotating the core. However, it must be recognized that orthogonal transformations also restrict the possible solution space as compared to oblique rotations. Thus, even though maximal superdiagonality is sought, this is under the added constraint of having orthogonal transformation matrices, and hence an intrinsic low-rank PARAFAC structure is unlikely to be deduced this way, because such PARAFAC structures mostly have nonorthogonal component matrices. Thus, the goal is not to obtain PARAFAC structure, but only to simplify the interpretation of the 'arbitrary' solution obtained from the Tucker3 algorithm.

The second choice pertains to the kind of simple structure that is desired. A predefined pattern of zeros can be postulated and a transformation can be sought that transforms the core-array as close as possible to this predefined structure (e.g. in a least squares sense).

Another possibility is to transform the core-array to a structure preferably having only nonzeros at the superdiagonal (assuming that $P = Q = R$). With such transformations it is possible to check how far an (R, R, R) Tucker3 model is from an R-component PARAFAC model. Trying to obtain diagonal frontal slices of the core-array is another possibility. If all frontal slices of the core-array can be diagonalized simultaneously, then the (R, R, R) Tucker3 model is equivalent to a PARAFAC model with R components [Kroonenberg 1983]. Finally, it is common to try to rotate the core to a simple structure in a more general sense [Kiers 1991b, Kiers 1997a]. Instead of rotating the core towards a predefined structure, the core-array is rotated such that as few elements as possible are of a high absolute value, and as many core elements as possible have a low absolute value. With this approach, no predefined structure is imposed such as e.g. a PARAFAC structure. It is merely sought to find a solution with few interactions. This can be an advantageous type of rotation to use as a one-shot general rotation tool.

The third aspect is whether a loss of fit is allowed or not. Sacrificing a certain amount of fit in order to introduce an extra zero in the core-array is a matter of choice for the data analyst. If a considerable number of entries in a core-array can be put to zero without loosing too much of the fit, then this may be preferred [Kiers 1994, Kiers 1998a].

As an example of obtaining a simpler core-array structure consider a (P, Q, R) Tucker3 model with $P = Q \times R$. The core-array $\mathbf{G}$ can be rearranged as a two-way $\mathbf{G}$ $(P \times P)$ and, hence, is square. This means that premultiplying $\mathbf{G}$ with $\mathbf{S} = \mathbf{G}^{-1}$ (if this inverse exists) produces a (matricized) core-array that is the identity matrix. The transformation is oblique, thereby destroying the orthogonality of the component matrices.

Usually, in core-arrays of considerable size, many elements can be forced to zero without loss of any fit by proper rotations. For that reason alone, transformations are worth considering [ten Berge & Kiers 1999]. Additionally, from an exploratory point of view, it is always helpful to reduce the complexity of the representation of a model. Several algorithms for core-rotations are given in the literature [Andersson & Henrion 1999, Henrion & Andersson 1999, Kiers 1992, Kiers 1993, Kiers 1994, Kiers 1997b, Kiers 1998a, ten Berge & Kiers 1999]. An algorithm for rotating towards such simple structure is provided in Chapter 6.

EXAMPLE 5.3

Rotating a Tucker3 core-array

A (3,3,3) Tucker3 model was fitted to a data set of five samples measured by fluorescence excitation–emission spectroscopy. The five samples contain different amounts of three fluorescent amino acids. The $3 \times 3 \times 3$ core $\mathbf{G}$ can be illustrated in terms of the percentage variance explained by each core element, i.e. the percentage of variance that the particular combination of factor loadings represent.

$$\mathbf{G}_{\%} = \begin{bmatrix} 0 & 0.2 & 2.9 & 0 & \mathbf{13.8} & \mathbf{20.2} & 0.4 & 1.9 & 0 \\ 1.8 & 0.6 & 3.7 & 3.8 & 0.3 & 1.7 & 0.3 & 0.5 & .2 \\ 2.7 & 0 & 5.0 & \mathbf{18.9} & 0.1 & \mathbf{17.8} & 1.7 & 0 & 1.5 \end{bmatrix}$$

As can be seen there are four elements marked with bold that describe most of the variation (71 % of the variance).

By rotating, a simpler structure may perhaps be obtained. Orthogonal rotations are chosen because efficient algorithms for such rotations are readily available. Because orthogonal rotations are used, it does *not* make sense to seek a superdiagonal core. Even though it is known that these data are very trilinear and hence a superdiagonal core would be appropriate when using (oblique) PARAFAC loadings, a superdiagonal core cannot be found using orthogonal rotations, because the Tucker3 loadings will then remain orthogonal.

Instead of maximal superdiagonality, it is sought to rotate orthogonally to a more general simple structure. This can be achieved by maximizing the variance of the squared core elements in the transformed core. As explained by Andersson and Henrion [1999] this variance is defined as

$$\sum_{p=1}^{P}\sum_{q=1}^{Q}\sum_{r=1}^{R}\left(g_{pqr}^2 - \frac{SS_g}{IJK}\right)^2$$

where SS_g is the sum of squared core elements. The solution is equivalent to a certain case of the so-called three-mode orthomax rotation where orthogonal transformations are used [Henrion & Andersson 1999, Kiers 1997b]. An algorithm for this rotation is given in Chapter 6. When the core is rotated orthogonally to maximize the variance of all squared core elements, the following core is obtained (expressed in terms of percentage variance explained by the core element)

$$\mathbf{G}_{\%}^{diag} = \begin{bmatrix} 0.6 & 1.1 & 0 & 0 & 0 & \mathbf{64.4} & 0.6 & 0 & 0 \\ 4.0 & 2.1 & 1.8 & 0 & 3.6 & 0 & 0 & 0.1 & 0 \\ 0.2 & 0.9 & 0.1 & \mathbf{19.9} & 0 & 0 & 0.2 & 0 & 0.5 \end{bmatrix}$$

As can be seen, a significant simplification has now been achieved with only two elements explaining 84 % of the variance! Thus, this core has a more parsimonious structure than the core above and is therefore preferred in terms of interpretation.

It is also possible to perform rotations focusing on achieving simple structure of the loadings rather than the core. This can simplify the interpretation of the loadings as such [Kiers 1993, Kiers 1997b, Kiers 1998a, Kiers 1998b, Kiers & ten Berge 1994a, Kiers & ten Berge 1994b].

A special case of simple core-array structures is when the positions of the zero elements in the core-array are known beforehand [Kiers 1992]. In such a situation, no rotations are necessary, because the simple structure is known beforehand. The simple structure is then imposed directly into the fitting of the model, rather than rotating the model after it has been fit. This is the situation in constrained Tucker3 models (see Section 5.1) where the model parameters are estimated by simply fixing the particular core-array elements to zero.

Uniqueness and partial uniqueness of PARAFAC models

UNIQUENESS OF PARAFAC MODELS

One of the most attractive features of the PARAFAC model is its uniqueness property. Many of the applications of PARAFAC models are based on this property, such as second-order calibration and resolution of spectral data [Leurgans & Ross 1992, Sanchez & Kowalski 1988]. It is, therefore, worthwhile to treat this topic in detail.

The term uniqueness is used in the sense discussed earlier, that is, trivial nonuniqueness due to scaling and permutation of the component matrices is not considered. These indeterminacies are usually handled in a problem-specific way. Suppose that $\underline{\mathbf{X}}$ is modeled with an R-component PARAFAC model, with component matrices $\mathbf{A}$, $\mathbf{B}$, $\mathbf{C}$, with elements a_{ir}, b_{jr} and c_{kr}, respectively. Hence, the elements x_{ijk} of $\underline{\mathbf{X}}$ can be written as

$$x_{ijk} = \sum_{r=1}^{R} a_{ir}b_{jr}c_{kr} + e_{ijk} \tag{5.18}$$

where e_{ijk} is an element of the residual array $\underline{\mathbf{E}}$ and Equation (5.18) holds for $i = 1, \ldots, I; j = 1, \ldots, J; k = 1, \ldots, K$. Now assume that there exist component matrices $\tilde{\mathbf{A}}, \tilde{\mathbf{B}}, \tilde{\mathbf{C}}$ with elements $\tilde{a}_{ir}, \tilde{b}_{jr}, \tilde{c}_{kr}$, respectively, such that

$$x_{ijk} = \sum_{r=1}^{R} \tilde{a}_{ir}\tilde{b}_{jr}\tilde{c}_{kr} + e_{ijk} \tag{5.19}$$

where x_{ijk} and e_{ijk} are *identical* to the corresponding values in Equation (5.18), then the uniqueness property states that $\mathbf{A},\mathbf{B},\mathbf{C}$ are equal to $\tilde{\mathbf{A}},\tilde{\mathbf{B}},\tilde{\mathbf{C}}$, respectively, up to scaling and permutation differences. In other words, it is not possible to rotate the PARAFAC component matrices without changing the fit. It is important to understand that the uniqueness property is in terms of the estimated model parameters. There is no guarantee that a PARAFAC model obtains a 'true' solution. However, if the model is correctly specified with the right number of components, and the trilinear PARAFAC structure is approximately valid, then the PARAFAC solution will provide estimates of the underlying parameters.

A sufficient condition for a PARAFAC model to give unique parameter estimates is given by Kruskal [1989] and later extended by Sidiropoulos and Bro [2000]. This condition is

$$k_\mathrm{A} + k_\mathrm{B} + k_\mathrm{C} \geq 2R + 2 \tag{5.20}$$

where k_A, k_B and k_C are the k-ranks of the component matrices $\mathbf{A}$, $\mathbf{B}$ and $\mathbf{C}$, respectively (see Section 2.5) and R is the number of components in the PARAFAC model. Furthermore, a one-component model is always unique regardless of the above even for a bilinear model. The Kruskal conditions are not necessary conditions except for R smaller than four [ten Berge & Sidiropoulos 2001]. The necessary conditions for uniqueness of PARAFAC models with more than three components are still unknown. Sanchez & Kowalski [1988] and Leurgans & Ross [1992] also give conditions for uniqueness, but these conditions are stronger than those of Equation (5.20). Hence, the Kruskal conditions are preferred. In many of the situations encountered in chemical practice, the condition of Equation (5.20) is met. In these cases, PARAFAC models give unique parameter estimates. When correctly specified, PARAFAC also provides estimates of the underlying physical parameters generating the data.

ELABORATION **5.2**

Uniqueness of PARAFAC models

A straightforward way of proving the uniqueness of PARAFAC models is outlined here (for details see Harshman [1972]). It proves uniqueness under the same conditions as those given by Sanchez & Kowalski [1988] and Leurgans & Ross [1992]. Suppose that the three-way array $\underline{\mathbf{X}}$ has two slices $\mathbf{X}_1$ and $\mathbf{X}_2$ and can be modeled with a PARAFAC model:

$$\mathbf{X}_1 = \mathbf{A}\mathbf{D}_1\mathbf{B}'$$
$$\mathbf{X}_2 = \mathbf{A}\mathbf{D}_2\mathbf{B}'$$

where $\mathbf{A} = [\mathbf{a}_1, \ldots, \mathbf{a}_R]$ and $\mathbf{B} = [\mathbf{b}_1, \ldots, \mathbf{b}_R]$ are $I \times R$ and $J \times R$ matrices with $I, J \geq R$ and of rank R; and $\mathbf{D}_1, \mathbf{D}_2$ are nonsingular diagonal matrices such that $(\mathbf{D}_1)(\mathbf{D}_2)^{-1} = \Delta$ where Δ has distinct diagonal elements. Let an alternative PARAFAC model for $\underline{\mathbf{X}}$ be represented by

$$\mathbf{X}_1 = \tilde{\mathbf{A}}\tilde{\mathbf{D}}_1\tilde{\mathbf{B}}'$$
$$\mathbf{X}_2 = \tilde{\mathbf{A}}\tilde{\mathbf{D}}_2\tilde{\mathbf{B}}'$$

Because $\mathbf{A}$ and $\tilde{\mathbf{A}}$ span the same subspace (i.e. the ranges of $\mathbf{X}_1$ and $\mathbf{X}_2$), there exists a nonsingular matrix $\mathbf{S}$ $(R \times R)$ such that

$$\tilde{\mathbf{A}} = \mathbf{A}\mathbf{S}$$

and, likewise, a nonsingular $\mathbf{T}$ $(R \times R)$ exists such that $\tilde{\mathbf{B}} = \mathbf{B}\mathbf{T}$. Hence,

$$\mathbf{A}\mathbf{D}_1\mathbf{B}' = \mathbf{A}\mathbf{S}\tilde{\mathbf{D}}_1\mathbf{T}'\mathbf{B}'$$
$$\mathbf{A}\mathbf{D}_2\mathbf{B}' = \mathbf{A}\mathbf{S}\tilde{\mathbf{D}}_2\mathbf{T}'\mathbf{B}'$$

which follows from combining the two alternative models. Premultiplying the above by $\mathbf{A}^+$ and postmultipying by $(\mathbf{B}^+)'$ gives

$$\mathbf{D}_1 = \mathbf{S}\tilde{\mathbf{D}}_1\mathbf{T}'$$
$$\mathbf{D}_2 = \mathbf{S}\tilde{\mathbf{D}}_2\mathbf{T}'$$

and, hence, $\mathbf{D}_2^{-1} = (\mathbf{T}')^{-1}(\tilde{\mathbf{D}}_2)^{-1}\mathbf{S}^{-1}$ Therefore, $\mathbf{D}_1\mathbf{D}_2^{-1} = \mathbf{S}\tilde{\mathbf{D}}_1\mathbf{T}'(\mathbf{T}')^{-1}(\tilde{\mathbf{D}}_2)^{-1}\mathbf{S}^{-1} = \mathbf{S}\tilde{\mathbf{D}}_1(\tilde{\mathbf{D}}_2)^{-1}\mathbf{S}^{-1}$. Written differently, this becomes $\Delta = \mathbf{S}\Gamma\mathbf{S}^{-1}$, where the diagonal matrices Δ and Γ are implicitly defined. This equation is actually an eigenvalue/eigenvector equation because it can be rewritten as

$$\Delta\mathbf{S} = \mathbf{S}\Gamma$$

and $\mathbf{S}$ contains the eigenvectors of Δ and Γ the corresponding eigenvalues in weakly descending order. The matrix Δ is diagonal, and it is known from matrix theory that the eigenvalues of a diagonal matrix are its diagonal elements. Hence,

$$\Gamma = \Pi\Delta\Pi'$$

where Π is a permutation matrix used to account for the fact that Δ is not necessarily in weakly descending order. Substituting $\Gamma = \Pi\Delta\Pi'$ in $\Delta\mathbf{S} = \mathbf{S}\Gamma$ gives:

$$\Delta\mathbf{S} = \mathbf{S}\Pi\Delta\Pi'$$

or, by postmultipying both sides by Π (which is an orthogonal matrix),

$$\Delta\mathbf{S}\Pi = \mathbf{S}\Pi\Delta$$

and this is again the eigenvalue/eigenvector equation of Δ. The matrix Δ has distinct diagonal elements and therefore distinct eigenvalues. Hence, its eigenvectors are uniquely determined up to their length. A solution to this problem is $S\Pi = \Lambda$, where, Λ is a diagonal matrix. Hence, $S\Pi = \Lambda$ is also the only solution and thus $S = \Lambda \Pi'$. This means that the only difference between $\tilde{A}$ and A are permutations and scale. Apart from these differences, A is unique. In a similar fashion, it can be shown that T is the product of the permutation matrix Π and a diagonal matrix. Together with $\tilde{B} = BT$ this establishes the essential uniqueness of the PARAFAC model.

It is clear from this proof that the condition of Δ to have distinct diagonal elements is essential. This comes down to having *sufficient variation* between the slices X_1 and X_2. Stated more precisely: the dyads (a_1, b_1) to (a_R, b_R) should express themselves in different proportions in the slices X_1 and X_2. At least two slices are needed to have this *sufficient variation*. Without the condition of distinct elements (and eigenvalues) on the diagonal of Δ, the eigenvalue/eigenvector system has no unique solution for those eigenvalues that are the same. Still, the eigenvectors corresponding to the distinct eigenvalues are unique and this means that the corresponding parts of S are unique (up to permutation and scale). This is called *partial uniqueness* by Harshman.

PARTIAL UNIQUENESS OF PARAFAC MODELS

In some cases, the Kruskal condition is not met and unique resolution of the data cannot be obtained. However, it may still be possible to obtain partial uniqueness in the sense that some parameters are uniquely determined. One often occurring example is in second-order calibration (Chapter 10) if the pure standard contains only the analyte of interest and the mixture contains the analyte and two or more interferents.

As an example, assume that two interferents are present in an unknown sample. The two measured matrices (standard and mixture) are stacked and three PARAFAC components are needed to model the data. The component matrix A in the sample direction has size 2×3 – two samples; three components – and the two columns in A associated with the interferents are a multiple of each other, because they both have a zero in the first row (concentrations of interferents in the standard). Hence, the k-rank of A is 1. Assuming that the other component matrices B and C have full rank (and thus full k-rank) gives a sum of k-ranks of $1 + 3 + 3 = 7$. This is less than $2 \times 3 + 2 = 8$, and the Kruskal condition is not met. Therefore, uniqueness is not guaranteed.

Harshman showed that uniqueness can still be obtained for those components that *do* have 'adequate' variation across all three modes [Harshman 1972]. Adequate variation, in this case, means that as the concentration of the analyte of interest is varying independently of the interferents, its parameters can be determined uniquely. Therefore, the analyte of interest can be quantified and its profiles recovered, whereas the profiles and concentrations of the interferents are subject to rotational indeterminacy.

Uniqueness and partial uniqueness of hybrid (constrained Tucker3) models

On the one hand, PARAFAC models give unique solutions under mild conditions and on the other hand, Tucker3 models have rotational indeterminacy. An interesting question is then whether hybrid (e.g. constrained Tucker3) models have uniqueness properties. This turns out to be a difficult question and only partial results are available [Kiers & Smilde 1995, Kiers & Smilde 1998, ten Berge & Smilde 2002].

EXAMPLE 5.4

Uniqueness of constrained Tucker3 models

Consider a (3,3,3) Tucker3 model with core-array $\underline{\mathbf{G}}$, which is matricized to $\mathbf{G}$:

$$\mathbf{G} = \begin{bmatrix} x & 0 & 0 & 0 & 0 & d & 0 & f & 0 \\ 0 & 0 & b & 0 & y & 0 & e & 0 & 0 \\ 0 & a & 0 & c & 0 & 0 & 0 & 0 & z \end{bmatrix}$$

where x,y,z,a,b,c,d,e,f are assumed to be nonzero. Kiers *et al.* [Kiers 1997b] show that under mild conditions the constrained Tucker3 model with this core-array gives unique estimates (apart from scaling and permutation differences). Illustrating the extravagance and complications arising in the study of uniqueness of such models: one of the conditions is that $(xyz + ade + bcf)^3 \neq 27abcdefxyz$. Needless to say, this condition is almost always met in practice.

The question of uniqueness of PARAFAC and constrained Tucker3 models is especially important in curve resolution and second-order calibration (Chapter 10). In second-order calibration, the response of a second-order instrument, e.g. liquid chromatography hyphenated with ultraviolet spectroscopy, is used to assess the concentration of an analyte in an unknown mixture. The possibility of doing so is called the second-order advantage. For simple second-order instruments, it can be shown that a PARAFAC model can be used for the calibration. In more complicated calibration situations a constrained Tucker3 model has to be used [Smilde *et al.* 1994a, Smilde *et al.* 1994b]. The question is then whether the calibration based on the constrained Tucker3 model retains the second-order advantage. Uniqueness for constrained Tucker3 models can only be assessed for each specific case [Kiers & Smilde 1998].

5.3 Properties of Tucker3 Models

Size of the Tucker3 core-array

In order to decide on the size of the Tucker3 core-array the following is a useful result. If (P, Q, R) is the dimensionality vector of $\underline{\mathbf{X}}$ (see Section 2.6), then $\underline{\mathbf{X}}$ can be modeled exactly with a (P, Q, R) Tucker3 model [Kruskal 1989]. This means that it is always useful to calculate the dimensionality vector, which can be done by matricizing $\underline{\mathbf{X}}$ $(I \times J \times K)$ in the three different two-way matrices $\mathbf{X}_{(I \times JK)}$, $\mathbf{X}_{(J \times IK)}$ and $\mathbf{X}_{(K \times IJ)}$ (see notation section) and subsequently use two-way tools for establishing the ranks of these matrices. Methods to establish the size of the Tucker3 core-array based on this principle are given by Timmerman and Kiers [2000].

An alternative procedure for establishing the size of the core-array is by extending the principle of cross-validation to three-way arrays which is explained in Chapter 7.

Rank of a Tucker3 core-array

A result which sheds light on the properties of a Tucker3 model is the following. Suppose that a (P, Q, R) Tucker3 models fits exactly a three-way array $\underline{\mathbf{X}}$ of rank S. Then the core-array $\underline{\mathbf{G}}$ of the Tucker3 model also has rank S. Hence, the rank of the three-way array $\underline{\mathbf{G}}$ inherits the rank of $\underline{\mathbf{X}}$ [Carroll *et al.* 1980].

5.4 Degeneracy Problem in PARAFAC Models

Degenerate solutions are sometimes encountered when fitting the PARAFAC model. Degeneracy is a situation in which the algorithm has difficulties in correctly fitting the model. The estimated model parameters are hence often unstable and unreliable. Degeneracy occurs when the model is simply inappropriate, for example, when the data are not appropriately modeled by a trilinear model. Some of these situations are referred to as *two-factor degeneracies*. When two factors are interrelated a Tucker3 model is appropriate and fitting PARAFAC models can then yield degenerate models that can be shown to converge completely only after infinitely many iterations, while the norm of the correlated loading vectors diverges. In such a case, the Tucker (or constrained versions) might be better [Smilde *et al.* 1994a].

Persistent degeneracies can also be caused by poor preprocessing. In such a case, degenerate solutions are observed even for a (too) low number of components. Extracting too many components can also give a degeneracy-like problem because the 'noise' components can be correlated. Some details about degeneracies can be found in Section 6.3.

A typical sign of a degeneracy is that two of the components become almost identical but with opposite sign or contribution to the model. The contribution of the components almost cancel each other. In PARAFAC each rank-one component can be expressed as the vectorized rank-one array obtained as

$$\mathbf{z}_r = \mathbf{c}_r \otimes \mathbf{b}_r \otimes \mathbf{a}_r \tag{5.21}$$

of size $IJK \times 1$. For the degenerate model, it would hold that for some r and p

$$\mathbf{z}_r \approx -\mathbf{z}_p \Rightarrow$$
$$\cos(\mathbf{z}_r, \mathbf{z}_p) = \frac{\mathbf{z}_r' \, \mathbf{z}_p}{\|\mathbf{z}_r\| \|\mathbf{z}_p\|} \approx -1 \tag{5.22}$$

This means that the loading vectors in component r and component p will be almost equal in shape, but negatively correlated.

An indication of degenerate solutions can thus be obtained by monitoring the correlation between all pairs of components. For three-way PARAFAC, the measure in Equation (5.22) can be calculated as

$$\begin{aligned} T_{rp} &= \cos(\mathbf{a}_r, \mathbf{a}_p)\cos(\mathbf{b}_r, \mathbf{b}_p)\cos(\mathbf{c}_r, \mathbf{c}_p) \\ &= \frac{\mathbf{a}_r' \, \mathbf{a}_p}{\|\mathbf{a}_r\| \|\mathbf{a}_p\|} \frac{\mathbf{b}_r' \, \mathbf{b}_p}{\|\mathbf{b}_r\| \|\mathbf{b}_p\|} \frac{\mathbf{c}_r' \, \mathbf{c}_p}{\|\mathbf{c}_r\| \|\mathbf{c}_p\|} \end{aligned} \tag{5.23}$$

where r and p indicate the rth and pth component, respectively. This measure is called Tucker's congruence coefficient and is also sometimes referred to as the uncorrected correlation coefficient [Mitchell & Burdick 1994]. It is easy to show that the symmetric $R \times R$ matrix **T** containing all possible triple cosines can be computed as

$$\mathbf{T} = (\mathbf{A}'\mathbf{A})^* (\mathbf{B}'\mathbf{B})^* (\mathbf{C}'\mathbf{C}) \tag{5.24}$$

for the three-way PARAFAC model if all loading vectors are scaled to length one.

In a situation with degeneracy, the more iterations are performed, the closer the correlation between the two components will come to minus one. However, the correlation will theoretically never reach minus one [Kruskal *et al.* 1989].

EXAMPLE 5.5

Example of a $2 \times 2 \times 2$ array with degeneracy

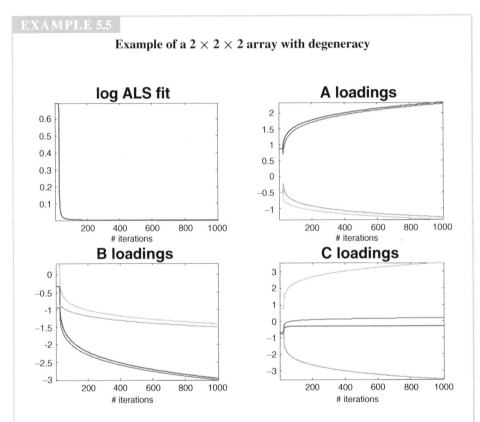

Figure 5.3. Fit values (upper left) and parameter values as a function of iteration number in a PARAFAC-ALS algorithm. For each mode, there are four parameters (2×2) which are shown as lines in the plots.

There exists a simple $2 \times 2 \times 2$ array which shows degeneracy problems and often serves as a benchmark data set to test algorithms [Kruskal *et al.* 1983, ten Berge & Kiers 1988]. This array $\underline{\mathbf{X}}$ has frontal slabs $\mathbf{X}_1$ and $\mathbf{X}_2$:

$$\mathbf{X}_1 = \begin{bmatrix} 1 & 0 \\ 0 & -1 \end{bmatrix}; \quad \mathbf{X}_2 = \begin{bmatrix} 0 & 1 \\ 1 & 0 \end{bmatrix}$$

Fitting a two-component PARAFAC model to this array will lead to the quickly converging fit values shown in Figure 5.3, whereas the parameters in **A**, **B**, and **C** do not converge at all as evidenced in the same figure.

5.5 Summary

This chapter discusses some properties of three-way component models. First of all, hierarchical relationships exist between Tucker, constrained Tucker and PARAFAC models. These relationships have consequences for fit values, model complexity and model selection. Moreover, if there are no hierarchical relationships between different component models, then comparing fit values becomes problematic.

The uniqueness properties of Tucker, constrained Tucker and PARAFAC models are discussed. A Tucker3 model finds unique subspaces, whereas a PARAFAC model finds unique axes. For constrained Tucker models, the situation is more complicated and no straightforward results are available.

Finally, some results are presented for Tucker3 models and the problem of degenerate PARAFAC solutions is discussed. Such degenerate solutions can be encountered in practice and sometimes have to do with postulating the wrong model.

6
ALGORITHMS

6.1 Introduction

The algorithms for estimating the parameters in the multi-way models are described in this chapter. First an introduction to optimization is given using the fitting of two-way models as an example. After describing the algorithms, it is described how to use fitted models on new data. The purpose may, for example, be to predict certain variables (e.g., calibration) or to conduct an outlier assessment (e.g., statistical process control). Usually the data are of such a structure that one mode is the observation/sample mode and the remaining modes are variable modes. Having the loading matrices and possibly the core array from a previously fit model, the scores of new samples are sought. How to predict the scores of new samples using a given model is discussed for each model individually. Finally, a core rotation algorithm is devised and it is discussed how to handle missing data and how to impose nonnegativity constraints in e.g. a PARAFAC algorithm.

6.2 Optimization Techniques

There are different routes to estimating the parameters of a model. Finding the parameters is an optimization problem, and in some situations, a directly computable solution may exist. In other situations an iterative algorithm has to be used. The two most important tools for fitting models in multi-way analysis are called *alternating least squares* and *eigenvalue* based solutions. Other approaches also exist, but these are beyond the scope of this book.

Consider the two-way problem of fitting a bilinear model of rank R to a given $I \times J$ matrix, $\mathbf{X}$. Thus, the parameters, $\mathbf{A}$ $(I \times R)$ and $\mathbf{B}$ $(J \times R)$, yielding the least squares solution

$$\min_{\mathbf{A},\mathbf{B}} \|\mathbf{X} - \mathbf{AB}'\|^2 \tag{6.1}$$

are sought ($\|\mathbf{E}\|$ is the Frobenius norm of $\mathbf{E}$). An attractive situation is to have a direct closed-form solution to the problem. In that case, the computational complexity of the

Multi-way Analysis With Applications in the Chemical Sciences.   A. Smilde, R. Bro and P. Geladi
© 2004 John Wiley & Sons, Ltd   ISBN: 0-471-98691-7

algorithm is known beforehand, and the algorithm can be more easily optimized. If a closed-form solution is not available, iterative algorithms have to be used. These work by gradually improving an interim estimate of the parameters until no further improvements are possible. The drawback of such an approach is that the complexity may vary from situation to situation [Gill *et al.* 1981].

For the stated two-way problem above, singular value decomposition of $\mathbf{X}$ or eigen-decomposition of $\mathbf{X}'\mathbf{X}$ can provide $\mathbf{A}$ and $\mathbf{B}$ directly. In the three-way case, similar 'closed-form' solutions also exist but they do not, as in the two-way case, provide a least-squares solution, but rather an approximate solution.

Iterative algorithms also exist for fitting the bilinear model of Equation (6.1). A well-known iterative algorithm for fitting the above model is the NIPALS or NILES algorithm developed by H. Wold [Wold 1966, Wold 1975]. The NIPALS algorithm can be viewed as based on the principle of alternating least squares.[1] In alternating least squares, the parameters are divided into several sets. Each set of parameters is estimated in a least squares sense conditionally on the interim values of the remaining parameters. The estimation of parameters is repeated iteratively until no change is observed e.g. in the fit of the model to the data. The reason for dividing the parameters into groups is to make it possible to use simpler algorithms for estimating the parameters.

Consider the bilinear model

$$\hat{\mathbf{X}} = \mathbf{A}\mathbf{B}' \tag{6.2}$$

If an initial estimate of $\mathbf{A}$, called $\tilde{\mathbf{A}}$, is provided, then estimating $\mathbf{B}$ given $\tilde{\mathbf{A}}$ is a regression problem with the solution

$$\tilde{\mathbf{B}} = \mathbf{X}'(\tilde{\mathbf{A}}^+)' \tag{6.3}$$

This does not solve the overall problem, but it improves any current estimate of $\mathbf{B}$. A new and better estimate of $\mathbf{A}$ given $\tilde{\mathbf{B}}$ can subsequently be determined as

$$\tilde{\mathbf{A}} = \mathbf{X}(\tilde{\mathbf{B}}^+)' \tag{6.4}$$

and a better estimate of $\mathbf{B}$ can then be obtained etc. This is the basic idea in alternating least squares algorithms: redefine the global problem (6.1) into usually linear subproblems that are easy to solve. Then iteratively solve these problems until convergence. As all estimates of parameters are least squares estimates, such an algorithm may only improve the fit or keep it the same if converged. Therefore the accompanying loss function value is nonincreasing, and in practice usually decreasing. Because the problem is a bounded-cost problem (the loss function cannot be lower than zero) convergence usually follows. Although convergence to a global minimum is not guaranteed, convergence is practical and attractive and one of the reasons for the widespread use of alternating least squares [de Leeuw *et al.* 1976]. Another benefit of alternating least squares algorithms is the simplicity of the involved substeps as compared to an algorithm working simultaneously on the entire problem and also the many variations available from the general theory of least squares regression, e.g. for handling restricted parameters or missing values. The drawback in difficult cases can be a rather slow convergence.

---

[1] NIPALS may also be viewed as a variant of the power method for calculating eigenvectors and -values [Hotelling 1936].

Consider an array $\underline{\mathbf{X}}$ and a generic model

$$\hat{\mathbf{X}} = \mathbf{M}(\mathbf{A}, \mathbf{B}, \mathbf{C}) \tag{6.5}$$

where the model of $\mathbf{X}$, $\mathbf{M}$, is a function of the parameters $\mathbf{A}$, $\mathbf{B}$ and $\mathbf{C}$ and is a matrix of the same size as the data held in $\mathbf{X}$. The number of parameter sets may vary from model to model. In PARAFAC there are three matrices, $\mathbf{A}$, $\mathbf{B}$ and $\mathbf{C}$, while in a bilinear model there are only two matrices. To estimate the parameters $\mathbf{A}$, $\mathbf{B}$, $\mathbf{C}$ of the model, an alternating least squares algorithm can be formulated

1. Initialize the parameters
2. $\tilde{\mathbf{A}} = \arg \min_{\mathbf{S}} \|\mathbf{X} - \mathbf{M}(\mathbf{S}, \mathbf{B}, \mathbf{C})\|^2$
3. $\tilde{\mathbf{B}} = \arg \min_{\mathbf{T}} \|\mathbf{X} - \mathbf{M}(\tilde{\mathbf{A}}, \mathbf{T}, \mathbf{C})\|^2$
4. $\tilde{\mathbf{C}} = \arg \min_{\mathbf{U}} \|\mathbf{M} - (\tilde{\mathbf{A}}, \tilde{\mathbf{B}}, \mathbf{U})\|^2$
5. Return to step 2 until convergence

The expression $\arg \min_{\mathbf{S}} \|\mathbf{X} - \mathbf{M}(\mathbf{S}, \mathbf{B}, \mathbf{C})\|^2$ means that the argument, $\mathbf{S}$, minimizing the loss function $\|\mathbf{X} - \mathbf{M}(\mathbf{S}, \mathbf{B}, \mathbf{C})\|^2$ provides the optimal update of the matrix $\mathbf{A}$. In the previous example with a bilinear model $\mathbf{M}(\mathbf{A}, \mathbf{B}) = \mathbf{A}\mathbf{B}'$, the updates given in Equations (6.3) and (6.4) provide the necessary steps in an alternating least squares algorithm.

The initialization of the parameters and check for convergence is described in detail under each model. Usually the iterative algorithm is stopped when either the parameters or the fit of the model does not change much.

In the next sections, it is shown how alternating least squares algorithms work for the PARAFAC model (Section 6.3) and Tucker models (Sections 6.4 and 6.5). The algorithm for N-PLS is given in Section 6.6. For N-PLS1, i.e. a situation with one dependent variable, the algorithm has a closed-form solution reminiscent of an eigenvalue solution, whereas for several dependent variables the algorithm is iterative. In Section 6.7, the algorithm for multi-way covariates regression models is described and in Section 6.8 core rotations are treated. After discussing the algorithms, it is shown how to handle missing data in Section 6.9 and how to impose nonnegativity on parameters (Section 6.10).

6.3 PARAFAC Algorithms

Algorithms for fitting the PARAFAC model are usually based on alternating least squares. This is advantageous because the algorithm is simple to implement, simple to incorporate constraints in, and because it guarantees convergence. However, it is also sometimes slow. An alternative type of algorithm based on a (generalized) eigenvalue problem gives an approximate solution and is described in Appendix 6.A.

Alternating least squares PARAFAC algorithm

The general three-way PARAFAC model is defined as

$$\mathbf{X}_{(I \times JK)} = \mathbf{A}(\mathbf{C} \odot \mathbf{B})' + \mathbf{E} \tag{6.6}$$

Array representation

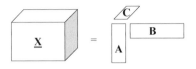

Matrix representation

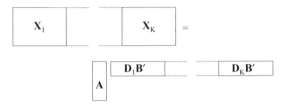

Figure 6.1. The structure of the PARAFAC model represented as a three-way array and matricized to a matrix.

and the corresponding least squares loss function is

$$\min_{A,B,C} \|X - A(C \odot B)'\|^2 \tag{6.7}$$

In order to fit the model using alternating least squares it is necessary to come up with an update for **A** given **B** and **C**; for **B** given **A** and **C** and for **C** given **A** and **B**. Due to the symmetry of the model, an update for one mode, e.g. **A**, is essentially identical to an update for any of the modes with the role of the different loading matrices shifted. To estimate **A** conditionally on **B** and **C** formulate the optimization problem as

$$\min_{A} \|X - AZ'\|^2 \tag{6.8}$$

where **Z** equals $C \odot B$ ($= [\mathbf{D}_1\mathbf{B}' \; \mathbf{D}_2\mathbf{B}' \dots \mathbf{D}_k\mathbf{B}']'$) and the diagonal matrix $\mathbf{D}_k$ holds the kth row of **C** in its diagonal. In Figure 6.1 it is seen that the matricized model, which is mathematically equivalent to the array representation, is a bilinear model in **A** and **Z** ($= C \odot B$). Thus, finding the optimal **A** amounts to a regression step where the data are regressed on $C \odot B$. It follows that for given **B** and **C**, the problem turns into a two-way problem of finding the least squares optimal **A** in the model $\mathbf{X} = \mathbf{AZ}' + \mathbf{E}$. The solution to this problem, when **Z** has full column-rank, is simply

$$A = X(Z')^+ = XZ(Z'Z)^{-1} \tag{6.9}$$

From the symmetry of the problem, it follows that **B** and **C** can be updated in similar ways. An algorithm follows directly as shown in Algorithm 6.1.

ALGORITHM **6.1**

PARAFAC-ALS

Given: $\underline{\mathbf{X}}$ of size $I \times J \times K$ and sought dimension R. Superscripts indicating current estimates are omitted for notational convenience

1. Initialize $\mathbf{B}$ and $\mathbf{C}$
2. $\mathbf{Z} = (\mathbf{C} \odot \mathbf{B})$
   $\mathbf{A} = \mathbf{X}_{(I \times JK)} \mathbf{Z} (\mathbf{Z}'\mathbf{Z})^{-1}$
3. $\mathbf{Z} = (\mathbf{C} \odot \mathbf{A})$
   $\mathbf{B} = \mathbf{X}_{(J \times IK)} \mathbf{Z} (\mathbf{Z}'\mathbf{Z})^{-1}$
4. $\mathbf{Z} = (\mathbf{B} \odot \mathbf{A})$
   $\mathbf{C} = \mathbf{X}_{(K \times JI)} \mathbf{Z} (\mathbf{Z}'\mathbf{Z})^{-1}$
5. Go to step 1 until relative change in fit is small

Calculating $\mathbf{Z}$ explicitly in the above problem can be computationally costly for large arrays. Also, the data array has to be continuously rearranged or kept in three versions ($\mathbf{X}_{(I \times JK)}$, $\mathbf{X}_{(J \times IK)}$ and $\mathbf{X}_{(K \times IJ)}$) which requires excessive memory. Harshman and Carroll and Chang [Carroll & Chang 1970, Harshman 1970] have observed that a simpler updating scheme is possible due to the structure of the problem. The algorithm can be formulated in terms of the frontal slices, $\mathbf{X}_k$, and therefore no rearrangements of the data array are necessary. The algorithm calculates the small matrices $\mathbf{XZ}$ and $\mathbf{Z}'\mathbf{Z}$ directly from $\mathbf{B}$, $\mathbf{C}$, and $\mathbf{X}$. This is beneficial because it is computationally less expensive and less memory demanding. For the update of $\mathbf{A}$, the matrix $\mathbf{X}$ ($I \times JK$) equals $[\mathbf{X}_1 \mathbf{X}_2 \ldots \mathbf{X}_K]$. Thus

$$\mathbf{XZ} = [\mathbf{X}_1 \mathbf{X}_2 \ldots \mathbf{X}_K][\mathbf{D}_1 \mathbf{B}' \mathbf{D}_2 \mathbf{B}' \ldots \mathbf{D}_K \mathbf{B}']' \tag{6.10}$$
$$= \mathbf{X}_1 \mathbf{B} \mathbf{D}_1 + \mathbf{X}_2 \mathbf{B} \mathbf{D}_2 + \cdots + \mathbf{X}_K \mathbf{B} \mathbf{D}_K$$

and further it follows from the properties of the Khatri–Rao product that

$$\mathbf{Z}'\mathbf{Z} = (\mathbf{C} \odot \mathbf{B})'(\mathbf{C} \odot \mathbf{B}) = (\mathbf{C}'\mathbf{C}) * (\mathbf{B}'\mathbf{B}) \tag{6.11}$$

as shown in Chapter 2. Thus, the update of $\mathbf{A}$ can be written as

$$\mathbf{A} = \mathbf{XZ}(\mathbf{Z}'\mathbf{Z})^{-1} = \left(\sum_{k=1}^{K} \mathbf{X}_k \mathbf{B} \mathbf{D}_k \right) \{(\mathbf{C}'\mathbf{C}) * (\mathbf{B}'\mathbf{B})\}^{-1} \tag{6.12}$$

For the two remaining modes, similar algorithmic shortcuts can be developed and it is customary to express all the three sub-problems in terms of the frontal slices of $\underline{\mathbf{X}}$ as above [Kiers & Krijnen 1991]. A more efficient algorithm for PARAFAC follows as shown in Algorithm 6.2, where it is noted that in step 4 the inverse need only be calculated once for all $k = 1, \ldots, K$. This algorithm is less transparent than that previously given, but it is computationally more efficient. The three updating steps for each of the modes are not identical. The reason for this apparent nonsymmetry of the algorithm is that the updates are all based on frontal slices of the data ($\mathbf{X}_k$). The model as such is still symmetric and equivalent to that in Algorithm 6.1.

ALGORITHM **6.2**

PARAFAC-ALS

Given: $\underline{\mathbf{X}}$ of size $I \times J \times K$ and the dimension R. Superscripts indicating current estimates are omitted for notational convenience

1. Initialize **B** and **C**

2. $\mathbf{A} = \left(\sum_{k=1}^{K} \mathbf{X}_k \mathbf{B} \mathbf{D}_k \right) \{ (\mathbf{C}'\mathbf{C}) * (\mathbf{B}'\mathbf{B}) \}^{-1}$

3. $\mathbf{B} = \left(\sum_{k=1}^{K} \mathbf{X}_k' \mathbf{A} \mathbf{D}_k \right) \{ (\mathbf{C}'\mathbf{C}) * (\mathbf{A}'\mathbf{A}) \}^{-1}$

4. $\mathbf{d}_k = \{ (\mathbf{B}'\mathbf{B}) * (\mathbf{A}'\mathbf{A}) \}^{-1} \{ (\mathbf{A}'\mathbf{X}_k\mathbf{B}) * \mathbf{I} \} \mathbf{1}, k = 1, \ldots, K$

5. Go to step 2 until relative change in fit is small

In step four, $\mathbf{d}_k$ is the kth row of **C**, and **I** is an $R \times R$ identity array and **1** an R vector of ones. The term $\{ (\mathbf{A}'\mathbf{X}_k\mathbf{B}) * \mathbf{I} \} \mathbf{1}$ is an R-column vector holding the diagonal of $\mathbf{A}'\mathbf{X}_k\mathbf{B}$.

Theoretically, convergence of fit does not necessarily imply convergence of parameters, but in practical situations this is usually the case. Therefore, the stopping criterion for the algorithm is usually a small relative change in the fit of the model. For example, the algorithm can be stopped when the improvement in fit from one iteration to the next is less than 0.0001 % of the prior fit. This criterion is usually sufficient for ensuring a good solution but in certain difficult cases[2] it may be observed that the solutions are different when fitted several times from different starting points. The reason can be that the algorithm does not, to a sufficient degree, approach the minimum. Although the same solution would be reached if the iterations had continued, the convergence criterion is not sufficiently small so that the solutions obtained are so far from this minimum that they are not similar. In such cases, lowering the convergence criterion usually helps achieving convergence to a minimum.

Calculating the fit after each iteration can be costly when the involved arrays are large. If the algorithm is monotonically convergent, as is the alternating least squares algorithm, then it is possible to use other simpler measures of convergence. One such measure, is to follow the relative change in the parameters instead of the model itself. Though this measure is not equivalent to using the change in fit, it will converge simultaneously with the fit and a low change in the estimated parameters is therefore indicative of convergence [Mitchell & Burdick 1994]. An exception occurs when the solution suffers from a *two-factor degeneracy* where the parameter estimates do not converge even though the fit converges at least within finite computational accuracy. Endless iterations in such a situation are usually avoided by allowing only a finite number of iterations.

Initializing PARAFAC-ALS

As with any optimization, good initial estimates of the parameters can speed up the PARAFAC-ALS algorithm and decrease the risk of converging to a local minimum. A good estimate is characterized as one that leads more directly to a global minimum. Random initialization does not always provide a good first estimate [Burdick *et al.* 1990, Harshman & Lundy 1984, Li & Gemperline 1993, Sanchez & Kowalski 1990, Sands & Young 1980]. Two problems can occur for difficult cases if inadequate initialization is used in PARAFAC-ALS. The algorithm produces local minima solutions or the computations are time-consuming.

---

[2] Difficult cases are usually situations in which some loading vectors are very correlated or situations in which too many components are extracted and noise therefore being modeled.

Both problems seem to be related to the same type of characteristics. The most common ones are discussed below. Conclusive evidence of problems and suitable remedies is, however, not yet available.

- *Some loading vectors are highly correlated*. This leads to a situation where the regression problems solved during the iterative fitting are ill-conditioned. The interim solutions are uncertain and apparently this leads to a situation where the fit only improves slowly. If the true least squares model parameters are intrinsically ill-conditioned, it is difficult to avoid long fitting times with PARAFAC-ALS. A poor initialization can also lead to an interim solution with highly correlated loading vectors. Even when the final solution does not involve highly correlated loadings, the interim solution bears the same characteristics as above. Thus the slow convergence is an interim problem; a so-called *swamp*. The difference between a swamp and the intrinsically correlated parameters is that a swamp may sometimes be avoided either by refitting the model from another starting point [Mitchell & Burdick 1994], or by using better initial values or by using regularized regression [Bro 1998, Kiers 1998, Rayens & Mitchell 1997]. An extreme situation of intrinsically correlated loadings is the two-factor degeneracy.
- *The model is not fully identified*. Unidentified models appear when the k-rank conditions (p.111) are not satisfied. Consider for example a second-order calibration problem in which two samples are modeled using PARAFAC. One sample may be an unknown sample with, say, R underlying components and the other sample a pure standard with only one underlying component. It follows that the true sample-mode loading ideally holding the estimated relative concentrations has the form

$$\mathbf{A} = \begin{bmatrix} a_{11} & 0 & \cdots & 0 \\ a_{21} & a_{22} & \cdots & a_{2R} \end{bmatrix}$$

As all columns but the first are perfectly correlated, the k-rank condition predicts that this model is likely not identifiable unless R equals two.[3] Even though some parameters are not identified,[4] it is still possible to fit the PARAFAC model. However, estimation can be time-consuming, because the orientation of some loading vectors is arbitrary, hence the error surface is flat.

A solution to some of the above problems except degeneracy is to use as good an initialization as possible. This is termed a *rational start*. For PARAFAC a rational start is usually based on an approximate solution obtained from the generalized rank annihilation method or direct trilinear decomposition. It may also be obtained from a traditional curve resolution method [Saurina *et al.* 1997]. The rational start has the advantage that it is likely to be close to the true solution, thereby saving computation time. A problem with the rational start is that of assessing whether convergence to a global minimum has been obtained, because refitting several times from the same starting point leads to the exact same parameters per definition. An approach similar to the rational start is the *semi-rational start*. It is usually based on

---

[3] For the k-rank conditions to hold, the inequality $k_A + k_B + k_C \geq 2R + 2$ must be valid. Assuming that the second and third mode loadings, **B** and **C**, have full rank R it must hold that $k_A + R + R \geq 2R + 2 \Rightarrow k_A \geq 2$. When there are two or more interferents in the sample, then **A** has more than one column reading $[0\,a]'$. Regardless that a may differ for different columns of **A** these columns are collinear and hence the k-rank of **A** is one. Therefore the PARAFAC model is likely not unique according to the Kruskal conditions.

[4] Harshman [1972] showed that even though all other factors are subject to rotational freedom, the factor of the analyte of interest can be uniquely recovered (partial uniqueness, see p. 111).

singular vectors. The singular vectors approximately span the right subspace and this seems to be sufficient for avoiding swamps in many situations. Both approaches generally work well, but sometimes they may lead to a local minimum. On the other hand, a *random start* may be used. If several different random starts are used it is possible to assess if local minima are encountered. If all refit models have the same fit to the data, it is an indication that the algorithm has correctly estimated the parameters. If not, then either local minima exist or the algorithm has been terminated prematurely.

A recommendable approach is to fit the model from a rational or semi-rational starting point as well as several random starting points. If the same solution is obtained for every refit model, it is likely not a local minimum.

Calculating scores of new samples

When a model has been fitted and scores of new samples are required, the basic problem to solve is a simple multiple linear regression problem. The PARAFAC model is given

$$\mathbf{X}_{(I \times JK)} = \mathbf{A}(\mathbf{C} \odot \mathbf{B})' + \mathbf{E} \tag{6.13}$$

for a three-way array matricized to a two-way matrix. A new sample can be held in a matrix $\mathbf{X}_{new}$ of size $J \times K$. Reshaping the data matrix according to the above model yields a $1 \times JK$ vector $(\text{vec } \mathbf{X}_{new})'$ and the model of this new sample given the above calculated loadings can be written as

$$(\text{vec } \mathbf{X}_{new})' = \mathbf{a}'(\mathbf{C} \odot \mathbf{B})' + \mathbf{E}_{new} \tag{6.14}$$

where the R-vector $\mathbf{a}$ holds the score values for the new sample (corresponding to one row of $\mathbf{A}$ above) and where $\mathbf{B}$ and $\mathbf{C}$ are known from the prior fit model. In Figure 6.2 the model of the new sample is depicted. When the first mode refers to samples, the problem

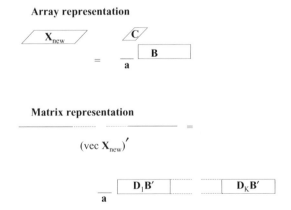

Figure 6.2. The structure of the PARAFAC model of one new sample represented as a two-way array and vectorized to a vector. In a prediction context, only the scores (**a**) are unknown. The noise part is left out of the plots for clarity.

of estimating the scores of a sample given by the data matrix $\mathbf{X}_{new}$ ($J \times K$) is then found in a least squares sense by finding

$$\min_{\mathbf{a}} \| \text{vec } \mathbf{X}_{new} - (\mathbf{C} \odot \mathbf{B})\mathbf{a} \|^2 \tag{6.15}$$

where $\mathbf{B}$ and $\mathbf{C}$ are given from a prior fit model. Therefore $\mathbf{a}$ is given by

$$\mathbf{a} = (\mathbf{C} \odot \mathbf{B})^+ \text{vec } \mathbf{X}_{new} \tag{6.16}$$

Similar expressions are derived if mode two or mode three parameters are sought.

6.4 Tucker3 Algorithms

Two different algorithms for fitting the Tucker3 model are given. First, an algorithm is given for noise-free data, i.e., for data that conform exactly to the model. As for PARAFAC, a simple closed-form algorithm exists and it also provides an approximate solution in the noisy case. This algorithm will be described and afterwards a least squares algorithm is given. Before discussing the algorithms, the calculation of the core array is first discussed. The core is implicitly given by the loading matrices and the original $\underline{\mathbf{X}}$. For a Tucker3 model with orthogonal loading matrices a compact expression for the core is given which is algorithmically helpful.

Calculating the core of the Tucker3 model

The core array $\underline{\mathbf{G}}$ of a Tucker3 model is implicitly given by the loading matrices. In the Tucker3 model there are interactions between every component vector in every mode. An individual element of the core array, g_{pqr}, gives the importance of the specified combination of components (Figure 6.3) as explained in Chapter 4. The core array of the Tucker3 model holds the coefficients of the array expressed within the truncated bases defined by the loading matrices. The core can be determined by regressing the data array onto these bases. For general nonorthogonal loading matrices, the regression can be expressed in the following way. If $\mathbf{A}$ ($I \times P$) holds the first mode loadings, then the coefficients of $\underline{\mathbf{X}}$ with respect to

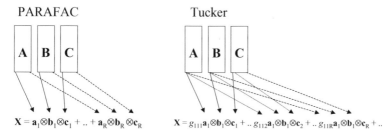

Figure 6.3. The difference between PARAFAC and Tucker3.

this basis can be found by solving

$$\mathbf{X}_{(I \times JK)} = \mathbf{AW}' + \mathbf{E}_{(I \times JK)} \Rightarrow \mathbf{W}' = (\mathbf{A}'\mathbf{A})^{-1}\mathbf{A}'\mathbf{X}_{(I \times JK)} \tag{6.17}$$

For the noiseless case, $\mathbf{X}_{(I \times JK)} = \mathbf{A}(\mathbf{A}'\mathbf{A})^{-1}\mathbf{A}'\mathbf{X}_{(I \times JK)}$ showing that $\mathbf{X}$ can be expressed on the truncated basis $\mathbf{A}$. The regression onto $\mathbf{A}$, $\mathbf{W}$, is an array of size $P \times JK$ which holds the coefficients of $\underline{\mathbf{X}}$ expressed with respect to the truncated basis $\mathbf{A}$ in the first mode. For all three modes considered simultaneously the coefficient matrix is the array expressed in terms of all the new bases; $\mathbf{A}$ in the first mode, $\mathbf{B}$ in the second mode, and $\mathbf{C}$ in the third mode. This array reads

$$\mathbf{G}_{(P \times QR)} = (\mathbf{A}'\mathbf{A})^{-1}\mathbf{A}'\mathbf{X}_{(I \times JK)} \left[(\mathbf{C}'\mathbf{C})^{-1}\mathbf{C}' \otimes (\mathbf{B}'\mathbf{B})^{-1}\mathbf{B}' \right] \tag{6.18}$$

which is the general equation for finding the core array given the bases. For orthogonal loadings the expression in Equation (6.19) follows due to the identity of the cross-product of the individual loading matrices:

$$\mathbf{G} = \mathbf{A}'\mathbf{X}(\mathbf{C} \otimes \mathbf{B}) \tag{6.19}$$

where $\mathbf{G}$ is the $P \times Q \times R$ core array $\underline{\mathbf{G}}$ matricized to a $P \times QR$ matrix and $\mathbf{X}$ is the data array matricized to an $I \times JK$ matrix. If the fit is perfect, then $\underline{\mathbf{G}}$ contains exactly the same information as $\underline{\mathbf{X}}$ merely expressed using different 'coordinates' (see Elaboration 4.4)

Closed-form solution for the Tucker3 model

The loss function for a (P, Q, R) Tucker3 model is

$$\min_{\mathbf{A},\mathbf{B},\mathbf{C},\mathbf{G}} \|\mathbf{X} - \mathbf{AG}(\mathbf{C} \otimes \mathbf{B})'\|^2 \tag{6.20}$$

subject to $\mathbf{A}$, $\mathbf{B}$, and $\mathbf{C}$ being orthogonal without loss of generality. The orthonormality constraints can be relaxed if, for example, nonnegative parameters are desired. In the following, only the orthonormal model is considered in detail. Assume that the data can be written as

$$\mathbf{X} = \mathbf{AG}(\mathbf{C} \otimes \mathbf{B})' \tag{6.21}$$

The column-space of $\mathbf{X}$ is completely within the range of $\mathbf{A}$, $\Re(\mathbf{A})$, because the presumed model is perfectly fitting. This subspace may be determined as the first P left singular vectors of a singular value decomposition of $\mathbf{X}$ thus immediately providing $\mathbf{A}$. Similarly, the matrix $\mathbf{B}$ can be found directly from a singular value decomposition of the data matricized in the second mode, i.e., by performing a singular value decomposition of $\mathbf{X}_{(J \times IK)}$ and from $\mathbf{X}_{(K \times IJ)}$ the matrix $\mathbf{C}$ can be found. The core array can subsequently be calculated as shown in Equation (6.19) yielding the Algorithm 6.3.

ALGORITHM **6.3**
Closed-form solution for orthogonal Tucker3

Given: $\underline{\mathbf{X}}$ of size $I \times J \times K$ and sought dimensions (P, Q, R).

1. $\mathbf{A}$ equals first P left singular vectors of $\mathbf{X}_{(I \times JK)}$
2. $\mathbf{B}$ equals first Q left singular vectors of $\mathbf{X}_{(J \times IK)}$

3. **C** equals first R left singular vectors of $\mathbf{X}_{(K \times IJ)}$
4. $\mathbf{G} = \mathbf{A}'\mathbf{X}(\mathbf{C} \otimes \mathbf{B})$

Least squares algorithm for the orthogonal Tucker3 model

Originally, the algorithms proposed for fitting the Tucker models were not least squares algorithms but rather of the kind shown in the preceding section. Later, least squares algorithms based on the principle of alternating least squares were introduced. The most important algorithm is called TUCKALS3 for fitting the model in a least squares sense with orthonormal loading vectors [Kroonenberg & de Leeuw 1980]. More efficient algorithms and algorithms handling higher order models have later appeared [Andersson & Bro 1998, Röhmel *et al.* 1983, ten Berge *et al.* 1987].

From the definition of $\underline{\mathbf{G}}$ in Equation (6.19) it follows that the Tucker3 model of $\underline{\mathbf{X}}$ with orthogonal loading matrices can be stated as

$$\mathbf{AG}(\mathbf{C}' \otimes \mathbf{B}') = \mathbf{AA}'\mathbf{X}(\mathbf{C} \otimes \mathbf{B})(\mathbf{C}' \otimes \mathbf{B}') \tag{6.22}$$

For **B** and **C** fixed it follows that finding the optimal **A** conditional on **B** and **C** is equal to minimizing the norm of $\mathbf{X} - \mathbf{AA}'\mathbf{X}(\mathbf{C} \otimes \mathbf{B})(\mathbf{C}' \otimes \mathbf{B}')$. This is equivalent to minimizing

$$\|\{\mathbf{X} - \mathbf{AA}'\mathbf{X}(\mathbf{C} \otimes \mathbf{B})(\mathbf{C}' \otimes \mathbf{B}')\}(\mathbf{C} \otimes \mathbf{B})\|^2 \tag{6.23}$$

because $(\mathbf{C} \otimes \mathbf{B})$ is orthogonal. This can be restated as

$$\min\|\mathbf{X}(\mathbf{C} \otimes \mathbf{B}) - \mathbf{AA}'\mathbf{X}(\mathbf{C} \otimes \mathbf{B})(\mathbf{C}' \otimes \mathbf{B}')(\mathbf{C} \otimes \mathbf{B})\|^2$$
$$= \min\|\mathbf{X}(\mathbf{C} \otimes \mathbf{B}) - \mathbf{AA}'\mathbf{X}(\mathbf{C} \otimes \mathbf{B})\|^2$$
$$= \min\|(\mathbf{I} - \mathbf{AA}')\mathbf{X}(\mathbf{C} \otimes \mathbf{B})\|^2 \tag{6.24}$$

The matrix $(\mathbf{I} - \mathbf{AA}')$ projects on the orthogonal complement of the column space of **A**; or stated otherwise; $(\mathbf{I} - \mathbf{AA}')$ produces the residuals after projection onto the column space of **A**. Hence, components have to be found such that, after projection of $\mathbf{X}(\mathbf{C} \otimes \mathbf{B})$ on these components, the residual variation is minimal in a least squares sense. This is what principal component analysis does (see Chapter 3) and a solution is to take the first P left singular vectors of $\mathbf{X}(\mathbf{C} \otimes \mathbf{B})$. The components found are automatically in the column space of $\mathbf{X}(\mathbf{C} \otimes \mathbf{B})$, and, therefore, in the column space of **X**.

For fixed **A** and **C**, a similar update can be derived for **B** by taking the first Q left singular vectors of the matrix $\mathbf{X}_{(J \times IK)}(\mathbf{C} \otimes \mathbf{A})$ and likewise for **C** from the matrix $\mathbf{X}_{(K \times IJ)}(\mathbf{B} \otimes \mathbf{A})$. Thus an alternating least squares algorithm for fitting the Tucker3 model with orthogonal loading matrices can be implemented as shown in Algorithm 6.4.

ALGORITHM **6.4**

Orthogonality constrained Tucker3 ALS

Given: $\underline{\mathbf{X}}$ of size $I \times J \times K$ and sought dimension $(P \times Q \times R)$. Superscripts indicating current estimates are omitted for notational convenience.

1. Initialize **B** and **C**
2. **A** equals first P left singular vectors of $\mathbf{X}_{(I \times JK)}(\mathbf{C} \otimes \mathbf{B})$

3. **B** equals first Q left singular vectors of $\mathbf{X}_{(J \times IK)}(\mathbf{C} \otimes \mathbf{A})$
4. **C** equals first R left singular vectors of $\mathbf{X}_{(K \times IJ)}(\mathbf{B} \otimes \mathbf{A})$
5. Go to step 2 until relative changes are small
6. $\mathbf{G} = \mathbf{A}'\mathbf{X}(\mathbf{C} \otimes \mathbf{B})$

Updating **A**, **B**, and **C** iteratively, implicitly updates **G** as well. Therefore, the core array need only be explicitly calculated once upon convergence. Usually, though, the change in fit is used as a measure of convergence. Fit is easily calculated from the sum of the squared elements of **G**, which represents the sum-of-squares explained by the interim model. Therefore steps five and six are usually exchanged in practical implementations. As for the PARAFAC algorithm, the convergence criterion commonly used is that the relative change in fit between two subsequent iterations is sufficiently small.

In the above algorithms, it has been assumed that the loading matrices are orthogonal. Having orthogonal loading matrices leads to efficient algorithms, but in some situations nonorthogonal loading matrices may be preferred. This can, for example, happen in the use of the so-called restricted Tucker3 models where the loading vectors typically have a physical meaning, and are mostly not to be orthogonal. An algorithm for unconstrained (nonorthogonal) loading parameters can be obtained by solving specifically for each component matrix in the model using standard regression techniques. For example, an update of **A** given the data and **B**, **C**, and **G**, is easily obtained from

$$\min\|\mathbf{X} - \mathbf{A}\mathbf{G}(\mathbf{C}' \otimes \mathbf{B}')\|^2 \tag{6.25}$$

from which the update of **A** follows as

$$\mathbf{A} = \mathbf{X}\{\mathbf{G}(\mathbf{C}' \otimes \mathbf{B}')\}^{+} \tag{6.26}$$

For the nonorthogonal model, updating the component matrix does not provide an indirect update of the core array, whose update must therefore be directly included in the iterative fitting. The core array may be found from posing the model as

$$\text{vec } \mathbf{X} = (\mathbf{C} \otimes \mathbf{B} \otimes \mathbf{A})\text{vec } \mathbf{G} + \text{vec } \mathbf{E} \tag{6.27}$$

Efficient algorithms, taking advantage of the Kronecker structure of the model, are usually implemented [Andersson & Bro 1998, Kiers *et al.* 1992, Kroonenberg *et al.* 1989, Paatero & Andersson 1999].

Initializing the Tucker3 algorithm

For initializing Algorithm 6.4 it is necessary to provide initial values of **B** and **C**. Usually, these are taken as the left singular vectors from a singular value decomposition of the array matricized to a matrix, where the given mode is kept as row-mode and the two additional modes are combined into the column-mode. That is, the solution from Algorithm 6.3 is usually taken as starting values for the least squares algorithm. For large arrays, this approach can be time-consuming and other approaches may be used instead, based on calculating singular value decompositions of smaller matrices than the raw data [Andersson & Bro 1998].

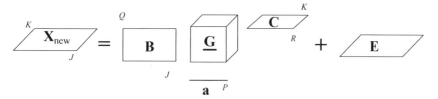

Figure 6.4. The model for determining the scores, **a** $(P \times 1)$, of a new sample, $\mathbf{X}_{\text{new}}$, given the parameters of a pre-estimated Tucker3 model.

Calculating scores of new samples

When a Tucker3 model is used for calculating the scores of a new sample, the loading parameters and the core parameters are usually considered fixed. This turns the problem into a multiple linear regression problem. Assume that the first mode is the sample mode and a new sample, $\mathbf{X}_{\text{new}}$, of size $J \times K$ is given. This sample matrix corresponds in size to one horizontal slice of the original array (Figure 6.4). The problem can be expressed

$$\min_{\mathbf{a}} \|\text{vec } \mathbf{X}_{\text{new}} - (\mathbf{C} \otimes \mathbf{B})\mathbf{G}'\mathbf{a}\|^2 \tag{6.28}$$

The matrices **B**, **C**, and **G** are known from the existing model. The least squares solution is

$$\mathbf{a} = ((\mathbf{C} \otimes \mathbf{B})\mathbf{G}')^+ \text{vec } \mathbf{X}_{\text{new}} \tag{6.29}$$

6.5 Tucker2 and Tucker1 Algorithms

An algorithm for fitting the Tucker2 model may be developed in line with the Tucker3 algorithms. The loss function for the Tucker2 model is

$$\min_{\mathbf{B},\mathbf{C},\mathbf{G}} \|\mathbf{X} - \mathbf{G}(\mathbf{C} \otimes \mathbf{B})'\|^2 \tag{6.30}$$

assuming that it is the first mode that is not reduced. The second mode loading matrix **B** is of size $J \times Q$, the third mode loading matrix **C** is of size $K \times R$ and the extended core array $\underline{\mathbf{G}}$ is of size $I \times Q \times R$. Both loading matrices are here assumed to be orthogonal though this is not mandatory. Similar expressions are obtained if the second or third mode is not reduced. An algorithm for fitting the Tucker2 model is shown in Algorithm 6.5.

ALGORITHM **6.5**

Orthogonality constrained Tucker2-ALS algorithm

Given: $\underline{\mathbf{X}}$ of size $I \times J \times K$ and sought dimensionality $(Q \times R)$. Superscripts indicating current estimates are omitted for notational convenience.

1. Initialize **B** and **C**
2. **B** equals first Q left singular vectors of $\mathbf{X}_{(J \times IK)}(\mathbf{C} \otimes \mathbf{I})$
3. **C** equals first R left singular vectors of $\mathbf{X}_{(K \times IJ)}(\mathbf{B} \otimes \mathbf{I})$

4. Go to step 1 until relative change in fit is small

5. $\mathbf{G} = \mathbf{X}(\mathbf{C} \otimes \mathbf{B})$

The Tucker1 model is equivalent to a bilinear model of the data matricized to a two-way array (Section 4.2). Thus, standard two-way algorithms may be used for fitting the model.

6.6 Multilinear Partial Least Squares Regression

The background for the extension of two-way partial least squares regression to multiway data (N-PLS) was provided by Bro [1996] and further elaborated on by Smilde [1997] and de Jong [1998]. An improved model of $\underline{\mathbf{X}}$ was later introduced which, however, maintains the same predictions as the original model [Bro *et al.* 2001]. Only the three-way version of N-PLS is considered here. It differs from the N-way ($N > 3$) partial least squares regression algorithm in that it has a closed-form solution for the situation with only one dependent variable.

Theoretical background

The basic outline of a traditional PLS regression algorithm was described in Chapter 3. The algorithm to be extended to multi-way regression differs from most PLS algorithms in that no extra loading vectors (usually called $\mathbf{p}$) appear. It has been shown that the introduction of the loading vector $\mathbf{p}$ is immaterial with respect to predictions for the two-way problem and it turns out that it is more natural to extend the version without additional loading vectors to multiway data. Loading vectors can be introduced as shown by Ståhle [1989]. Even though the predictions remain the same [de Jong 1998], it complicates the interpretation of the model, because the model of the data, in fact, becomes a two-way matricized model.

Partial least squares regression is a sequential algorithm where one component is computed at a time [Burnham *et al.* 1996, ter Braak & de Jong 1998]. Each component is calculated by first finding the variable weights ($\mathbf{w}$) of that component. By regressing the data onto these weight vectors, a score vector is found in the $\mathbf{X}$-space. This score vector has the attractive property of providing a conditional least squares model of the $\mathbf{X}$-data (or its residuals) [Smilde 1997]. Furthermore, by choosing the weights such that the covariance between $\mathbf{X}$ and $\mathbf{y}$ is maximized, a predictive model is obtained as elaborated on in Chapter 3 and 4.

In extending the PLS regression algorithm to three-way data, the only thing needed is to change the bilinear model of $\mathbf{X}$ to a trilinear model of $\underline{\mathbf{X}}$. For example, the first component in a bilinear two-way model

$$\mathbf{X}_{(I \times J)} = \mathbf{t}\mathbf{w}' + \mathbf{E_X} \tag{6.31}$$

is exchanged with

$$\mathbf{X}_{(I \times JK)} = \mathbf{t}(\mathbf{w}^K \otimes \mathbf{w}^J)' + \mathbf{E_X} \tag{6.32}$$

where the data are now three-way data and the model is a trilinear rank-one PARAFAC-like model with scores held in $\mathbf{t}$ ($I \times 1$), loadings in the second mode held in $\mathbf{w}^J$ ($J \times 1$)

and loadings in the third mode held in $\mathbf{w}^K (K \times 1)$. The aim is still the same; namely to determine loadings for the $\underline{\mathbf{X}}$ and $\mathbf{Y}$ models such that the scores have maximal covariance. Details for higher order models can be found in the literature [Bro 1998, Nilsson *et al.* 1998, Nilsson 1998].

The partial least squares regression algorithm

In PLS regression the weight vectors are normally defined to be of length one. Under this constraint, it is easy to show that for the two-way case with one dependent variable, the first loading vector is given by the vector

$$\mathbf{w} = \frac{\mathbf{X}'\mathbf{y}}{\|\mathbf{X}'\mathbf{y}\|} \tag{6.33}$$

The regression of $\mathbf{X}$ onto $\mathbf{w}$ with $\|\mathbf{w}\| = 1$ equals

$$\mathbf{t} = \mathbf{X}\mathbf{w}/(\mathbf{w}'\mathbf{w}) = \mathbf{X}\mathbf{w} \tag{6.34}$$

and leads to a score vector which has maximal covariance with $\mathbf{y}$ [Bro 1996, de Jong & Phatak 1997, Höskuldsson 1988].

For the three-way case, it is appropriate to use a shorthand notation introduced by Smilde [1997] before discussing the details of the algorithm. For a one-component three-way model with loading vectors $\mathbf{w}^J$ and $\mathbf{w}^K$ of length one, the notation $\mathbf{w}$ is used for the Kronecker product of the two:

$$\mathbf{w} = \mathbf{w}^K \otimes \mathbf{w}^J \tag{6.35}$$

When $\mathbf{w}^J$ and $\mathbf{w}^K$ are of length one it can be shown that $\mathbf{w}$ will also be of length one. Calculating the score of a set of samples, $\mathbf{X}_{(I \times JK)}$, for given $\mathbf{w}^J$ and $\mathbf{w}^K$ is equivalent to solving for $\mathbf{t}$ in $\min\|\mathbf{X}_{(I \times JK)} - \mathbf{t}(\mathbf{w}^K \otimes \mathbf{w}^J)'\|^2 = \|\mathbf{X}_{(I \times JK)} - \mathbf{t}\mathbf{w}^T\|^2$. As $\mathbf{w}$ has length one, it follows that the least squares score vector $\mathbf{t}$ given $\mathbf{w}$ can be found as $\mathbf{X}_{(I \times JK)}\mathbf{w}$. For an individual sample, $\mathbf{X}_i$ $(J \times K)$, the score can also be calculated by left-multiplying the $J \times K$ data matrix with $\mathbf{w}^J$ and right-multiplying with $\mathbf{w}^K$. Thus, $t_i = (\mathbf{w}^J)'\mathbf{X}_i\mathbf{w}^K$.

For a situation with one dependent variable, a score vector is sought which has maximal covariance with $\mathbf{y}$. For the two-way case the loading vector is found directly as $\mathbf{X}'\mathbf{y}/\|\mathbf{X}'\mathbf{y}\|$. For the three-way case the weights can also be found directly as the first left and right singular vectors from a singular value decomposition of a matrix $\mathbf{Z}$ $(J \times K)$ which is obtained as the inner product of $\underline{\mathbf{X}}$ and $\mathbf{y}$ (see proof in Appendix 6.B). Each element in $\mathbf{Z}$ is given by

$$z_{jk} = \sum_{i=1}^{I} y_i x_{ijk} \tag{6.36}$$

From the nonorthogonal two-way PLS1 algorithm (Chapter 3) and the above property it follows that an algorithm for three-way PLS1 can be formulated as in Algorithm 6.6, where it is assumed that the data have been properly preprocessed (centered across the first mode and possibly scaled). The three-way PLS1 algorithm is called tri-PLS1. The Arabic indicates the order of the dependent data, whereas the number before PLS indicates the order of the independent data. The algorithm provided is identical to the original algorithm whereas the algorithm with extended modeling of $\underline{\mathbf{X}}$ can be found in the literature [Bro *et al.* 2001].

ALGORITHM **6.6**

Tri-PLS1 algorithm

Given: Independent data $\underline{\mathbf{X}}$ ($I \times J \times K$) and dependent data $\mathbf{y}$ ($I \times 1$)
Let $f = 1$

1. Calculate $\mathbf{Z}$ from Equation (6.36).
2. Determine $\mathbf{w}_f^J$ and $\mathbf{w}_f^K$ as the first left and right singular vectors of $\mathbf{Z}$.
3. Calculate $\mathbf{t}_f$ from Equation (6.34) and define $\mathbf{T} = [\mathbf{t}_1 \, \mathbf{t}_2 \dots \mathbf{t}_f]$.
4. $\mathbf{b}_f = (\mathbf{T}'\mathbf{T})^{-1}\mathbf{T}'\mathbf{y}$.
5. Each sample, $\mathbf{X}_i$ is replaced with its residuals $\mathbf{X}_i - \mathbf{t}_{if}\mathbf{w}_f^J(\mathbf{w}_f^K)'$ and $\mathbf{y} = \mathbf{y} - \mathbf{Tb}_f$.
6. $f = f + 1$. Continue from 1 until proper description of original $\mathbf{y}$.

In step four, the score vectors from different components are not orthogonal. Hence, the calculation of the regression coefficients in step four has to be done taking formerly calculated score vectors into account. In step five, the deflation of $\mathbf{X}$ can be skipped without consequences for the predictions [de Jong 1998]. However, residuals, e.g. for diagnostic purposes, are then not available. Alternatively, deflation of $\mathbf{y}$ can be skipped while maintaining the deflation of $\mathbf{X}$.

The algorithm outlined above corresponds to the bilinear PLS1, in that there is only one dependent variable, $\mathbf{y}$. If several dependent variables are present, it is possible to model each dependent variable separately with this algorithm. It is also possible to model all ys simultaneously as in the PLS2 algorithm. This takes advantage of eventual/existing collinearity in the dependent variables by using a bilinear model of these (or even a multi-way model in case of multi-way structure in the dependent variables [Bro 1996]). The algorithm for several dependent variables, which is iterative, is shown in Algorithm 6.7. The algorithm follows directly from the trilinear tri-PLS1 and the ordinary two-way algorithm denoted bi-PLS2. The PLS1 algorithm is equivalent to the PLS2 algorithm for one dependent variable. The check for convergence in step 6 is usually performed by checking if the $\|\mathbf{t} - \mathbf{t}^{\text{old}}\| / \|\mathbf{t}^{\text{old}}\|$ is less than e.g. 10^{-6}. The score $\mathbf{t}$ is the current estimate and $\mathbf{t}^{\text{old}}$ is the prior estimate.

ALGORITHM **6.7**

Tri-PLS2 algorithm

Given: Independent data $\underline{\mathbf{X}}$ ($I \times J \times K$) and dependent data $\mathbf{Y}$ ($I \times M$)
Let $\mathbf{u}_1$ equal a column in $\mathbf{Y}$ and $f = 1$

1. Calculate the matrix $\mathbf{Z}$ using $\mathbf{X}$ and $\mathbf{u}_f$ analogous to Equation (6.36).
2. Determine $\mathbf{w}_f^J$ and $\mathbf{w}_f^K$ as the first left and right singular vectors of $\mathbf{Z}$.
3. Calculate $\mathbf{t}_f = \mathbf{X}(\mathbf{w}_f^K \otimes \mathbf{w}_f^J)$.
4. $\mathbf{q}_f = \mathbf{Y}'\mathbf{t}_f / \|\mathbf{Y}'\mathbf{t}_f\|$.
5. $\mathbf{u}_f = \mathbf{Yq}_f$.

6. If **u** has converged, continue, else step 1.
7. $\mathbf{b}_f = (\mathbf{T}^{\mathrm{T}}\mathbf{T})^{-1}\mathbf{T}^{\mathrm{T}}\mathbf{u}_f$.
8. $\mathbf{X}_i - t_{if}\mathbf{w}_f^J(\mathbf{w}_f^K)'$ and $\mathbf{Y} = \mathbf{Y} - \mathbf{T}\mathbf{b}_f\mathbf{q}_f'$.
9. $f = f + 1$. Continue from 1 until proper description of **Y**.

Bro *et al.* [2001] developed a modification of *N*-PLS in which the same predictions are obtained, but the model of $\underline{\mathbf{X}}$ is modified using a Tucker3 core array. This provides a better-fitting model of $\underline{\mathbf{X}}$ and also enables post-rotations of scores and weights, e.g. for jack-knife based assessment of parameters.

Predicting responses of new samples

Depending on what kind of information is needed, different possibilities exist for using a multilinear partial least squares regression model on new data. If only the prediction of **y** is wanted, it is possible to obtain a set of regression coefficients that relates **X** directly to **y**. This has been shown in detail by Smilde and de Jong for the multilinear PLS1 model [de Jong 1998, Smilde 1997] as follows. The first score vector $\mathbf{t}_1$ is

$$\mathbf{t}_1 = \mathbf{X}\mathbf{w}_1 \tag{6.37}$$

where **X** is the array $\underline{\mathbf{X}}$ matricized to an $I \times JK$ matrix and

$$\mathbf{w}_1 = \mathbf{w}_1^K \otimes \mathbf{w}_1^J \tag{6.38}$$

The second score vector can be found from regressing the residual $\mathbf{X} - \mathbf{t}_1\mathbf{w}_1'$ onto

$$\mathbf{w}_2 = \mathbf{w}_2^K \otimes \mathbf{w}_2^J \tag{6.39}$$

i.e.,

$$\begin{aligned}\mathbf{t}_2 &= (\mathbf{X} - \mathbf{t}_1\mathbf{w}_1')\mathbf{w}_2 \\ &= (\mathbf{X} - \mathbf{X}\mathbf{w}_1\mathbf{w}_1')\mathbf{w}_2 \\ &= \mathbf{X}(\mathbf{I} - \mathbf{w}_1\mathbf{w}_1')\mathbf{w}_2\end{aligned} \tag{6.40}$$

etc. because the vectorized weight vectors are of length one. This derivation leads to a general formula for a matrix

$$\mathbf{R} = \left[\mathbf{w}_1 \; (\mathbf{I} - \mathbf{w}_1\mathbf{w}_1')\mathbf{w}_2 \cdots \prod_{r=1}^{R-1}(\mathbf{I} - \mathbf{w}_r\mathbf{w}_r')\mathbf{w}_R \right] \tag{6.41}$$

for which it holds that

$$\mathbf{T} = \mathbf{X}\mathbf{R} \tag{6.42}$$

Thus, the R scores of a new sample, $\mathbf{x}'$ ($1 \times JK$) can be found as

$$\mathbf{t}' = \mathbf{x}'\mathbf{R} \tag{6.43}$$

For *N*-PLS1, the predictions of y for a new sample can be obtained from the scores in the following way. Let $\mathbf{t}_r = [t_1 \; t_2 \ldots t_r]$ and let $\mathbf{b}_r$ be an r-vector holding the inner relation regression coefficients for the rth component. Recall that for the rth component, the contribution to the predicted y is dependent on all the first r components. Hence, the contribution from the first component is $t_1 b_1$ (b_1 is 1×1); the contribution from the second component is $[t_1\ t_2] \mathbf{b}_2$ where $\mathbf{b}_2$ is of size 2×1 etc. Let $\mathbf{B}$ be an $R \times R$ matrix holding $\mathbf{b}_s$ in the first s elements of the sth column of $\mathbf{B}$ and zeros elsewhere. Then

$$\hat{\mathbf{y}} = t_1 b_1 + [t_1\ t_2]\mathbf{b}_2 + \cdots + [t_1\ t_2 \ldots t_R]\mathbf{b}_R = \mathbf{TB1} = \mathbf{Tb} \tag{6.44}$$

where $\mathbf{b} = \mathbf{B1}$ and where $\mathbf{1}$ is a vector of ones. Thus $\mathbf{b}$ is simply the sum of $\mathbf{B}$ summed across the rows.

It follows that

$$\mathbf{b}_{\text{pls}} = \mathbf{Rb} \tag{6.45}$$

holds the regression coefficients that directly compute $\hat{\mathbf{y}}$ from $\mathbf{X}$ or more correctly, predicts the centered/scaled $\mathbf{y}$ from the centered/scaled $\mathbf{X}$. Hence for a new sample $\mathbf{x}'_{\text{new}}$ ($1 \times JK$) the predicted y-variable is

$$\hat{y} = \mathbf{x}'_{\text{new}} \mathbf{b}_{\text{pls}} \tag{6.46}$$

In case residuals are needed for diagnostic purposes, each component must be calculated specifically. This can be done in the following way. The scores for the prediction samples can be found from Equation (6.42). Residuals in $\mathbf{x}'$ can then be determined as

$$\mathbf{e}_X = \mathbf{x}' - \mathbf{t}'\mathbf{W}' \tag{6.47}$$

where $\mathbf{W} = [\mathbf{w}_1 \mathbf{w}_2 \ldots \mathbf{w}_R]$ with the $\mathbf{w}$s defined as in e.g. Equation (6.38).

6.7 Multi-way Covariates Regression Models

The algorithm for multi-way covariates regression is explained using the example of regressing $\mathbf{y}$ ($I \times 1$) on $\underline{\mathbf{X}}$ ($I \times J \times K$), where a Tucker3 structure for $\underline{\mathbf{X}}$ is assumed, with P, Q and R components in the I, J and K direction, respectively. Generalizations to multivariate (and multi-way) $\mathbf{Y}$ and other three-way structures for $\underline{\mathbf{X}}$ can be inferred from this example.

The formal problem is

$$\min_{\mathbf{W}} \left[\alpha \|\mathbf{X} - \mathbf{XWP}'_X\|^2 + (1 - \alpha)\|\mathbf{y} - \mathbf{XWp}_y\|^2 \right] \tag{6.48}$$

where $0 \leq \alpha \leq 1$ and this value is chosen a priori.[5] For illustration purposes, the loading matrix $\mathbf{P}_X$ has a Tucker3 structure $\mathbf{G}(\mathbf{C} \otimes \mathbf{B})'$ where $\mathbf{B}$ ($J \times Q$) and $\mathbf{C}$ ($K \times R$) are loading matrices for the second and third direction of $\underline{\mathbf{X}}$, respectively and $\mathbf{G}$ ($P \times QR$) is the matricized version of the three-way core-array $\underline{\mathbf{G}}$ ($P \times Q \times R$). The components $\mathbf{A}$ ($I \times P$) are chosen to be in the column-space of $\mathbf{X}$ ($I \times JK$) by using the matrix of weights $\mathbf{W}$ ($JK \times R$) resulting in the scores being defined as $\mathbf{A} = \mathbf{XW}$. The vector $\mathbf{p}_y$ ($R \times 1$) is a vector of regression coefficients for predicting $\mathbf{y}$ from the scores $\mathbf{A} = \mathbf{XW}$.

---

[5] In practice α is estimated, e.g. through cross-validation, together with other meta-parameters such as the number of components.

An algorithm for solving Equation (6.48) can be based on alternating least squares. Starting values are chosen and then iterations are performed to minimize the loss function between brackets in Equation (6.48). The full algorithm is described in Algorithm 6.8. Several parts of the algorithm are explained below.

Assuming that all other parameters are given, $\mathbf{W}$ can be solved by rewriting the loss function of Equation (6.48) as

$$\alpha \left\| \mathbf{X} - \mathbf{XWP}'_{\mathbf{X}} \right\|^2 + (1 - \alpha) \left\| \mathbf{y} - \mathbf{XWp}_{\mathbf{y}} \right\|^2$$
$$= \left\| \sqrt{\alpha}(\mathbf{X} - \mathbf{XWP}'_{\mathbf{X}}) \right\|^2 + \left\| \sqrt{(1 - \alpha)}(\mathbf{y} - \mathbf{XWp}_{\mathbf{y}}) \right\|^2$$
$$= \left\| \sqrt{\alpha}\mathbf{X} - \sqrt{\alpha}\mathbf{XWP}'_{\mathbf{X}} \; \sqrt{(1 - \alpha)}\mathbf{y} - \sqrt{(1 - \alpha)}\mathbf{XWp}_{\mathbf{y}}) \right\|^2$$
$$= \left\| \sqrt{\alpha}\mathbf{X} - \mathbf{XW}\sqrt{\alpha}\mathbf{P}'_{\mathbf{X}} \; \sqrt{(1 - \alpha)}\mathbf{y} - \mathbf{XW}\sqrt{(1 - \alpha)}\mathbf{p}_{\mathbf{y}}) \right\|^2$$
$$= \left\| [\sqrt{\alpha}\mathbf{X} \; \sqrt{(1 - \alpha)}\mathbf{y}] - \mathbf{XW}[\sqrt{\alpha}\mathbf{P}'_{\mathbf{X}} \; \sqrt{(1 - \alpha)}\mathbf{p}_{\mathbf{y}}] \right\|^2$$
$$= \left\| \mathbf{Z} - \mathbf{XWU}' \right\|^2 \tag{6.49}$$

where the matrices $\mathbf{Z}$ and $\mathbf{U}$ are defined implicitly. The problem of Equation (6.49) is known as the Penrose problem [Penrose 1956] and the solution is $\mathbf{W} = \mathbf{X}^+\mathbf{Z}(\mathbf{U}^+)'$.

The vector $\mathbf{p}_{\mathbf{y}}$ only appears in the second term of the loss function. Given the other parameters, $\mathbf{p}_{\mathbf{y}}$ can be estimated as $\mathbf{p}'_{\mathbf{y}} = (\mathbf{XW})^+\mathbf{y}$. This is a simple least squares step. The matrix $\mathbf{P}_{\mathbf{X}}$ consists of three Tucker-type matrices $\mathbf{B}$, $\mathbf{C}$ and $\mathbf{G}$. Given the other parameters, these matrices $\mathbf{B}$, $\mathbf{C}$ and $\mathbf{G}$ can be updated by proper Tucker3 steps (see Algorithm 6.4) where the score matrix $\mathbf{A} = \mathbf{XW}$ is kept constant.

ALGORITHM **6.8**

Multi-way covariates regression–ALS

Given: Independent data $\mathbf{X}$ ($I \times J \times K$) and dependent data $\mathbf{y}$ ($I \times 1$). Superscripts indicating current estimates are omitted for notational convenience

1. Initialize $\mathbf{W}$, $\mathbf{B}$, $\mathbf{C}$.
2. Update $\mathbf{G}$ and $\mathbf{p}_{\mathbf{y}}$.
3. Update $\mathbf{W}$ by solving $\min\|\mathbf{Z} - \mathbf{XWU}'\|^2$.
4. Update $\mathbf{B}$, $\mathbf{C}$, and $\mathbf{G}$ using one iteration of Algorithm 6.4 but skipping the update for $\mathbf{A}$.
5. Go to step 3 until relative change in fit is small.

There are different possibilities for the initialization step. One solution is to take a semi-rational start using appropriate eigenvectors, e.g., the starting $\mathbf{A}$ contains the first P eigenvectors of $\mathbf{XX}'$ and the resulting starting value for $\mathbf{W} = \mathbf{X}^+\mathbf{A}$.

The algorithm can easily be adapted to accommodate e.g. a PARAFAC structure of $\underline{\mathbf{X}}$. Then the $\mathbf{G}$ matrix disappears and, hence, updating in step 4 is only over $\mathbf{B}$ and $\mathbf{C}$. These have to be updated with PARAFAC steps using Algorithm 6.2 but keeping $\mathbf{A}$ unchanged. If a Tucker1 structure of $\underline{\mathbf{X}}$ is assumed, then the problem reduces to two-way principal covariates regression on the properly matricized $\mathbf{X}$. For this problem, a noniterative solution exists [de Jong & Kiers 1992]. The algorithm can be extended to accommodate any numbers of ways for $\underline{\mathbf{X}}$ and $\underline{\mathbf{Y}}$.

Predicting responses of new samples

When data from new samples are to be evaluated with an existing multi-way covariates regression model, the principle follows almost immediately from the procedures used in other decomposition models. The scores of size $P \times 1$ of a new sample can be found by from the preprocessed data, $\mathbf{X}_{new}$ ($J \times K$) and the weights, $\mathbf{W}$ ($JK \times R$)

$$\mathbf{a} = \mathbf{W}' \text{ vec } \mathbf{X}_{new} \tag{6.50}$$

From these new scores, the predicted values can be calculated as

$$\hat{y} = \mathbf{a}\mathbf{p}_y \tag{6.51}$$

possibly followed by appropriate postprocessing (adding constants and inverting/undoing scaling).

6.8   Core Rotation in Tucker3 Models

For Tucker3 models, there is rotational freedom in the parameters as discussed earlier (Chapter 5). It is therefore possible to rotate the solution much in line with how principal component and factor analysis models can be rotated. The rotations can aim at either simplifying the structure of the loading matrices or simplifying the structure of the core array. An algorithm will be described here for rotating the Tucker3 core to what has been termed a maximal variance structure. Maximal variance structure means that the core elements are rotated such that there are many elements with low absolute values and few with high absolute values, meaning that the variance is concentrated on as few core elements as possible. This is achieved by rotating the core in such a way that the variance of the squared core elements is maximized [Andersson & Henrion 1999, Henrion & Andersson 1999]. This criterion can also be viewed in the light of more traditional methods for rotating to simple structure. Specifically it is identical to the quartimax (orthomax with weight zero and orthogonal transformations) criterion [Kruskal 1988]. An efficient algorithm for this problem can be found as a special case of three-mode orthomax rotation [Kiers 1997a]. Three rotation matrices, $\mathbf{S}$ ($P \times P$), $\mathbf{T}$ ($Q \times Q$) and $\mathbf{U}$ ($R \times R$) are sought, that define the rotations of the Tucker3 model. In this algorithm, the rotation matrices are constrained to be orthogonal, hence $\mathbf{S}^{-1} = \mathbf{S}'$, etc. This provides significant computational advantages, but also sacrifices the degree to which the core can be rotated. The loss-less rotation of the model can be written

$$\begin{aligned}
&\|\mathbf{X} - \mathbf{AG}(\mathbf{C} \otimes \mathbf{B})'\|^2 \\
&= \|\mathbf{X} - \mathbf{AS}'\mathbf{SG}(\mathbf{CU}'\mathbf{U} \otimes \mathbf{BT}'\mathbf{T})'\|^2 \\
&= \|\mathbf{X} - \mathbf{AS}'\mathbf{SG}(\mathbf{U} \otimes \mathbf{T})'(\mathbf{CU}' \otimes \mathbf{BT}')'\|^2 \\
&= \|\mathbf{X} - \tilde{\mathbf{A}}\tilde{\mathbf{G}}(\tilde{\mathbf{C}} \otimes \tilde{\mathbf{B}})'\|^2
\end{aligned} \tag{6.52}$$

where $\tilde{\mathbf{G}} = \mathbf{SG}(\mathbf{U} \otimes \mathbf{T})'$ is the rotated core and $\tilde{\mathbf{A}} = \mathbf{AS}'$, $\tilde{\mathbf{B}} = \mathbf{BT}'$, and $\tilde{\mathbf{C}} = \mathbf{CU}'$ are the counter-rotated component matrices. The three rotation matrices are found to optimize var(vec $\mathbf{G} * $ vec $\mathbf{G}$)), i.e. the sums of squares of the squared core elements. The algorithm is based on alternating least squares and works by updating $\mathbf{S}$, $\mathbf{T}$ and $\mathbf{U}$ in turn until convergence as shown in Algorithm 6.9.

ALGORITHM **6.9**

Rotating core to simple structure

Given: A Tucker3 core array $\mathbf{G}$ ($P \times Q \times R$)

Initialize $\mathbf{S} = \mathbf{I}_{(I \times I)}$, $\mathbf{T} = \mathbf{I}_{(J \times J)}$, $\mathbf{U} = \mathbf{I}_{(K \times K)}$

1. $\mathbf{S}^{\text{new}} = \text{quartimax}(\mathbf{G}_{(QR \times P)})$,   update $\mathbf{G}$ as $\mathbf{G}_{(QR \times P)}\mathbf{S}^{\text{new}}$ and $\mathbf{S}$ as $(\mathbf{S}^{\text{new}})'\mathbf{S}$.
2. $\mathbf{T}^{\text{new}} = \text{quartimax}(\mathbf{G}_{(PR \times Q)})$,   update $\mathbf{G}$ as $\mathbf{G}_{(PR \times Q)}\mathbf{T}^{\text{new}}$ and $\mathbf{T}$ as $(\mathbf{T}^{\text{new}})'\mathbf{T}$.
3. $\mathbf{U}^{\text{new}} = \text{quartimax}(\mathbf{G}_{(PQ \times R)})$,   update $\mathbf{G}$ as $\mathbf{G}_{(PQ \times R)}\mathbf{U}^{\text{new}}$ and $\mathbf{U}$ as $(\mathbf{U}^{\text{new}})'\mathbf{U}$.
4. If change in var(vec $\mathbf{G}$∗vec $\mathbf{G}$)) is low stop, else go to step 1.

In each step 1 to 3, the quartimax algorithm [Carroll 1953, Ferguson 1954, Neuhaus & Wrigley 1954] is applied to the matricized core to provide an update of one of the core rotation matrices. The quartimax algorithm is available in most statistical packages.

6.9 Handling Missing Data

Missing values should be treated with care in any model. It is important to realize that missing *information* cannot be compensated for. But if the data are redundant, missing values should ideally not pose problems for the analysis. Ad hoc approaches such as setting the missing values to zero are hazardous and often give completely incorrect results. The implicit rationale behind setting missing values to zero is that a zero value means 'nothing' and thus has no influence on fitting the model. However, that is not correct in a modeling situation. Consider the situation depicted in Figure 6.5. The left-most data matrix without missing values poses a rank one problem because there is only one spectral phenomenon. The data set shown to the right is identical but now there are some missing values in the spectrum with the lowest intensities. In principle, the rank must still be one as there is still only one spectral phenomenon. However, the missing values have been set to zero and clearly, it is not possible to model the data with zero-values by a one-component model.

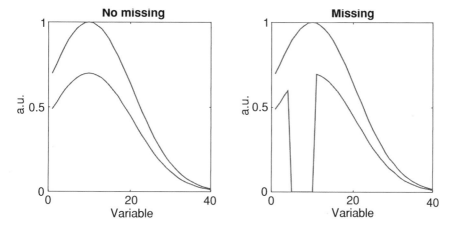

Figure 6.5. Data set of two synthetic spectra shown to the left. To the right the same data set is shown but now with missing values in one spectrum. The missing values have been set to zero.

Another approach for handling missing data is to impute the missing elements from neighboring elements in case that these elements are correlated. This approach can provide reasonable results if only a few data elements are missing, but it does not work satisfactorily in critical situations with many missing elements. In two-way PCA and two- and multi-way PLS regression fit through NIPALS-like algorithms, the approach normally advocated for in chemometrics [Martens & Næs 1989] is to simply skip the missing elements in the appropriate inner products of the algorithm. In these sequential algorithms, each component is estimated from the yet unexplained part of the data. In e.g. PCA, with missing elements, the components of such a procedure are still optimal according to the component-wise criterion. That is, the first component is an optimal one-component representation of the data, the second component is an optimal one-component representation of the residual etc. However, the overall model does not have any well-defined optimality. For example, orthogonality of scores and loadings is lost, and a least squares solution is not obtained. The approach has been shown to work well for a small amount of randomly missing data, but also to be suboptimal in some cases [Grung & Manne 1998, Nelson *et al.* 1996]. This is also illustrated in practice in Example 6.1 where the NIPALS approach is compared to a least squares method to be discussed next.

EXAMPLE 6.1

Inadequacy of NIPALS with large amounts of missing data

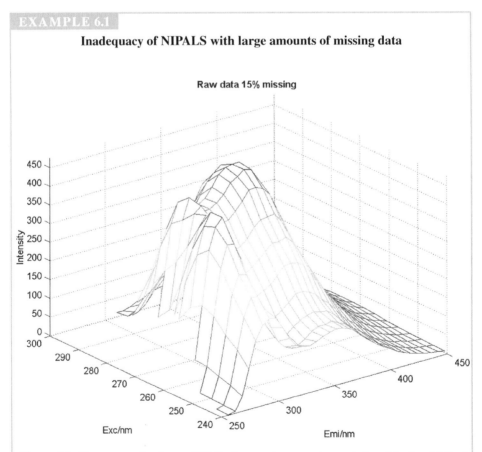

Figure 6.6. Fluorescence landscape (EEM) of tryptophan, tyrosine, and phenylalanine. 15 % of the data are missing in the left corner primarily because of Rayleigh scatter.

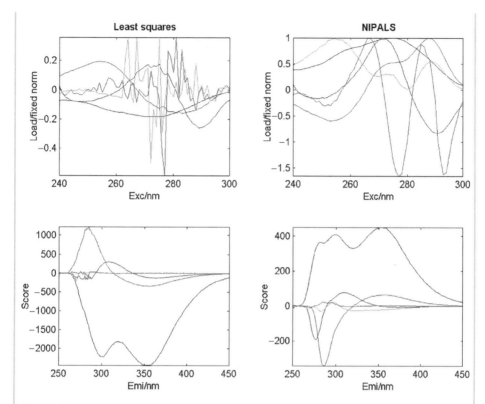

Figure 6.7. Left: scores and loadings from a five-component least squares PCA model. Right: similar result using the NIPALS method for handling missing data.

Fluorescence excitation–emission measurements from a single sample containing three amino acids are used for illustrating the possible problems with the NIPALS approach for handling missing data (Figure 6.6). Ideally the matrix has rank three, one for each amino acid in the sample. The data matrix contains the fluorescence excitation–emission matrix from one single sample with approximate rank three and 15 % missing data. In Figure 6.7 it is seen that the results using the NIPALS approach are useless compared to the least squares approach where the bilinear PCA model is fitted in a least squares sense to the nonmissing part of the matrix. Three components should be sufficient for fitting the data, but NIPALS overestimates the number of components by far whereas the least squares approach provides noisy loadings and low-variance scores after the third component as expected.

In most modeling situations it is not a priori known how many components to choose. The main difficulty in using NIPALS is therefore to be able to *detect* whether the found model is real, or, as in this case, just an artifact reflecting the position of the missing data rather than the variation in present data.

A natural way to handle missing data is to fit the model *only* to the nonmissing data. Thus, only optimizing the loss function over nonmissing elements. The loss function for any model of incomplete data can thus be stated

$$\min_{\mathbf{M}} \| \mathbf{W}*(\mathbf{X} - \mathbf{M}) \|^2 \qquad (6.53)$$

where $\mathbf{X}$ is a matrix containing the data, $\mathbf{M}$ contains the model and $*$ denotes the Hadamard (elementwise) product. The structure and constraints of $\mathbf{M}$ are given by the specific model being fit, e.g. $\mathbf{AB}'$ in case of a bilinear model. The matrix $\mathbf{W}$ contains weights that are either one if corresponding to an existing element or zero if corresponding to a missing element. The natural way to fit the model with missing data is thus by a weighted regression approach with the weights being a binary array [Kiers 1997b, Paatero 1997]. This approach works for *all* models that have a well-defined overall optimization criterion; e.g. PARAFAC or Tucker. For partial least squares regression this approach cannot be used directly, and mostly a NIPALS-like approach is adopted, where the missing data are handled separately in the calculation of each component.

Yet another approach for handling incomplete data is to impute the missing data iteratively during the estimation of the model parameters [Dempster *et al.* 1977, Rubin *et al.* 1997]. The missing data are initially replaced with either sensible or random elements. A standard algorithm is used for estimating the model parameters using the now complete, but incorrect, data set. After each iteration the model, $\hat{\mathbf{X}}$, of $\mathbf{X}$ is calculated, and the missing elements are replaced with the model estimates. The iterations and replacements are continued until no changes occur in the estimates of the missing elements and the overall convergence criterion is fulfilled. When the algorithm has converged, the elements replacing the missing elements have zero residual.

How then, do these two approaches–weighted regression and imputation–compare? Kiers [1997b] has shown that the two approaches give identical results, which can also be realized by considering data imputation more closely. As residuals corresponding to missing elements are zero, they do not influence the parameters of the model, which is the same as saying they have zero weight in the loss function. Algorithmically, however, there are some differences. Consider two competing algorithms for fitting a model of data with missing elements; one where the parameters are updated using weighted least squares regression with zero weights for missing elements and one where ordinary least squares regression and data imputation is used. Using weighted least squares regression instead of ordinary least squares regression is computationally more costly per iteration, and therefore slows down the algorithm. Using iterative data imputation on the other hand often requires more iterations due to the data imputation (typically 30–100 % more iterations). This is so, because, when the missing data are imputed from the interim model, the original loss function is not directly optimized (because the imputed values are incorrect). The closer the interim model comes to the true least squares model, the closer the imputed missing values get to the correct imputations. Only upon convergence are the imputed elements correct. It is difficult to say which method is preferable, as this is dependent on implementation, size of the data, and size of the computer memory available. Data imputation has the advantage of being easy to implement, also for problems that are otherwise difficult to estimate under a weighted loss function.

Louwerse *et al.* [1999] also compared the imputation approach to another method for handling missing data specifically designed for cross-validation.

6.10 Imposing Nonnegativity

The algorithms as presented in the previous sections allow the loadings to contain positive and negative values. This is mathematically correct, but not always in line with prevailing chemical knowledge. Sometimes, chemical knowledge imposes nonnegativity constraints: concentrations cannot become negative, correctly measured spectra can usually not be negative and chromatograms can usually not be negative. This is important background knowledge that, on occasion, can be included directly in the algorithms. Fitting models subject to nonnegativity constraints is thus of practical importance. Further, Lawton and Sylvestre [1971] conjectured that using nonnegativity significantly reduces the feasible space of the parameters to be estimated. Thereby the estimated components may sometimes be determined up to a smaller rotational ambiguity.

Nonnegativity is a lower bounded problem, where each parameter is bound to be above or equal to zero. Such a bounded problem can be efficiently solved with an active set algorithm [Gill *et al.* 1981]. How this algorithm can be implemented in an alternating least squares algorithm, e.g., for fitting the PARAFAC model is explained.

Consider a PARAFAC model of an $I \times J \times K$ array $\underline{\mathbf{X}}$. The model is usually fitted by minimizing the loss function

$$\min_{\mathbf{A},\mathbf{B},\mathbf{C}} \|\mathbf{X} - \mathbf{A}(\mathbf{C} \odot \mathbf{B})'\|^2 \qquad (6.54)$$

If the parameters of the model are known to be nonnegative (e.g. reflecting concentrations), then it is possible to fit a constrained version of this model where nonnegativity is imposed. This problem can be written

$$\min_{\mathbf{A},\mathbf{B},\mathbf{C}} \|\mathbf{X} - \mathbf{A}(\mathbf{C} \odot \mathbf{B})'\|^2 \qquad (6.55)$$

subject to $a_{ir} \geq 0, b_{jr} \geq 0, c_{kr} \geq 0$

$(i = 1, \ldots, I; j = 1, \ldots, J; k = 1, \ldots, K; r = 1, \ldots, R)$

The elements a_{ir} constitute the elements of the first mode loadings in the $I \times R$ matrix $\mathbf{A}$, and b_{jr} and c_{kr} are elements of $\mathbf{B}$ and $\mathbf{C}$ respectively. The problem is different from the unconstrained problem, because additionally all parameters must be nonnegative. An algorithm for fitting this model can be devised using alternating least squares much in line with the ordinary unconstrained algorithm. The loading matrices, $\mathbf{A}$, $\mathbf{B}$ and $\mathbf{C}$ are updated iteratively considering the other parameters fixed. For updating, e.g., the first mode parameters the problem is

$$\min_{\mathbf{A}} \|\mathbf{X} - \mathbf{A}(\mathbf{C} \odot \mathbf{B})'\|^2 \qquad (6.56)$$

subject to $a_{ir} \geq 0, i = 1, \ldots, I; r = 1, \ldots, R$

Because the update of the parameters for a given i (a given row of $\mathbf{A}$) is independent of the other parameters in $\mathbf{A}$, each row may be updated separately[6] still providing the global least squares solution to Equation (6.56). Thus, instead of considering the problem in

---

[6] To see that this is so, observe that the problem of finding the optimal $\mathbf{A}$ may be written $\min_{\mathbf{A}} \|\mathbf{X} - \mathbf{A}\mathbf{Z}'\|^2 = \min_{\mathbf{A}} \|\mathbf{X}' - \mathbf{Z}\mathbf{A}'\|^2$ where $\mathbf{Z} = (\mathbf{C} \odot \mathbf{B})$. Thus, the problem is a multivariate linear regression problem and hence each row of $\mathbf{A}$ may be updated independently.

Equation (6.56) it suffices to consider the problem

$$\min_{\mathbf{a}_i} \|\text{vec } \mathbf{X}_i - (\mathbf{C} \odot \mathbf{B})\mathbf{a}_i\|^2 \tag{6.57}$$

subject to $a_{if} \geq 0$, $f = 1, \ldots, F$

where vec $\mathbf{X}_i$ is the ith horizontal slab (sample i) rearranged into a vector. The vector $\mathbf{a}_i$ contains the parameters in the ith row of $\mathbf{A}$ (see also Figure 6.2). Solving this problem for all $i(i = 1, \ldots, I)$ provides the least squares update for $\mathbf{A}$. The problem is equivalent to a standard multiple linear regression problem and forms the basis for one of the most used algorithms for solving for nonnegativity constrained parameters.

The solution to the nonnegativity constrained problem in Equation (6.57) can be found by the algorithm NNLS (nonnegative least squares) which is an *active set algorithm* and is based on finding the solution iteratively. In each step, the interim solution is improved by adjusting the parameters such that these do not turn negative. After a finite number of iterations, the solution is found by simple linear regression on the unconstrained subset of the variables and the parameters of the constrained variables are set to zero. Detailed descriptions of active set algorithms for nonnegativity constrained regression can be found in the literature [Bro & de Jong 1997, Hanson & Lawson 1974].

6.11 Summary

In this chapter, algorithms for the models described in Chapter 4 are given. For PARAFAC and Tucker models, both exact least squares and faster approximate algorithms are provided. The least squares models are based on the principle of alternating least squares whereas the fast methods are based on solving a fixed number of eigenvalue problems. Implementation issues pertaining to efficient algorithms that take advantage of the structural model and initialization are also described. For each type of model, it has been shown how to use a model for finding scores of new available data. Finally, important issues such as handling missing data and imposing nonnegativity are discussed.

Appendix 6.A: Closed-form Solution for the PARAFAC Model

The earliest closed-form solution for the PARAFAC model is called the generalized rank annihilation method or GRAM. It is originally based on a calibration method called rank annihilation factor analysis–RAFA [Ho *et al.* 1980, Ho *et al.* 1981] which is a method for predicting the concentration of an analyte in an unknown sample [McCue & Malinowski 1983, Nørgaard & Ridder 1994]. While rank annihilation factor analysis does not directly fit a PARAFAC model, it is based on the assumption that the data (or at least the analyte of interest) follows a trilinear model. For historical as well as practical reasons it is interesting to have an idea of how rank annihilation factor analysis works. A short description is provided in Elaboration 6.1. The amazing property that rank annihilation factor analysis is able to quantitate analytes without specific knowledge of the interferents has later been coined the *second-order advantage*. This name is used because the advantage is obtained by using the second-order (two-way) structure of the individual samples instead of merely rearranging the two-way data of each sample to a long vector or first order structure.

ELABORATION **6.1**

Rank annihilation factor analysis

Rank annihilation factor analysis treats the special situation in which there are only two slices, where each slice contains the second-order measurements of one sample. One sample is usually a pure standard containing only the analyte of interest in a known concentration and the other is an unknown sample containing unknown interferents as well as the analyte of interest. The purpose in rank annihilation factor analysis is to estimate the concentration of the analyte of interest in the unknown sample. It is assumed that the signal from the analyte of interest corresponds to a rank-one component with the intensity proportional to the concentration. Thus the standard sample, say X_1, can be described

$$X_1 = c_{11} a_1 b_1'$$

ignoring the noise part. The vector a_1 holds the pure profile for the first mode (for example emission spectrum of the analyte) and b_1 is the pure profile in the second mode (for example the excitation spectrum of the analyte). The scalar c_{11} is the concentration of the analyte. The index 11 refers to the fact that c is the concentration of the first analyte in the first sample. For simplicity, assume that also the interferents in the unknown sample can be modeled as rank-one bilinear components. Then a PARAFAC model applies and the model of the two samples can be written

$$X_1 = AD_1B' \quad \text{and} \quad X_2 = AD_2B'$$

where D_1 is a diagonal matrix with c_{11} as the first element and zeros on the other diagonal elements because the interferents are absent in the standard sample. The matrix D_2 is a diagonal matrix holding the unknown concentrations of the analyte and interferents of sample two.

The idea behind rank annihilation factor analysis is based on reducing the rank of the data matrix of the unknown sample by subtracting the contribution from the analyte of interest. If the signal from the analyte of interest is subtracted from the sample data, then the rank of this matrix decreases by one, as the contribution of the analyte of interest to the rank is one. Thus, the rank of $X_2 - \gamma X_1$ is calculated. If the scalar γ is equal to the ratio of the concentration of the analyte in the unknown and standard sample ($\gamma = c_{21}/c_{11}$) then the rank, $r(X_2 - \gamma X_1)$, is lowered by one because the contribution of the analyte to X_2 is removed (annihilated), whereas in all other cases, the rank will be the constant. If the data are noisy, the significant eigenvalues are usually monitored, and the point where the smallest significant eigenvalue is at its minimum (as a function of γ) is usually taken to be the estimated ratio. An example of rank annihilation factor analysis is illustrated in Figure 6.8, where it is seen that for the value 0.6 the smallest significant singular value (third from top) clearly reaches a minimum indicating that the contribution of the standard has been minimized in the unknown sample; hence this is the most likely concentration

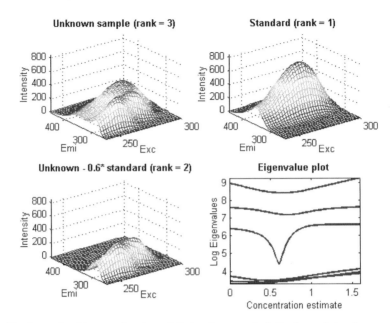

Figure 6.8. Example of the use of rank annihilation factor analysis for determining the concentration of tryptophane using fluorescence excitation–emission spectroscopy. In the top left plot the unknown sample is shown. It contains three different analytes. The standard sample (only tryptophane) is shown in the top right plot. In the lower right plot, it is shown that the smallest significant (third from top) singular value of the analyte-corrected unknown sample matrix reaches a clear minimum at the value 0.6. In the lower left plot the unknown sample is shown with 0.6 times the standard sample subtracted. It is evident that the contribution from the analyte is practically absent.

estimate under the premises. Theoretically, the third singular value will decrease to values in the neighborhood of the remaining noise singular values. This does not happen, though, for most real data due to the existence of nonlinearities etc. The true reference concentration in the unknown sample is 0.592 relative to the concentration in the standard sample!

However, the second-order advantage is more naturally seen as a manifestation of the uniqueness of the PARAFAC model. In that context, it also provides means for separating not only concentration-related parameters but also, for example, separating a spectral part of a data set into the pure underlying spectra.

The rank annihilation factor analysis method is intuitively appealing but somewhat complicated and slow. The solution, however, can be found directly by expressing the sought reduction in rank as a generalized eigenvalue problem. This was realized by Lorber [1984, 1985] and Sanchez & Kowalski coined the term generalized rank annihilation method–GRAM–for this automated version of rank annihilation factor analysis [Sanchez *et al.* 1987, Sanchez & Kowalski 1986, Wilson *et al.* 1989] which also has the advantage that several components may be present or absent in both calibration and standard sample.

The basic theory behind the generalized rank annihilation method is that the rank reduction can be re-expressed and automated. A scalar γ (relative concentration of the analyte in the unknown sample) is sought such that the matrix pencil

$$\mathbf{X}_2 - \gamma\mathbf{X}_1 \qquad (6.58)$$

has reduced rank. Here $\mathbf{X}_2$ contains the noiseless response of the mixture sample and has rank R; and $\mathbf{X}_1$ contains the noiseless response of the standard sample which is here assumed to contain only the analyte of interest and hence is of (pseudo-) rank one. Generalization to situations with different numbers and amounts of analyte in the two samples can be made but is not pursued here for brevity [Kiers & Smilde 1995].

There is one value of γ that will lead to a rank-reduction of the expression in Equation (6.58); namely the one for which γ equals the concentration of the analyte in $\mathbf{X}_2$ relative to the concentration in $\mathbf{X}_1$. It is assumed that $\mathbf{X}_1$ and $\mathbf{X}_2$ are square ($R \times R$) matrices. Then seeking a scalar γ with the property that the rank of $\mathbf{X}_2 - \gamma\mathbf{X}_1$ is $R - 1$ can be formalized in

$$|\mathbf{X}_2 - \gamma\mathbf{X}_1| = 0 \qquad (6.59)$$

where $|\mathbf{X}|$ means the determinant of the matrix $\mathbf{X}$. Stated otherwise, the system of equations $(\mathbf{X}_2 - \gamma\mathbf{X}_1)\mathbf{v} = 0$ should have a nontrivial solution for $\mathbf{v}$ (i.e. $\mathbf{v} \neq 0$). Such a nontrivial solution can be found by solving a generalized eigenvalue problem

$$\mathbf{X}_2\mathbf{v} = \mathbf{X}_1\mathbf{v}\gamma \qquad (6.60)$$

Algorithms are available to solve Equation (6.60), e.g, using the QZ algorithm [Golub & van Loan 1989].

The generalized eigenvalue problem looks for common directions in the column- and row-spaces of the two matrices involved. This is also the basic idea behind PARAFAC models. Therefore, it is not surprising that there is an intimate relationship between PARAFAC models and the generalized eigenvalue problem. The generalized rank annihilation method is based on the fact that the generalized eigenvalue problem approach can be used to estimate the parameters in a PARAFAC model. This will be illustrated for a case in which a three-way array is given with only two slices in the third mode. Generalizations to more than two slices will be made later. Assume that a three-way array $\underline{\mathbf{X}}$ with two frontal slices can be modeled as

$$\mathbf{X}_k = \mathbf{A}\mathbf{D}_k\mathbf{B}', k = 1,2 \qquad (6.61)$$

i.e., the data follow a PARAFAC structure. No errors are assumed for now and $\mathbf{X}_k(k = 1,2)$ are ($R \times R$) matrices where R is the number of components. If the original data in $\mathbf{X}_k$ is not of size $R \times R$, the data can be transformed into such size by replacing $\mathbf{X}_k$ ($I \times J$) with $\mathbf{U}'\mathbf{X}_k\mathbf{V}$ where $\mathbf{U}$ ($I \times R$) and $\mathbf{V}$ ($V \times R$) are orthonormal matrices usually found from a singular value decomposition of $\sum_{k=1}^{2} \mathbf{X}_k$. Thus, the compressed data $\mathbf{U}'\mathbf{X}_k\mathbf{V}$ will be of size $R \times R$ and a PARAFAC model fitted to this compressed array can be transformed back to the original domain by pre-multiplying $\mathbf{A}$ and $\mathbf{B}$ with $\mathbf{U}$ and $\mathbf{V}$ (see Elaboration 6.2). For noisy data, the transformation back and forth between subspaces is not lossless (Elaboration 6.3), but in the following explanation of GRAM, this issue will not be addressed further.

ELABORATION **6.2**

Expressing PARAFAC models in subspaces

Suppose that an R-component PARAFAC model is sought of the data $\underline{\mathbf{X}}$ ($I \times J \times K$) where I and J are larger than R and where K equals two. Select $\mathbf{U}$ ($I \times R$) as an R-dimensional orthogonal basis for the column-space of $\mathbf{X}_k$ ($k = 1,2$) and $\mathbf{V}$ as an R-dimensional orthogonal basis for the row-space of $\mathbf{X}_k$, ($k = 1,2$). Equation (6.61) ensures that $\mathbf{X}_1$ and $\mathbf{X}_2$ have a common column- and row-space. The components collected in $\mathbf{A}$ *will* be in the column-space of both the $\mathbf{X}_1$ and $\mathbf{X}_2$ matrix and $\mathbf{A}$ is a basis of this column-space. Thus, a matrix $\mathbf{P}$ ($R \times R$) exists such that $\mathbf{A} = \mathbf{UP}$. Likewise, there exists a matrix $\mathbf{Q}$ ($R \times R$) such that $\mathbf{B} = \mathbf{VQ}$. This means that $\mathbf{U}$ and $\mathbf{V}$ span the same spaces as $\mathbf{A}$ an $\mathbf{B}$ respectively and $\mathbf{P}$ and $\mathbf{Q}$ are nonsingular matrices that transform $\mathbf{U}$ and $\mathbf{V}$ into $\mathbf{A}$ and $\mathbf{B}$. Therefore, fitting the model $\mathbf{X}_k = \mathbf{A}\mathbf{D}_k\mathbf{B}'$, $k = 1,2$ or fitting the model $\mathbf{U}'\mathbf{X}_k\mathbf{V} = \mathbf{P}\mathbf{D}_k\mathbf{Q}'$, $k = 1,2$ is equivalent. Elaboration 6.3 shows different ways of estimating $\mathbf{U}$ and $\mathbf{V}$ from real data where additional noise is present.

ELABORATION **6.3**

Finding suitable subspaces from real data

For real data, the data matrices will usually be of full rank due to inherent noise and model deviations. Finding suitable bases for the row- and column-spaces is therefore a nontrivial task. The bases, denoted $\mathbf{U}$ and $\mathbf{V}$ above, must span the common variation in the data. In practice, $\mathbf{U}$ is often found from a singular value decomposition of $[\mathbf{X}_1 \ \mathbf{X}_2]$ and $\mathbf{V}$ from a singular value decomposition of $[\mathbf{X}_1' \ \mathbf{X}_2']'$ or some similarly derived matrices reflecting the systematic variation in both samples. Another approach is to simply take the R most significant left and right singular vectors from a singular value decomposition of the matrix $\mathbf{X}_1 + \mathbf{X}_2$. Little has been done to investigate the influence of the choice of method for finding the bases [Booksh *et al.* 1994]. The generalized rank annihilation method can be viewed as a special case of the direct trilinear decomposition as will be shown in the next section. Viewed as such, it can be shown that the following set of bases are natural. A Tucker2 model (p. 78) is fitted to the three-way array with frontal slices $\mathbf{X}_1$ and $\mathbf{X}_2$. This model fitted using ($R, R, 2$) components can be written as

$$\mathbf{X} = [\mathbf{X}_1 \ \mathbf{X}_2] = \mathbf{UG}(\mathbf{I} \otimes \mathbf{V})' + [\mathbf{E}_1 \ \mathbf{E}_2] = \mathbf{U}[\mathbf{G}_1 \ \mathbf{G}_2](\mathbf{I} \otimes \mathbf{V})' + [\mathbf{E}_1 \ \mathbf{E}_2]$$

or equivalently

$$\mathbf{X}_1 = \mathbf{UG}_1\mathbf{V}' + \mathbf{E}_1, \ \mathbf{X}_2 = \mathbf{UG}_2\mathbf{V}' + \mathbf{E}_2$$

where $\mathbf{G}_k = \mathbf{U}'\mathbf{X}_k\mathbf{V}$. As seen from this, a bilinear model is found for both slices with the same loadings but different $\mathbf{G}$ matrices. The first mode loadings can be seen to provide a best-spanning basis for the first mode in a least squares sense and likewise for the second mode loading matrix $\mathbf{V}$. The core array, $\mathbf{G}$, directly provides compressed slices $\mathbf{G}_1$ and $\mathbf{G}_2$. These two slices are of size $R \times R$ (rank R) and can be used subsequently in the algorithm of the generalized rank annihilation method. When fitting the PARAFAC model to the $\mathbf{G}$ matrices, the solution can be converted back to the original domain through $\mathbf{U}$ and $\mathbf{V}$.

Equation (6.61) can be rewritten as

$$
\begin{aligned}
\mathbf{X}_1 &= \mathbf{AD}_1\mathbf{B}' \Rightarrow \mathbf{X}_1(\mathbf{B}')^{-1} = \mathbf{AD}_1 \\
\mathbf{X}_2 &= \mathbf{AD}_2\mathbf{B}' \Rightarrow \mathbf{X}_2(\mathbf{B}')^{-1} = \mathbf{AD}_2 \Rightarrow \mathbf{X}_2(\mathbf{B}')^{-1}\mathbf{D}_2^{-1}\mathbf{D}_1 = \mathbf{AD}_1 \\
\mathbf{X}_1(\mathbf{B}')^{-1} &= \mathbf{X}_2(\mathbf{B}')^{-1}\mathbf{D}_2^{-1}\mathbf{D}_1
\end{aligned}
\tag{6.62}
$$

where it is assumed that $\mathbf{B}$ $(R \times R)$ and $\mathbf{D}_2$ are nonsingular. The last line of Equation (6.62) can be recognized as a generalized eigenvalue problem (6.60). The generalized eigenvalue problem of Equation (6.62) can be formulated in general terms as

$$
\mathbf{X}_1\Psi = \mathbf{X}_2\Psi\Lambda
\tag{6.63}
$$

where Λ is a diagonal matrix containing the generalized eigenvalues and Ψ contains the generalized eigenvectors. On the assumption that Λ has distinct values on its diagonal, the generalized eigenvalue problem gives a unique solution. Hence, this solution must be the one solving the last equation in Equation (6.62). This gives the following equations

$$
\begin{aligned}
\Psi &= (\mathbf{B}')^{-1} \\
\Lambda &= \mathbf{D}_2^{-1}\mathbf{D}_1
\end{aligned}
\tag{6.64}
$$

from which $\mathbf{B}$ and $\mathbf{D}_2^{-1}\mathbf{D}_1$ can be obtained. Moreover, $\mathbf{A}$ can be obtained by calculating $\mathbf{A} = \mathbf{X}_2(\mathbf{B}')^{-1}\mathbf{D}_2^{-1}$. If the original data were compressed, the components in $\mathbf{A}$ and $\mathbf{B}$ can be transformed back into the original domain by the use of $\mathbf{U}$ and $\mathbf{V}$.

In second-order calibration, the concentration of the analyte in the standard is known. This is held in one element of the diagonal of $\mathbf{D}_1$ whereas the remaining diagonal elements are zero, when the standard only contains the analyte of interest. The matrix $\mathbf{D}_2$ then contains the concentrations of all the chemical analytes in the unknown mixture. From the known $\mathbf{D}_1$ and the estimated $\mathbf{D}_2^{-1}\mathbf{D}_1$, the value in $\mathbf{D}_1$ corresponding to the analyte can be obtained.

The above derivation of the generalized rank annihilation method assumes noise-free data but an algorithm for approximating the solution in case of data that do not conform exactly to the model is easily implemented as in Algorithm 6.10 where the transformation of the raw data into the subspaces is explicitly written. In the algorithm, singular value decomposition is used for finding suitable subspaces. This is the approach normally adopted in the literature although many versions of generalized rank annihilation can be found [Booksh *et al.* 1994, Booksh & Kowalski 1994, Faber *et al.* 1994, Gerritsen *et al.* 1992, Li *et al.* 1992, Sanchez *et al.* 1987, Sanchez & Kowalski 1986, Wilson *et al.* 1989]. Although generalized rank annihilation is a method for finding the parameters in the PARAFAC model, it is less general than other PARAFAC algorithms. It requires $\mathbf{A}$ and $\mathbf{B}$ to be of full rank and it only handles the three-way case and only in the situation with two samples. For example, a three-component solution can be fitted to a $2 \times 2 \times 2$ array with a general PARAFAC algorithm while this is not possible with the generalized rank annihilation method. It is also important to be aware that the solution is not a least squares solution, although for well-behaved data, the difference is usually small.

The algorithm and derivations above also assume that $\mathbf{A}, \mathbf{B}$, and $\mathbf{D}_2$ are of full rank and that $\mathbf{D}_2^{-1}\mathbf{D}_1$ has distinct diagonal elements. However, when the standard only contains the analyte of interest, $\mathbf{D}_2^{-1}\mathbf{D}_1$ will not have distinct elements except when there is only one interferent. It can be shown that this does not exclude having a unique solution for the analyte, but the parameters pertaining to the remaining interferents will not be uniquely recovered [Harshman 1972, Kiers & Smilde 1995].

A practical numerical issue in the use of the generalized rank annihilation method is that, at certain times, complex-valued solutions may arise. Means have been provided, though, for eliminating this problem by simple similarity transformations [Faber 1997, Li *et al.* 1992].

ALGORITHM **6.10**

GRAM–generalized rank annihilation method

Given: $\underline{\mathbf{X}}$ of size $I \times J \times 2$ and sought dimension R

1. Let $\mathbf{U}$ be the first R left singular vectors of $[\mathbf{X}_1|\mathbf{X}_2]$ and $\mathbf{V}$ the first R right singular vectors of $[\mathbf{X}_1'|\mathbf{X}_2']$.
2. Let $\mathbf{G}_1 = \mathbf{U}'\mathbf{X}_1\mathbf{V}$ and $\mathbf{G}_2 = \mathbf{U}'\mathbf{X}_2\mathbf{V}$ be matrices of size $R \times R$.
3. Find the generalized eigenvectors (Ψ), and eigenvalues (Λ) such that $\mathbf{G}_1\Psi = \mathbf{G}_2\Psi\Lambda$, where Ψ holds the generalized eigenvectors and Λ is a diagonal matrix holding the eigenvalues in the diagonal.
4. Then $\mathbf{A} = \mathbf{U}\mathbf{G}_2\Psi\mathbf{D}_2^{-1}$ and $\mathbf{B} = \mathbf{V}(\Psi')^{-1}$.
5. The third mode loadings (concentration estimates) are indirectly given as $\Lambda = \mathbf{D}_2^{-1}\mathbf{D}_1$.

Several methods similar to the generalized rank annihilation method have been described [Leurgans *et al.* 1993, Mitchell & Burdick 1993, Sands & Young 1980] and compared with a least squares PARAFAC algorithm for their use in curve resolution. Generalized rank annihilation is mostly found to be inferior to the least squares PARAFAC model with respect to accuracy of the parameter estimates, but generalized rank annihilation is typically much faster than least squares PARAFAC algorithms. If the data have a high signal-to-noise ratio and the structural model is appropriate and correctly specified, the solutions from the two methods are often close [Windig & Antalek 1997] and hence in favor of generalized rank annihilation. Generalized rank annihilation is also often used for initialization of PARAFAC. Some theoretical aspects of the differences between different ways of using the generalized rank annihilation method for fitting the PARAFAC model have been discussed by e.g. Faber and Kiers & Smilde [Faber *et al.* 1994, Faber *et al.* 1997, Kiers & Smilde 1995].

The generalized rank annihilation method is based on the trilinear model and fits the model using a (generalized) eigenvalue problem. The method is restricted by one mode having maximal dimension two (i.e., two slices). If there are more than two slices, it is possible to pick two of the slices, say $\mathbf{X}_1$ and $\mathbf{X}_2$, and find the component matrices $\mathbf{A}$ and $\mathbf{B}$ using the generalized rank annihilation method. As $\mathbf{A}$ and $\mathbf{B}$ are then known, the loadings in the third mode can be found by a simple regression step. However, the problem of finding two suitable slices for applying the generalized rank annihilation method is nontrivial and has received considerable attention (see Elaboration 6.4 for details). When there are more than two slices, the algorithm is called the direct trilinear decomposition (DTLD–sometimes abbreviated TLD) and the algorithm follows as shown in Algorithm 6.11.

ELABORATION **6.4**

Finding two slices for direct trilinear decomposition

In direct trilinear decomposition, two slices have to be used for fitting the PARAFAC model using the generalized rank annihilation method. Instead of looking for two suitable slices of the array it is common and sensible to define a set of two synthetic slices based on linear combinations of all the original slices. This can be done in a variety of ways. Originally, Sanchez and Kowalski suggested using the Tucker3 model [Sanchez & Kowalski 1990] described in Section 6.4. Others have suggested alternative but similar approaches [Tu & Burdick 1992]. A general approach in line with the original approach is provided here.

Given an array $\underline{\mathbf{X}}$ of size $I \times J \times K$, two slices in, say, the third mode are needed in order to be able to use the generalized rank annihilation method. These may be formed as weighted averages of all the slices. A sensible way to define two such samples is to determine two slices that preserve the total variation in $\underline{\mathbf{X}}$ maximally in a least squares sense. Additionally these two slices must be within an R-dimensional subspace (R is the number of components in the PARAFAC model) in the first and second mode in order to maximize directly the appropriateness of the span of the data matrices. Thus, two slices $\mathbf{G}_1$ and $\mathbf{G}_2$ of size $R \times R$ are sought such that these are representative of the variation in $\underline{\mathbf{X}}$. This may be accomplished in a least squares sense by fitting a Tucker3 model with dimensionality $R \times R \times 2$,

$$\mathbf{X} = \mathbf{UG}(\mathbf{Z} \otimes \mathbf{V})' + \mathbf{E}$$

where $\mathbf{U}\,(I \times R)$, $\mathbf{V}\,(J \times R)$ and $\mathbf{Z}\,(K \times 2)$ are the orthogonal loading matrices of the model and $\mathbf{G}$ is the $R \times (2R)$ matricized array containing $\mathbf{G}_1$ in the first slice and $\mathbf{G}_2$ in the second.

ALGORITHM **6.11**

Direct trilinear decomposition

Given: $\underline{\mathbf{X}}$ of size $I \times J \times K$ and the dimension R of the PARAFAC model.

1. Determine rank-reduced bases for three modes, $\mathbf{U}\,(I \times R)$ $\mathbf{V}\,(J \times R)$ and $\mathbf{Z}\,(K \times 2)$. See Elaboration 6.4 or simply determine $\mathbf{U}$ as the first R left singular vectors from a singular value decomposition of $\mathbf{X}_{(I \times JK)}$, $\mathbf{V}$ from $\mathbf{X}_{(J \times IK)}$ and $\mathbf{Z}$ from $\mathbf{X}_c$.
2. Determine $\mathbf{G}_1$ and $\mathbf{G}_2$ of size $R \times R$. $[\mathbf{G}_1 \mathbf{G}_2] = \mathbf{U}'\mathbf{X}_{(I \times JK)}(\mathbf{Z} \otimes \mathbf{V})$.
3. Solve the generalized eigenproblem $\mathbf{G}_1\Psi = \mathbf{G}_2\Psi\Lambda$.
4. $\mathbf{A} = \mathbf{UG}_2\Psi\mathbf{D}_2^{-1}$, and $\mathbf{B} = \mathbf{V}(\Psi')^{-1}$.
5. Determine $\mathbf{C}$ as the least squares solution given $\mathbf{A}$ and $\mathbf{B}$ (see Algorithm 6.1). $\mathbf{C} = \mathbf{X}_{(K \times IJ)}(\mathbf{B} \odot \mathbf{A})^+$.

The basic principle of the generalized rank annihilation method has been (re-)invented several times, e.g., in signal processing under the name ESPRIT [Roy & Kailath 1989] and in psychology [Schönemann 1972] where it was based on the same idea [Cattell 1944] that Harshman generalized to the PARAFAC model and algorithm.

Appendix 6.B: Proof That the Weights in Trilinear PLS1 can be Obtained from a Singular Value Decomposition

A set of unit-length vectors $\mathbf{w}^J$ and $\mathbf{w}^K$ are sought that satisfy

$$\max_{\mathbf{w}^J \mathbf{w}^K} [\mathrm{cov}(\mathbf{t}, \mathbf{y}) | t_i = (\mathbf{w}^J)' \mathbf{X}_i \mathbf{w}^K] \tag{6.65}$$

that is, a set of weight vectors $\mathbf{w}^J$ and $\mathbf{w}^K$ that through a least squares model of $\underline{\mathbf{X}}$ produces a score vector with maximal covariance with $\mathbf{y}$. The covariance criterion can equivalently be posed as maximizing $\mathbf{t}'\mathbf{y}$ which written in scalar notation leads to

$$\max_{\mathbf{w}^J \mathbf{w}^K} \left[\sum_{i=1}^{I} t_i y_i \Big| t_i = (\mathbf{w}^J)' \mathbf{X}_i \mathbf{w}^K \right]$$

$$= \max_{\mathbf{w}^J \mathbf{w}^K} \left[\sum_{i=1}^{I} y_i (\mathbf{w}^J)' \mathbf{X}_i \mathbf{w}^K \right]$$

$$= \max_{\mathbf{w}^J \mathbf{w}^K} \left[(\mathbf{w}^J)' \left(\sum_{i=1}^{I} y_i \mathbf{X}_i \right) \mathbf{w}^K \right]$$

$$= \max_{\mathbf{w}^J \mathbf{w}^K} \left[(\mathbf{w}^J)' \mathbf{Z} \mathbf{w}^K \right] \tag{6.66}$$

where $\mathbf{Z}$ is a matrix of size $J \times K$ containing in its jkth element the inner product of $\mathbf{y}$ and the jkth column of $\underline{\mathbf{X}}$, i.e. the vector with typical elements $x_{ijk}, i = 1, \ldots, I$. This last expression is maximized by performing a singular value decomposition of $\mathbf{Z}$. The first left singular vector equals $\mathbf{w}^J$ and the first right singular vector equals $\mathbf{w}^K$. This follows directly from the properties of the singular value decomposition.

7

VALIDATION AND DIAGNOSTICS

7.1 What is Validation?

Data analysis is often based on evaluation of fitted models. These models provide approximations and interpolations of the real system under study. There is often a wide choice of models and the quality of these depends on a number of factors, because mathematical and statistical models based on real data have properties depending not only on the system under study but also on the sampling, the properties of the model, the algorithm etc. Validation can be seen as the part of the analysis where it is investigated if valid conclusions can be drawn from a model. Does the model *generalize* the data in a parsimonious way, i.e., express the main variation in the data in a way that is as simple as possible? Does the model predict efficiently? Has the model been influenced by certain samples or variables in such a way that these, rather than the data in general, are described? Has the algorithm converged etc.? Harshman [1984] divides validation into four topics

- theoretical appropriateness;
- computational correctness;
- statistical reliability;
- explanatory validity.

  Theoretical appropriateness of a model has to do with the choice of the model. Is a latent variable structure expected on theoretical grounds? Is this structure bilinear or trilinear in nature? Are offsets expected? Are linearizing transformations or preprocessing needed etc.

  Computational correctness is mainly concerned with whether the algorithm has converged to the desired global minimum or to unintended local minima and explaining why this has happened. Also rounding-off error is important, as is reproducibility of the calculations. These are algorithmic problems. For a data analyst such aspects should ideally not be of concern, but unfortunately, some three-way models are not numerically simple to fit.

Multi-way Analysis With Applications in the Chemical Sciences.   A. Smilde, R. Bro and P. Geladi
© 2004 John Wiley & Sons, Ltd ISBN: 0-471-98691-7

Statistical reliability is related to appropriateness of distributional assumptions, the stability of solutions to resampling, choice of dimensionality and confidence intervals of the model parameters. The statistical reliability is often difficult to quantify in practical data analysis, e.g., because of small sample sets or poor distributional knowledge of the system.

Explanatory validation is concerned with how well the model reflects the real phenomena under investigation, i.e., the appropriateness of the model. Do the residuals indicate that further systematic variation is present? Do other results confirm the conclusions drawn from the model etc. Is any knowledge about a physical model gained from studying the model of the data?

An important part of validation is the use of diagnostics for characterizing the quality of the solution. Diagnostics do not belong to three-way analysis per se, but are useful measures calculated in addition to the three-way model under study. Many diagnostics are modified versions of their two-way counterparts [Kroonenberg 1983]. Most diagnostics can also be plotted (Chapter 8), which makes them easier to explore and examine. The field of three-way analysis is not very old and many opportunities for diagnostic checking have not been tested thoroughly, since emphasis so far in this development has usually been on three-way models, uniqueness, convergence and fast algorithms. Harshman [1984] divides diagnostics into four levels: zero-fit diagnostics, one-fit diagnostics, many-fit (single data set) diagnostics, and many-fit (multiple data sets) diagnostics.

Zero-fit diagnostics are those related to the data *before* any model has been fit. Zero-fit diagnostics includes check for gross outliers and reliability of the data through replicates. It also includes initial judging of which model to use, perhaps guided by initial analyses of the data matricized to different two-way forms, etc.

One-fit diagnostics are those used after fitting a specific model and are useful for validating this particular model. This is not related to comparison to other competing models, but rather checking the internal statistical consistency of the estimated model itself. The diagnostics can be divided into those relating to the parameters and to the residuals. *Estimated parameters:* investigation of a model includes monitoring loadings to detect unusual variables or objects and constant factors, which can perhaps be eliminated by proper preprocessing. Also checking for signs of nonlinearities, interpretability, convergence etc. are important. *Residuals:* checking for heteroscedasticity, outliers, or systematic variation. Comparing goodness-of-fit with estimated data reliability etc.

Many-fit (single data set) diagnostics are useful for comparing either models estimated from different starting points, or alternative model forms. Diagnostics at this stage can again be divided into those related to parameters and those related to residuals. *Estimated parameters*: study parameters across successive iterations, different solutions, or different dimensionalities to investigate which seem sensible, interpretable, robust, converged etc. This is performed by comparing to the overall knowledge of the three-way problem. *Residuals*: similar to one-fit related diagnostics, but here the focus is on whether the algorithms have converged, if one model is significantly better than another according to fit and similar criteria, etc.

Many-fit (multiple data sets) diagnostics are used when several data sets are available for describing the same problem, or when the data set is so large that it can be split into several sets. In the first case, test-set validation is possible and in the latter case, the theory of resampling is applicable, making possible bootstrapping, jackknifing, cross-validation, split-half analyses etc.

This general description of practical validation is now substantiated by treating the following specific problems all of which are useful in different parts of the above-mentioned stages of analysis:

- cross-validation and test-set validation;
- selecting a model;
- determining the number of components;
- influence and residual analysis.

As for ordinary multivariate analysis these issues are all strongly connected. In-depth knowledge of the problem at hand is what makes the difference in most situations. Subjects not specifically related to three-way analysis are not treated in detail as information on these subjects can be found in most textbooks on standard multivariate analysis [Martens & Næs 1989, Rousseeuw & Leroy 1987, Weisberg 1985].

7.2 Test-set and Cross-validation

General aspects

The first treated is test-set and cross-validation which are general tools useful for many purposes in validation. For example, cross-validation is often used in two-way analysis for establishing the number of components in PCA or PLS models. In essence, test-set and cross-validation simply provide methods for obtaining more realistic residuals than those obtained by ordinary residuals from fitted models. Thus, test-set and cross-validation are of general interest when residuals are used for reporting explained variance, assessing outliers etc.

In certain cases, especially when overfitting occurs, fitted residuals suffer from not being similar to the residuals that will be obtained when the model is used on new data. The magnitude and structure of the fitted residuals may therefore be misleading even after correcting for degrees of freedom. Test-set or cross-validation can reduce this problem by providing residuals that more correctly reflect residuals that are obtainable on new data. Such residuals or sums of squares of these can be used for a number of purposes: (1) assessing which model to use, (2) assessing the number of components to use, (3) assessing the prediction ability of a regression model, (4) assessing overfitting by comparing cross-validated and fitted residuals, and (5) residual analysis. Hence, validated residuals are useful in the same situations where fitted residuals are used, but they provide fundamentally different and often more objective information with respect to the future use of the model.

Test-set validation is performed by fitting the current model to new data. How this is done is explained in Chapter 6 for the different models. The residuals are simply obtained by subtracting the model of the new data from the actual data. Test-set validation is natural because it specifically simulates the practical use of the model on new data in the future. If the test-set is made independent of the calibration data (future samples measured by future technicians or in a future location etc.) then the test-set validation is the definitive validation because the new data are, in fact, made as a drawing from the future total population of all possible measurements. Small test-sets may provide uncertain results merely because of the small sample size.

In component models, one practical problem associated with the test-set validation method is that the residuals obtained are not independent of the data used for fitting the model. Consider a two-way PCA model as a simple example of a component model. If the loadings $\mathbf{P}$ ($J \times R$) are found from a calibration set and a PCA model is to be tested on either a test-set or on left-out samples during cross-validation, then the residuals are found by first calculating the scores for the new data $\mathbf{X}_{\text{test}}$ as

$$\mathbf{T}_{\text{test}} = \mathbf{X}_{\text{test}}\mathbf{P} \tag{7.1}$$

assuming that $\mathbf{X}_{\text{test}}$ has been properly preprocessed (similar to the calibration data). The residual sum of squares for the test-set is then given as

$$\|\mathbf{X}_{\text{test}} - \hat{\mathbf{M}}\|^2 = \|\mathbf{X}_{\text{test}} - \mathbf{T}_{\text{test}}\mathbf{P}'\|^2 = \|\mathbf{X}_{\text{test}} - \mathbf{X}_{\text{test}}\mathbf{P}\mathbf{P}'\|^2 \tag{7.2}$$

As can be clearly seen, the model values ($\mathbf{X}_{\text{test}}\mathbf{P}\mathbf{P}'$) are not independent of the data ($\mathbf{X}_{\text{test}}$) and thus overfitting can be a problem even for a test set. Thus, to compare different models on the basis of residual sum of squares calculated in this way, some correction for degrees of freedom is necessary [Martens & Næs 1989]. Such degrees of freedom are not easily defined, especially for multi-way models and hence can, at most, be considered as reasonable approximations. For example, the approximate degrees of freedom can be set to the number of elements minus the number of parameters or minus the number of parameters corrected for scaling and rotational indeterminacy (Chapter 5 or [Bro 1998, Durell *et al.* 1990, Liu & Sidiropoulos 2001, Louwerse *et al.* 1999]).

The same problem does not exist in validation of regression models where the relevant residual sum of squares $\|\mathbf{y}_{\text{test}} - \hat{\mathbf{y}}\|^2$ is based on a model that yields predictions that are independent of the $\mathbf{y}$ data because they are solely based on the independent variables and on a model derived from prior data. Thus, $\mathbf{y}$ does not appear in the calculation of the prediction and therefore each residual has one degree of freedom.

An alternative to test-set validation is cross-validation. Cross-validation can be viewed as a simulation of test-set validation where a test-set validation is repeated using different objects or data points in the test-set. In this respect, cross-validation is particularly useful when a limited number of samples is available. Also, certain types of cross-validation are useful in validating component models because the above mentioned problem of dependent residuals is eliminated by construction.

Cross-validation is an internal resampling method much like the older Jackknife and Bootstrap methods [Efron 1982, Efron & Gong 1983, Efron & Tibshirani 1993, Wehrens *et al.* 2000]. The principle of cross-validation goes back to Stone [1974] and Geisser [1974] and the basic idea is simple:

(i) Leave out part of the data values.
(ii) Build the model without these data values.
(iii) Use the model to predict these data values.
(iv) Calculate the corresponding residual error.
(v) Cycle through this procedure such that each data value has been left out once and sum all the squared prediction errors (PRESS).

When cross-validation is used for selecting the number of components, the number yielding a minimum PRESS value is often chosen. In practice, a minimum in the PRESS values may not always be present or sometimes the minimum is only marginally better than simpler models. In such cases, it becomes more difficult to decide on the number of components and prior knowledge as well as other model results and the overall aim of the analysis have to be taken into account. When choosing which data values to leave out, it has to be borne in mind that different approaches will yield different results. For example, leaving out only replicates, a repeatability error will be obtained whereas if the left-out samples are from new experiments performed by new laboratories, a reproducibility error is obtained.

The PRESS is used for obtaining estimates of variance or for deciding on the number of components to use. If only the individual residuals are sought for residual analysis, the actual calculation of PRESS can be skipped.

The PRESS will be of the form

$$\text{PRESS} = \sum_{i=1}^{I}(y_i - \hat{y}_{(i)})^2 \tag{7.3}$$

where the term $\hat{y}_{(i)}$ is the prediction of y_i using a model that is fitted without using the ith object itself to estimate the model. This 'object' may also be a segment of several individual objects leading to the so-called segmented cross-validation. Further extending this segmentation, the segments may be independent test-sets, in which case, the cross-validation will yield results that are similar to the results from test-set validation.

As shown above, the residuals will not be independent of the data in the case of component modeling when individual samples are left out during the cross-validation. In regression though, the y_i and $\hat{y}_{(i)}$ are statistically independent and Equation (7.3) is an estimator of the summed squared prediction error of y using a certain model.

In order to obtain independent model estimates in component models a different cross-validation technique has to be used. Instead of leaving out a complete sample/row, it is possible to leave out only one (or a few) elements. Using an algorithm that handles missing data, it is possible to estimate the relevant component model without the left-out element. The estimate of this element is obtained from the model of the whole data array ($\sum_{r=1}^{R} t_{ir}p_{jr}$ in case of leaving out element x_{ij} in PCA) where there is no dependence between the left-out element and the model. This is the basis for the cross-validation routines in several papers [Louwerse *et al.* 1999].

There are many interesting and intricate details of test-set and cross-validation that can influence the quality and adequacy of the results. For example, data from designed experiments cannot easily be split if the design is close to saturated, measurements with correlated errors can give too optimistic results if not properly accounted for in the splitting of the samples etc. These issues are not specific to three-way analysis and will not be described in detail here [Duchesne & MacGregor 2001, Lorber & Kowalski 1990, Martens & Dardenne 1998, Næs & Ellekjær 1993, Osten 1988, Rivals & Personnaz 1999, Scarponi *et al.* 1990, Shao 1993, Stone 1974, Wold 1978]. Test-set and cross-validation of regression models can be performed exactly as in two-way analysis, but cross-validation of component models requires some special attention. This will be treated in the following.

Cross-validation of multi-way component models

In multi-way analysis, a distinction is made between cross-validation of regression models and cross-validation of component models. In regression models, the purpose is to predict a **y** and cross-validation can then be used to find the best regression model to achieve this. In component models cross-validation can be used to find the most parsimonious but adequate model of **X**. In this section, cross-validation of component models is described and in the next section regression models are treated.

Cross-validation of multi-way component models is described by Louwerse *et al.* [1999]. There are several alternatives for performing cross-validation of two-way component models such as PCA. The main approaches are the one of Eastment & Krzanowski [1982] and the one of Wold [1978]. The approach of Eastment & Krzanowski has been generalized and is called the 'leave-bar-out' method. The approach of Wold relies heavily on the special way the NIPALS algorithm deals with missing data and cannot be directly generalized for multi-way component models. A similar approach based on weighted regression [Kiers 1997b] is available and is called the expectation maximization method [Louwerse *et al.* 1999]. Both cross-validation methods are explained using a three-way array $\underline{\mathbf{X}}$ ($I \times J \times K$) where the problem is to find a proper Tucker3 model for this array, that is, finding P, Q and R in

$$\mathbf{X} = \mathbf{AG}(\mathbf{C} \otimes \mathbf{B})' + \mathbf{E} \qquad (7.4)$$

with loading matrices **A** ($I \times P$), **B** ($J \times Q$), **C** ($K \times R$) and core-array **G** ($P \times QR$).

LEAVE-BAR-OUT METHOD

For the Tucker3 model with a certain P, Q and R, the following steps are schematically taken:

(i) Leave out the ith slice (see Figure 7.1) and call the remaining array $\underline{\mathbf{X}}^{(i)}$. Build a Tucker3 ($P,Q,R$) model on $\underline{\mathbf{X}}^{(i)}$. This results in $\mathbf{A}^{(i)}$, $\mathbf{B}^{(i)}$, $\mathbf{C}^{(i)}$ and $\mathbf{G}^{(i)}$.

(ii) Leave out the jth slice of the original full array and call the remaining array $\underline{\mathbf{X}}^{(j)}$. Build a Tucker3 ($P,Q,R$) model on $\underline{\mathbf{X}}^{(j)}$. This results in $\mathbf{A}^{(j)}$, $\mathbf{B}^{(j)}$, $\mathbf{C}^{(j)}$ and $\mathbf{G}^{(j)}$.

(iii) Leave out the kth slice of the original full array and call the remaining array $\underline{\mathbf{X}}^{(k)}$. Build a Tucker3 ($P,Q,R$) model on $\underline{\mathbf{X}}^{(k)}$. This results in $\mathbf{A}^{(k)}$, $\mathbf{B}^{(k)}$, $\mathbf{C}^{(k)}$ and $\mathbf{G}^{(k)}$.

(iv) Predict the left out element (i,j,k) using the Tucker3 model based on the loading matrices $\mathbf{A}^{(i)}$, $\mathbf{B}^{(j)}$ and $\mathbf{C}^{(k)}$ and the geometric average of core-arrays above and calculate $(x_{ijk} - \hat{x}_{ijk})^2$.

(v) Repeat the whole procedure until each element is left out once. Calculate the prediction error sum of squares (PRESS).

The whole procedure can be repeated for different values of P, Q and R, and then the Tucker3 model with the lowest PRESS can be selected. By progressing through steps (i) to (iii) the block or element indicated in Figure 7.1 (shaded box) is never part of any model. Hence, this part can be independently predicted, which is crucial for proper cross-validation. It is a matter of choice how to leave out slices; one slice at a time, more slices at a time etc. This can also be different for the different modes of $\underline{\mathbf{X}}$. When the array is preprocessed (e.g. centered and/or scaled), then this preprocessing has to be repeated for every step again, that is, the data are left out from the original array.

The procedure sketched above is a simplification. There are several intricate details not described here, e.g. how to combine the different calculated loading matrices and core-arrays in step (iv) to predict the value of x_{ijk}, and how to deal with the sign and permutation indeterminacy problem of estimating comparable loading matrices from different submodels.

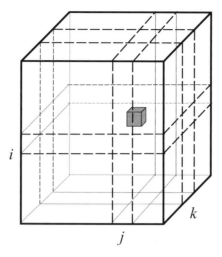

Figure 7.1. Illustration of the leave-bar-out method of cross-validating a three-way component model.

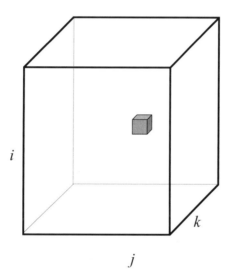

Figure 7.2. Illustration of the expectation maximization cross-validation of a three-way model.

EXPECTATION MAXIMIZATION

The second generalization of cross-validation is based on using weighted regression for dealing with missing data. This method of dealing with missing data is equivalent to the expectation maximization approach of handling missing data [Dempster *et al.* 1977, Kiers 1997b, Little & Rubin 1987, Zhang *et al.* 1997]. It works by leaving out only one specific element (or group of elements) at a time as illustrated in Figure 7.2. The cross-validation proceeds as follows:

(i) Leave out a part of $\underline{\mathbf{X}}$ (indicated with a box in Figure 7.2), and build a Tucker3 (P,Q,R) model giving weight zero to the left out part.

(ii) Use the calculated model parameters of step (i) to predict the values of the left out X-elements, calculate the prediction error.

(iii) Repeat steps (i) and (ii) until each element has been left out once.

(iv) Calculate the PRESS value.

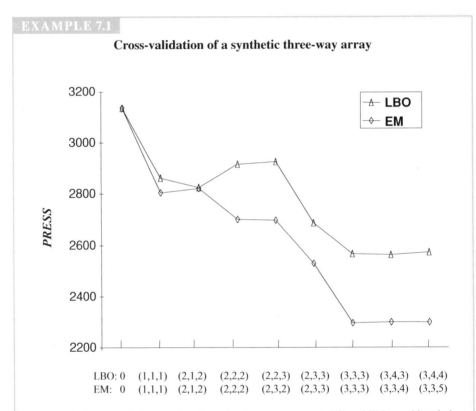

EXAMPLE 7.1

Cross-validation of a synthetic three-way array

LBO: 0   (1,1,1)   (2,1,2)   (2,2,2)   (2,2,3)   (2,3,3)   (3,3,3)   (3,4,3)   (3,4,4)
EM: 0   (1,1,1)   (2,1,2)   (2,2,2)   (2,3,2)   (2,3,3)   (3,3,3)   (3,3,4)   (3,3,5)

Figure 7.3. Cross-validation results of a noisy three-way array. LBO and EM are abbreviations of leave-bar-out and expectation maximization, respectively. The numbers between parentheses correspond to the size of the Tucker3 model, e.g. (2,2,2) means a Tucker3 (2,2,2) model.

A noisy synthetic array $(100 \times 20 \times 10)$ with a Tucker3 structure is generated. The core size is $(3 \times 3 \times 3)$. Elements of the unfolded true core (3×9) are:

$$\begin{bmatrix} 1 & 0 & 0 & 0 & 1 & 0 & 0 & 0 & 1 \\ 0 & 1 & 0 & 0 & 0 & 1 & 1 & 0 & 0 \\ 0 & 0 & 1 & 1 & 0 & 0 & 0 & 1 & 0 \end{bmatrix}$$

Normally distributed random numbers, with zero mean and unit variance, are used for $\mathbf{A}$, $\mathbf{B}$ and $\mathbf{C}$; 25 % of the array variation was described by this model and 75 % by normally distributed white independent identically distributed noise. Even though the data are very noisy, a (3,3,3) Tucker3 model will be adequate for describing the systematic part of the data. A bar size of $(10 \times 2 \times 1)$ was used for the leave-bar-out cross-validation; 20 data points, eliminated at random without replacement, were used for the expectation maximization cross-validation. Hence, in both cases 0.1 % of the data was left out in each cross-validation step. Two-way cross-validation of the matricized arrays, based on minimum two-way PRESS values, wrongly indicated a (3,3,1) core size. This indicates that in cases with a low signal to noise ratio the approach of using two-way cross-validation on the three matricized arrays does not work. Both types of three-way cross-validation indicated that a Tucker3 (3,3,3) model is appropriate (see Figure 7.3). The (3,3,3) model is preferred over e.g. the (3,3,4) model. Even though the (3,3,4) model is equally good with respect to the PRESS value, it uses more parameters and is thus more complicated without providing a better model.

There is yet only limited experience with these three-way cross-validation methods. The expectation maximization cross-validation method can be implemented in software without many difficulties. The leave-bar-out cross-validation method is more complex to implement. In general, the expectation maximization method outperforms the leave-bar-out method. In most cases the PRESS plots show a more distinct minimum and it is thus easier to identify the best model. However, the expectation maximization method requires much longer computation time [Louwerse *et al.* 1999] and this can be a problem for large arrays.

Cross-validation of multi-way regression models

Cross-validation of multi-way regression models does not essentially differ from cross-validating two-way regression models. Suppose that a three-way $\underline{\mathbf{X}}$ ($I \times J \times K$) is available and a univariate $\mathbf{y}$ ($I \times 1$). Then cross-validation can be performed in the following way:

(i) Leave out samples.
(ii) Build the multiway regression model on the remaining samples relating $\underline{\mathbf{X}}$ to the available part of $\mathbf{y}$.
(iii) Predict the y values of the left-out samples.
(iv) Calculate the prediction error.
(v) Repeat until each y value has been left out once.

In step (i) a selection has to be made on how many of the y values to leave out in each step. There are different alternatives: leave one out or leave more out. In the leave-more-out approach there are again different alternatives: leave out at random, or leave out in a pattern. Which one is the best depends on the situation and aim of the analysis. If the objects are not in random order, but patterned in a certain way, more thought has to go into this selection of what to leave out.

In step (ii) any multi-way regression model may be used and tested. Usually, different model types (e.g. *N*-PLS and Tucker3-based regression on scores model), or models with a different number of components (e.g. a two-component *N*-PLS model and a three-component *N*-PLS model) are tested. To have complete independence of $\hat{y}_{(i)}$ and y_i, the matrices involved in building the model have to be preprocessed based on interim calibration data each time step (ii) is entered.

7.3 Selecting Which Model to Use

Depending on the purpose of the data analysis, several different models may be appropriate. Choosing which model is better is therefore part of the overall validation procedure. For calibration problems, different models can be compared in terms of how well they *predict* the dependent variables. Sometimes a priori knowledge of the structure of the data is available (e.g., that fluorescence data can be well approximated by a trilinear model), but often this is not the case. If no external information is available on which comparison of different models can be based, other approaches have to be used. In the following, a discussion is given of how to assess the appropriateness of the model based on the mathematical properties of the data and the model. No exact rules will be given, but rather some guidelines that may be helpful for the data analyst.

Model hierarchy

It is important to have a feeling for the hierarchy of the different possible models (Chapter 5). Kiers [1991] showed that the PARAFAC model can be considered a constrained version of a Tucker3 model, and Tucker3 a constrained version of PCA on the matricized array (Tucker1). Any data set that can be modeled adequately by an R-component PARAFAC model can thus also be modeled by an (R,R,R) Tucker3 or an R-component Tucker1 model, but PARAFAC uses fewer parameters. A Tucker1 model always *fits* data better than a Tucker3 model, which again will *fit* better than a PARAFAC model, all except for extreme cases where the models may fit equally well. If a PARAFAC model is adequate, Tucker3 and Tucker1 models will tend to use the excess parameters to model noise or model the systematic variation in a redundant way. Therefore, it is generally preferable to use the simplest adequate model. The Tucker1 model can be considered the most complex and flexible model, as it uses most parameters, while PARAFAC is the most simple and restricted model. This principle of using the simplest possible model is old, in fact dating back to the fourteenth century (Ockham's razor), and is now also known as the law or principle of parsimony [Box & Jenkins 1976, Ferguson 1954, Harris & Judge 1998, Seasholtz & Kowalski 1993].

How then, do these observations help in choosing the most appropriate model? The sum of squares of a least squares modeled data set can be written as $SS_{tot} = SS_{mod} + SS_{res}$ where SS_{tot} is the sum of squares of the data array elements, SS_{mod} is the corresponding sum of squares of the model elements and SS_{res} of the residual elements (see Section 7.5). The following always holds:

$$SS_{mod}(\text{Tucker1}) \geq SS_{mod}(\text{Tucker3}) \geq SS_{mod}(\text{PARAFAC}) \qquad (7.5)$$

For this ranking to hold, it is assumed that the same number of components is used in all three models as well as the same number of components in each mode of the Tucker3 model. Suppose that for a given data set, a PARAFAC model is valid. From the above discussion and the inequality above, it follows that most likely SS_{mod} of a Tucker1, Tucker3, and a PARAFAC model of equivalent dimension will be similar (though lowest for the PARAFAC model). This is so, because they will all be capable of modeling the systematic variation. Even though the Tucker1 and the Tucker3 models have additional modeling power, this will only describe the (smaller) random variation. Hence, the models will be almost equally good with respect to SS_{mod}. Another extreme occurs in a situation where no three-way structure is appropriate but the Tucker1 model is. Then SS_{mod}(Tucker1) will be significantly higher than SS_{mod}(Tucker3) and SS_{mod}(PARAFAC). Much in line with how a scree-plot is used, the appropriate complexity of the model may be determined by scrutinizing the fit values of different models this way [Eastment & Krzanowski 1982, Timmerman & Kiers 2000]. Instead of using the fit residuals, a clearer picture can usually be obtained by using the sum of squares of errors in modeling cross-validated samples as demonstrated in Example 7.2.

Choosing sensory model from fit and cross-validated fit

Table 7.1. Percentage variation explained for PARAFAC, Tucker3 and Tucker1 fit to sensory data centered across the first sample mode and using different number of components. Percentages are defined as $100(1 - SS_{mod}/SS_{total})$

Number of Components	PARAFAC Fit	PARAFAC Cross-val.	Tucker3 Fit	Tucker3 Cross-val.	Tucker1 Fit	Tucker1 Cross-val.
1	35.3	14.5	35.3	14.5	44.6	13.2
2	49.2	26.2	49.2	26.2	65.8	26.5
3	57.4	32.9	57.7	31.6	74.3	18.6
4	62.7	34.4	64.6	19.6	80.7	<0
5	67.2	33.0	72.7	24.6	86.2	<0

In this data set ten breads (five breads baked in replicates) were assessed by a sensory panel of eight people using a fixed vocabulary of eleven different attributes. The data are arranged in a $10 \times 11 \times 8$ array and a model is sought of these data in order to be able to assess the differences between the breads from a perception point of view. One- to four-component models were fitted from the data centered across the first mode. This was done using PARAFAC, Tucker3, and Tucker1. The percentage of variation explained was calculated from the sum of squares of residuals relative to the sum of squares of the centered data. The percentage variation explained is given as both the cross-validated percentage and as the fit percentage. The cross-validated residuals were found by fitting the models using a full cross-validation routine with expectation maximization. In each segment, one element of the data set was set to missing, and the model (including offsets) was fit to the remaining part of the data. The cross-validated residuals were then determined as the discrepancy between fit and real values of the elements set to missing. The number of segments were also chosen to be fewer, such that more elements were left out at a time. This helps speeding up the algorithm but if too many elements are left out, the remaining elements will not be sufficient to estimate the model reliably. In this case, it was found, though, that even leaving out up to 100 elements at a time gave similar results except for Tucker1 which has less structure and hence is more influenced by the high amount of missing data. The elements that were left out in each segment were chosen to be systematically spread to avoid problems such as all data missing for any particular variable, assessor or bread.

The results are shown in Table 7.1 where the quality of the model is given in terms of fit rather than PRESS. Therefore, the values are increasing rather than decreasing until the optimal model complexity. It is evident that in terms of cross-validated fit there is no gain in going from the simple PARAFAC model to either Tucker3 or Tucker1. The fit will increase because of the higher flexibility of the models, but this increase in fit does not manifest itself in the cross-validated residuals. Hence, the increase in fit must be due to noise being fitted and is therefore simply overfit. This is seen, for example, for the three-component models where the fit increases from 57.4 to 74.3 % in going from PARAFAC to Tucker1 and at the same time the cross-validated fit decreases from 32.9 to 18.6 %! It is also clear from the above example, that the Tucker3 model is only moderately flexible and much closer to the PARAFAC than the Tucker1 model, both with respect to fit and to cross-validated errors. Finally, it is noted that fitted residuals can provide very misleading conclusions especially for noisy data.

Tucker3 core analysis

An exploratory way of finding the appropriate structural model is to investigate obtainable simplicity from a Tucker3 core array from a model using a high number of components. As the Tucker3 model has rotational freedom, it is possible to rotate the core by counter-rotating the loading matrices accordingly without changing the fit of the model as explained in Chapter 5. It can be sought either to rotate the core to a hypothesized structure such as superdiagonality or to a general simple structure in some well-defined way. If the same number of components is used in each mode and the core can be rotated approximately to a superdiagonal, this is clear evidence that a PARAFAC model is appropriate. However, oblique rotations must be used in order to allow for nonorthogonal PARAFAC component matrices. Another possibility is to seek a transformation of the core-array with as many zeros as possible, either by rotation preserving the fit, or by simply constraining elements to be zero (for references on core rotations see [Andersson & Henrion 1999, Brouwer & Kroonenberg 1991, Henrion 1993, Kiers 1992, Kiers 1997a, Kiers 1998a, Kiers 1998b, Murakami *et al.* 1998]). There are few examples in the literature on this approach and clear-cut solutions are seldom obtained. For more information on rotations see Chapter 5.

7.4   Selecting the Number of Components

It is necessary to decide the appropriate number of components to use in a model. The appropriate dimensionality of a model may even change depending on what the specified purpose of the model is. Hence, appropriate dimensionality of, e.g., a PARAFAC model, is *not* necessarily identical to the three-way pseudo-rank of the data array. Appropriate model dimensionality is not only a function of the data but also a function of the context and aim of the analysis. Hence, a suitable PARAFAC model for exploring a data set may have a rank different from a PARAFAC model where the scores are used for a subsequent regression model.

There are a number of tools for choosing the number of components. Some are based on statistical assumptions and significance testing, some based on empirical rules and some based on technological aspects. In general, it is not advisable to rely solely on one specific rule for choosing dimensionality. Every rule has its own assumptions and no real data set will meet these assumptions exactly. For unique models such as PARAFAC, choosing the proper number of components is often much easier than for, e.g., principal component analysis because the uniqueness of the model gives additional room for validating whether the model is reasonable.

In the following, some tools for determining the proper number of components are given. These are primarily scree plots, split-half analysis, cross-validation and visually checking residuals. The focus here is only on the data analytical validation tools. For specific applications, knowledge of the underlying phenomena in the data can often be helpful in assessing specific models. Spectra of certain analytes may be known, the shape of chromatographic profiles may be known or the nonnegativity of certain phenomena may be known. Such facts can be informative when comparing different models. The description will be divided into methods applicable for component models and methods applicable for regression models.

Complexity of component models

Several approaches are possible for determining the number of components. The general rule, is that all these methods can at most give good indications of suitable dimensionality. The definitive choice is the responsibility of the data analyst and follows from assessing one or better several diagnostics as well as combining these with background knowledge of the problem and purpose.

SCREE PLOT

In the following, the so-called scree plot for determining the number of components in PCA will be described. Afterwards, it will be shown how this method can be modified for finding the appropriate number of components in PARAFAC and in Tucker3 models.

One of the most frequently used tools for choosing the number of components in two-way analysis (e.g. PCA) is the *scree plot* [Cattell 1966]. The original scree plot gives the eigenvalues of the cross-product matrix sorted by size as a function of the number of components. It allows an interpretation of which components are well enough above the noise level for being useful. The first eigenvalue equals the sum of squares explained by the first component, etc. and the sum of all eigenvalues equals the trace of the cross-product matrix. Scree is a geological term referring to the debris that collects on the lower part of a rocky slope. Singular values from a singular value decomposition of the data can also be used instead of the eigenvalues. The singular values equal the square root of the corresponding eigenvalues. Eigenvalues, λ (or singular values or the percentages of the sum of squares left in the residual) are shown on the vertical axis as a function of component number on the horizontal axis (Figure 7.4). The scree plot with eigenvalues in PCA can also be shown cumulatively because the eigenvalues are independent and add up to 100 % of the sum of squares (%SS). Examples of many types of scree plots are shown in Chapter 8.

Eigenvalue

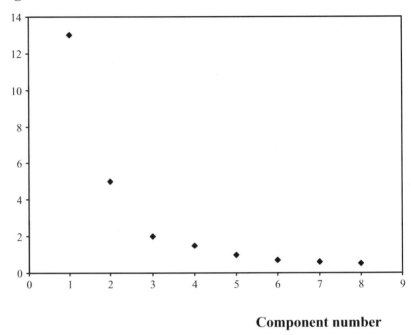

Component number

Figure 7.4. Example of a hypothetical scree plot.

A certain cutoff in the scree plot is used to determine which components are too small to be used. For exploratory purposes, it may suffice to choose a cutoff value that leads to, e.g., 80 % of the variation explained for noisy data, but for more quantitative purposes it is useful to have a more elaborate determination of the appropriate cutoff value. Usually the number of components is chosen where the plot levels off to a linear decreasing pattern (see also Horn [1965]). Thus, no more than the number of factors to the left of this point should be retained.

A plot similar to a scree plot can be made for PARAFAC, by plotting the sum of squares of the individual components. However, in this case, cumulative plots cannot be made directly because the variances of the individual factors are not additive due to the obliqueness of the factors. Furthermore, the sum of squares of the one-component model may not equal the size of the largest component in a two-component model. Hence, the scree plot is not directly useful for PARAFAC models. The cumulative scree plot for PARAFAC models, on the other hand, can be constructed by plotting the explained or residual sum of squares for a one-component model, a two-component model, etc. This will provide similar information to the ordinary two-way cumulative scree plot, with the exception that the factors change for every model, since PARAFAC is not sequentially fit. The basic principle is retained though, as the appropriate number of components to use is chosen as the number of components for which the decrease in the residual variation levels off to a linear trend (see Example 7.3).

EXAMPLE 7.3

Using scree plots for determining dimensionality of PARAFAC model

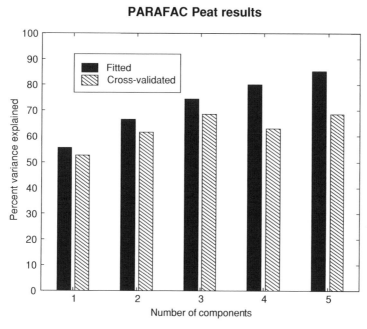

Figure 7.5. The scree plot shows percentage of the total sum of squares explained by the PARAFAC model with increasing number of components. Results from fit and (expectation maximization) cross-validated residuals are shown.

For five major industrial peat types (called S, I, M, A and N), seven treatments for better aggregation of colloidal particles in slurry are applied. The results of automated digital image analysis of the particles under a microscope give a histogram of particle areas in 21 bins. The array is peat type × treatment × particle area ($5 \times 7 \times 21$). PARAFAC models with increasing number of components are fit to the array. Figure 7.5 shows the fit percentage of variation explained by a PARAFAC model as a function of the complexity of the model (the number of components extracted). From the fit percentage, it is very difficult to see what the appropriate model dimensionality is. The main observation to be made is how much the magnitude of the residual variation decreases as a function of the rank of the model. With the specific background information of the expected noise in the peat data, a model that explains 75–80 % of the total sum of squares is satisfactory. See also Table 7.2 which illustrates that additivity of the sum of squares of individual PARAFAC components does not apply due to the nonorthogonality of the components.

As in ordinary two-way analysis, the scree plot based on fit values can sometimes be misleading, if the model is overfitting. Exchanging fit residuals with cross-validated

Table 7.2. Sum of squares as percentage of the total sum of squares for the peat example. Left column is calculated by adding the sum of squares for the individual components whereas the right column is the sum of squares explained by the total model

Number of Components	Sum of Components SS%	Model Sum of Squares (%)
1	55.4	55.4
2	84.3	66.7
3	102.3	74.4
4	106.2	80.0
5	398.0	85.3

values as shown in Figure 7.5, can often provide a clearer picture of the degree of overfit. In this case, the cross-validated variation explained suggests that no more than three components should be used. The cross-validation used was based on expectation maximization and several different numbers of segments were investigated ranging from leaving out one element at a time to leaving out 70 elements randomly at a time. Similar results were obtained in all cases.

Timmerman & Kiers [2000] developed a diagnostic based on the ideas discussed above. Rather than plotting the sum of squares of the model with a given number of components, the difference between, e.g., an $(R - 1)$-component model and an R-component model was plotted. In this way, the point where the relative improvement in fit is small is easier to detect. The technique was developed specifically for Tucker3 models, but is equally applicable to other types of models. Other ad hoc methods can also be used, either based on current two-way practice or based on the characteristics of the three-way model (e.g. Example 7.4).

EXAMPLE 7.4

**Determining number of PARAFAC components from the norm
of the components**

All PARAFAC loadings may and do often change when the model rank is increased. At the point where the correct number of components is used, these changes are often not very dramatic. This provides an empirical tool for selecting the number of components to use [Geladi *et al.* 2000]. This is shown in Figure 7.6 for the peat example of Example 7.3. Figure 7.6 shows a bar-plot of sum of squares of the ordered PARAFAC components as model rank increases. The models remain stable up to rank four and some of the components increase dramatically in variation in the rank five model. This is an indication

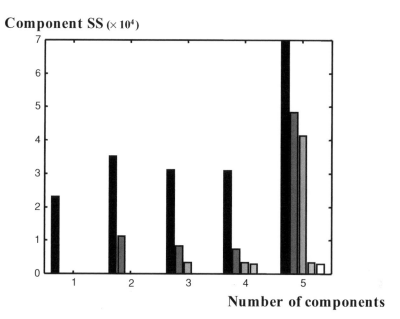

Figure 7.6. The ordered component sizes of the PARAFAC components as a function of the number of components in the model.

that the rank five model is unstable. The plot, therefore, suggests that a rank of more than four should not be used.

When fitting Tucker3 models, scree plots similar to the PARAFAC plots in Figure 7.5 can be made. However, as there are three numbers of components to set in the Tucker3 model, it is less straightforward to interpret such a plot. A general problem in validating the number of components for Tucker3 models is that there is no strict hierarchical relationship between different Tucker3 models, e.g. a Tucker3 (3,3,3) model has no clear hierarchical relationship with a Tucker3 (2,4,3) model. Methods to solve those problems are described in detail in the literature [Louwerse *et al.* 1999]. A commonly used simplification is to only look at the best-fitting of all combinations with the same number of components. Consider a series of Tucker3 models fit to a data set: (1,1,1), (1,1,2), (1,2,1), (2,1,1), (1,2,2), (2,1,2), (2,2,1), etc. Each model has an associated fit. Instead of plotting all the models, it is customary to plot the fit versus the *total* number of components. There are, for example, three different models with a total of five component vectors; namely (1,2,2), (2,2,1), and (2,1,2). *Only* the best-fitting of these three models is chosen. In Figure 7.7 the results from a cross-validation are shown. Note how the fit increases whereas the cross-validated variation definitely ceases to improve after the (3,4,3) model. Note also that there are no models with a total of four components. Having e.g. a (1,2,1) model is redundant and will fit exactly as a (1,1,1) model according to Equation (2.24).

USING RESIDUALS FOR DETERMINING THE NUMBER OF COMPONENTS

Residuals can be used for visually assessing the validity of a model. For example, the distribution of the residuals can be used for examining whether systematic variation is left

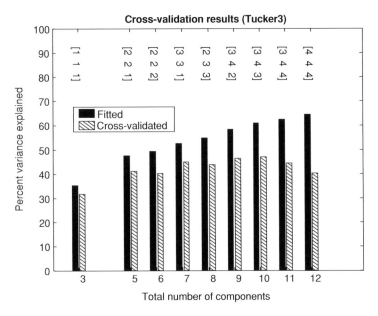

Figure 7.7. Scree-like plot for a Tucker3 model of the sensory bread data centered across the bread-mode. All combinations of components up to (4,4,4) are shown. For each total number of components, the model with best cross-validated fit is shown.

in the residuals or to what degree the distributional assumptions of the model are fulfilled. As discussed earlier, the size of the variation in the residuals can be used for determining the number of components to be used. Yet another aspect is to look for outlying samples in the residuals. Exploring the data is a problem-specific task. For spectral data, the spectral shape of the residuals can be interpreted. Residuals can also be used for more formal statistical tests. As in bilinear models, the characteristics of a model can be judged by the residuals of the model: if systematic variation is left in the residuals, it is an indication that more components can be extracted (Example 7.5). If the residual variance is larger than the known experimental error, it is indicative of more systematic variation in the data. To calculate variance-like estimators, Durell *et al.* [1990] suggest using the number of parameters for correcting mean squared errors. Such corrections might be used for exploratory purposes, but they are not to be taken as statistically correct numbers of degrees of freedom (see also Elaboration 5.1). Such are currently not available. Another way to obtain variance estimators is to fit hypothesized distributions to the residuals obtained either from fitted models or from predicted samples. This has been pursued in statistical process monitoring based on the work of Box, Nomikos and MacGregor and others [Box 1954, Jackson & Mudholkar 1979, Louwerse & Smilde 2000, Nomikos & MacGregor 1995].

EXAMPLE 7.5

Using residuals for assessing the number of components

A fluorescence excitation–emission data set of five samples with varying amounts of three amino acids was modeled using the PARAFAC model. In Figure 7.8 the residuals

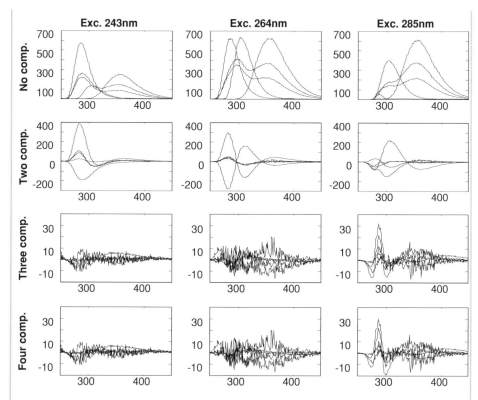

Figure 7.8. Residual plots of fluorescence data set. The plots are slabs of emission spectra obtained by fixing the excitation as indicated at the top. The top row figures show the raw data, and the subsequent three plots show the residuals after two, three, and four components respectively. Note the different vertical scales.

after zero, two, three, and four components are shown for three arbitrary excitation slabs of the three-way array. There are five samples in total, each giving rise to one line in each of the plots. The zero-component model holds the raw data. After two components, huge variations are still present in the residuals as compared to the raw data. After three and four components, the residual variation is of a much smaller size (note the scale) and more noisy. Since the residual variation does not decrease significantly going from three to four components, the three-component model must be the appropriate model as is also known from the chemistry of the data. Although these data are very close to being trilinear, there is still some minor systematic variation in the residuals after three components. This is expected due to deviations in the data from the ideal trilinear model. It illustrates that it may be problematic to rely completely on statistically based tests, since these explore the space of the residuals mainly based on assumptions that these contain no systematic variation. Before such tests are conducted, it is advisable to ensure that the distribution of the residuals complies with the underlying assumptions.

SPLIT-HALF ANALYSIS

Harshman advocates the use of split-half analysis for determining the proper rank of models with unique axes, that is, models with no rotational freedom [Harshman 1984, Harshman & De Sarbo 1984]. In split-half analysis different subsets of the data are analyzed independently. Due to the uniqueness of the PARAFAC model, the same result – same loadings – will be obtained in the nonsplit modes from models of any suitable subset of the data, if the correct number of components is chosen. To judge whether two models are equal, the indeterminacy in trilinear models has to be remembered: the order and scale of components may change if not fixed algorithmically. If too many or too few components are chosen, the model parameters will differ if a different data set is used for fitting the model.

If a model is stable in a split-half sense it is a clear indication that the model is 'real'; that it captures variation that not only pertains to the specific samples. If, on the other hand, some components are not stable in a split-half sense, it indicates that they reflect noise or that too few components are extracted. Hence, the model is not valid. It may also happen, though, that the phenomenon reflected in the nonstable component is simply only present in specific subsets. By including in the split-half analysis a model of the whole data set, this can be investigated by verifying if all components in the total model have significant scores in all subsets.

When doing a split-half experiment it has to be decided which mode to split. Splitting must be performed in a mode with a sufficient number of *disjoint* variables/samples, in such a way that each subset will span the whole domain and hence that all the parameters can be estimated from the individual subsets. For example, the conditions for uniqueness must apply to all the subsets. With a high-dimensional spectral mode, an obvious idea is to use this spectral mode for splitting, but the collinearity of the variables impedes sound results and split-half analysis is meaningless. Therefore, it is most common to split the data in the sample mode.

To avoid an unfortunate splitting of the samples causing some phenomena to be absent in certain groups, the following approach is often adopted. The samples are divided into two groups: A and B (Figure 7.9 and Example 7.6). If the samples are presumed to have some kind of correlation in time the sets are constructed contiguously, i.e., A consists of the first half of the samples and B of the last half. Accidentally, it may happen that one of these sets

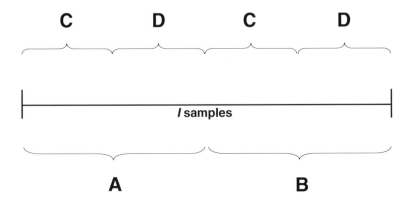

Figure 7.9. Partition of *I* samples in a split-half analysis. The samples are divided into two groups A and B which are disjoint, and into two other groups, C and D, which are also disjoint.

does not contain information on all latent phenomena. To assure or at least increase the possibility that the sets to be analyzed cover the complete variation, two more sets are generated, C and D. The set C is made from the first half of A and B and the set D consists of the last half of the samples in A and B. These sets are pairwise disjoint. A model is fit to each of the data sets, and if the solution replicates over set A and B *or* over set C and D, correctness of the solution is empirically verified. Replication of the solution over both pairs (A/B or C/D) is the most fortunate solution, but if the solution replicates over only one of the pairs this is sufficient, because it is unlikely that the solution accidentally replicates if the samples are independent and the solution random. The reason that the solution does not replicate over the other pair of sets can be that the number of samples is too low. Then it may simply not be possible to generate four different sets all containing all the latent phenomena.

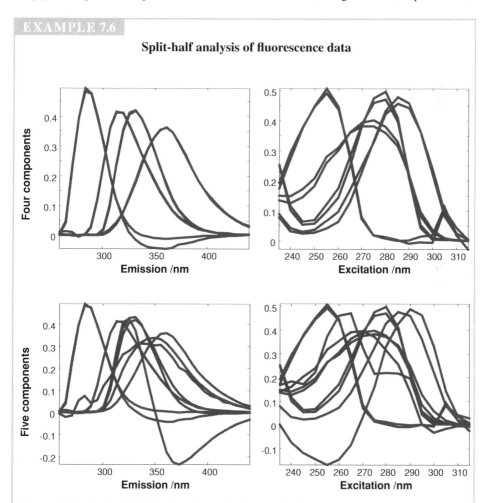

EXAMPLE 7.6

Split-half analysis of fluorescence data

Figure 7.10. Result of split-half analysis of a four- and a five-component PARAFAC model of fluorescence data. In the left-most plots, the emission loadings are shown and in the right-most the excitation loadings. Upper plots show the results from the four-component model and the lower ones for the five-component model. The profiles from set A and B are shown superimposed. It is seen that in the upper plots, the profiles of set A and B coincide quite closely whereas this is not the case in the lower plots.

Consider a data set of 27 samples containing four fluorophores which are measured by fluorescence excitation–emission spectroscopy. In order to find the appropriate number of PARAFAC components, a split-half analysis is conducted as described above dividing the 27 samples into two sets: A and B. An additional split (C and D) was also made, but the results are not shown here. These subsets were used for fitting a PARAFAC model with four components and with five components. The resulting loading vectors are given in Figure 7.10. Evidently, five components cannot be supported by the result of the split-half analysis, as some of the loading vectors change from subset to subset. Four, on the other hand, seems like an appropriate number of components to use. In Figure 7.10 it is seen that regardless of whether the samples given by subset A or B are used, the estimated profiles in the two modes shown are almost the same. The deviations between the two sets of estimates are not huge and may simply be caused by the noise and the small sample sizes. It is possible to quantify the similarity by calculating correlations or other similarity measures between different splits.

Number of components in regression models

Determining the number of components in regression models is usually performed in a similar manner to how this is done in two-way analysis. Thus, cross-validation using the root mean square error of cross-validation residuals as a diagnostic is one of the main tools for assessing regression models. Since determining the number of components in regression models has been described in numerous places, this will not be dealt with in detail here [Beebe & Kowalski 1987, Lorber & Kowalski 1990, Martens & Næs 1989, Osten 1988, Wakeling & Morris 1993, Wehrens & van der Linden 1997].

The scree plots described above are also used in regression models of the type $\mathbf{y} = \mathbf{Xb} + \mathbf{e}_y$ where $\mathbf{y}$ is decomposed into a model $\mathbf{Xb}$ and a residual $\mathbf{e}_y$. Plots of the percentage variation explained as a function of the number of PCA or PLS components are used to study the fit of the regression model. With results from cross-validation or a test-set, a similar plot can be used to select the rank of the regression model.

7.5 Residual and Influence Analysis

In this paragraph, influence and residual analysis is described. The focus is mainly on aspects particular to three-way models, whereas ordinary two-way influence and residual analysis can be found in the standard literature.

Residual analysis

It was explained in Chapters 2–4 that all models whether they are two-way or three-way, are built up as follows:

$$\underline{\mathbf{X}} = \underline{\hat{\mathbf{X}}} + \underline{\mathbf{E}} \text{ or } \mathbf{X} = \hat{\mathbf{X}} + \mathbf{E} \tag{7.6}$$

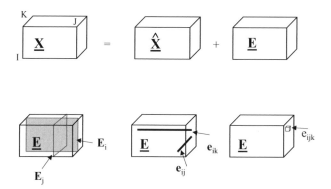

Figure 7.11. Model, $\hat{\underline{X}}$, of three-way array $\underline{X}$ with associated residual array $\underline{E}$. Below the model, different subarrays of the residual array are shown. Left, individual submatrices are shown in the first and second mode. In the middle, vectors are shown for two of the three possible directions while to the right, an individual element is shown.

The main goal of the data analysis is usually to find $\hat{\underline{X}}$, but the residual $\underline{E}$ can give important clues to the quality of this model. Possibly, residuals obtained from a test-set or from cross-validation can be used instead of fitted residuals. Random noise or some symmetrical type of distribution for the elements of $\underline{E}$ is normally expected and this can be verified from plotting the residuals and by the use of diagnostics. A good description of the use of residuals in three-way analysis is given by Kroonenberg [1983].

The fact that $\underline{E}$ is an $I \times J \times K$ three-way array when $\underline{X}$ is such an array, makes it possible to extract subarrays, as explained in Chapter 2, e.g. for investigating the residual variation of one particular sample or variable as compared to the remaining ones. This is shown in Figure 7.11. These subarrays can be summarized in a number of ways. The first subarray is $\underline{E}$ itself, and the sum of squares of $\underline{E}$ is used in scree plots, as explained earlier (Section 7.4 and Chapter 8). The array $\underline{E}$ can also be partitioned into matrices E_i ($i = 1,\ldots, I$), E_j ($j = 1,\ldots, J$) and E_k ($k = 1,\ldots, K$). Again, these matrices are studied using their sum of squares, or sometimes their variances, standard deviations, biases etc. (Table 7.3). The sum of squares, for example, can be calculated from these three matrices yielding three vectors of values: s_I ($I \times 1$), s_J ($J \times 1$) and s_K ($K \times 1$), where the ith element of s_I contains the sum of squares of E_i etc. These vectors can be studied in different ways, focusing on outliers, randomness or remaining structure. Figure 7.12 illustrates some practical interpration of such s vectors.

If $\underline{X}$ has e.g. I objects, J variables and K observation times, the study of s_I will indicate whether there are outliers among the objects, or whether there is a trend among the objects. In the same way, outliers among the variables and among e.g. the times may be found using s_J and s_K. If the variables are homogeneous, such as spectra, there may be discontinuities in certain parts of s_J that indicate a chemical or physical artefact. A trend in the s_K representing a time mode may point to a drift in time of an instrument.

Sum of squares of subelements of $\underline{E}$ can also be calculated for rows, columns and tubes: e_{ij}, e_{jk} and e_{ki} (Figure 7.11). The sum of squares of all the rows e_{jk} form a matrix S_{JK} of size ($J \times K$). It may be interesting to study the elements of S_{JK} as a matrix, or after a vec operation as a bar plot, or it may even be necessary to make a contour plot, especially if

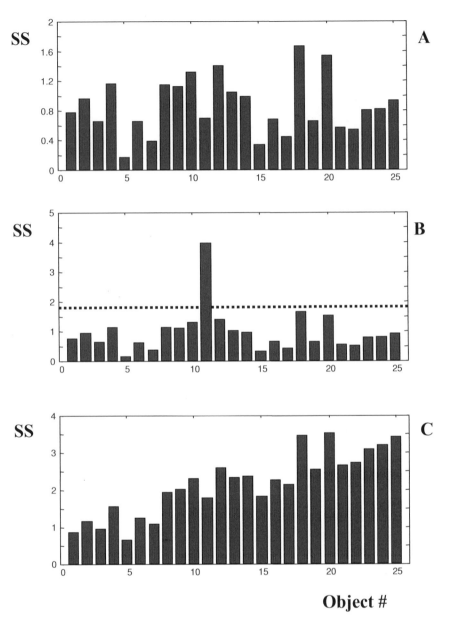

Figure 7.12. Different examples of sum of squares. (A) The sum of squares (SS) of the object-wise slices of **E** show a random distribution as expected. (B) There is a possible outlier (object 11) among the objects in mode (B). (C) There is a trend in the objects that is not captured by the model. The sum of squares may be replaced by a standard deviation, a bias or a variance in order to inspect different aspects of the residual variation.

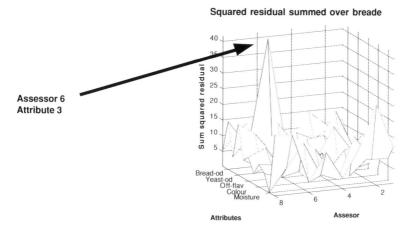

Figure 7.13. Sum of squared residuals from a model of a sensory data set investigating different breads. The residuals are summed across breads. Thus, each element corresponds to one specific combination of assessor and attribute.

J and K are large. An example where this would be useful is in fluorescence where S_{JK} forms an excitation–emission-like landscape that shows the residual sum of squares for each pair of excitation and emission wavelength. In multivariate image analysis, residual images can be made that by their intensity or color coding indicate how well a model fits to the image. The three possible types of S matrices S_{IJ}, S_{JK} and S_{KI} are not independent because they are calculated from the same elements. In Figure 7.13, an example is shown for the bread sensory data (Example 7.2) where it is observed from a plot of S_{JK}, that one assessor has difficulties in assessing one particular attribute. In some cases, it is necessary to study the individual elements of $\underline{E}$. This is not often done because there may be many individual elements. If needed it is possible to check each individual e_{ijk} against an a priori or a posteriori defined reference value, e.g. through the use of normal probability plots. Table 7.3 shows possible ways of summarizing subarrays of $\underline{E}$.

Table 7.4 presents different levels of studying residuals. Some of the plotting methods are described in Chapter 8. The first part of Table 7.4 is about visualizing the residuals and

Table 7.3. Ways of summarizing the residual $\underline{E}$. Subarrays, their summaries in terms of sum of squares. Other possible summaries can be derived similarly such as a variance, standard deviation, bias, skewness etc.

Subarrays	Summary
$\underline{E}$	Sum of squares (SS) a scalar s
$E_i, i = 1, \ldots, I$	SS vector $s_{(I \times 1)}$
$E_j, j = 1, \ldots, J$	SS vector $s_{(J \times 1)}$
$E_k, k = 1, \ldots, K$	SS vector $s_{(K \times 1)}$
$e_{ij}, i = 1, \ldots, I; j = 1, \ldots, J$	SS matrix $S_{(I \times J)}$
$e_{jk}, j = 1, \ldots, J; k = 1, \ldots, K$	SS matrix $S_{(J \times K)}$
$e_{ik}, i = 1, \ldots, I; k = 1, \ldots, K$	SS matrix $S_{(I \times K)}$
$e_{ijk}, i = 1, \ldots, I; j = 1, \ldots, J; k = 1, \ldots, K$	Three-way array $\underline{E}$

Table 7.4. Different levels of studying residuals. The right-most techniques in the table are more demanding because external knowledge of the residual distributions is required

Level 1 Visualization and Subjective Study	Level 2 A Posteriori	Level 3 A Priori
• Outlier detection • Detection of systematic structure especially of homogeneous variable • Detection of kinetics • Detection of drift in time, blocking effects • Study of contour plots/ distance to model/local features	• Study of first and second moments • Study of higher moments • Goodness of fit studies	• Study of first and second moments • Study of higher moments • Hypothesis testing

to look for problem-related information: spectral remainders, reaction kinetics, drift and blocking effects, outliers etc. (see e.g. Figure 7.8).

The middle part of Table 7.4 is about more formal statistical tests, where the distributional properties of the residual data are calculated from the residuals themselves. This is more demanding and often requires knowledge of degrees of freedom, which is a difficult subject for three-way residuals. If the residuals are centered around zero, then the sum of squared residuals is approximately χ^2 distributed [Box 1954]. The degrees of freedom can be estimated from the data [Box 1954, Jackson & Mudholkar 1979]. The resulting statistics are used in multivariate statistical process control (see Chapter 10).

The most demanding part of Table 7.4 is the a priori part, where a priori distributional knowledge is required (e.g. external knowledge of noise variance). This knowledge often comes from previous experiments and not from the data themselves. In multi-way analysis, levels 2 and 3 have not been studied extensively.

Residuals are also used for regression models (Chapters 3 and 4). Residuals of the dependent variables are obtained as

$$\mathbf{y} = \mathbf{Xb} + \mathbf{e}_y \quad \text{or} \quad \mathbf{Y} = \mathbf{XB} + \mathbf{E}_Y \tag{7.7}$$

The residuals may also be obtained from cross-validation. With the methods described in this book, both the $\mathbf{X}$ part and the $\mathbf{y}$ part of Equation (7.7) are modeled and hence have associated residuals. The residual $\mathbf{e}_y$ of $\mathbf{y}$ is usually summed to a number of regression statistics such as percentage variance explained, the coefficient of determination, root mean squared error of prediction etc. Diagnostics based on y-residuals are well covered in standard regression literature [Atkinson 1985, Beebe *et al.* 1998, Cook 1996, Cook & Weisberg 1980, Cook & Weisberg 1982, Martens & Næs 1989, Weisberg 1985].

One difference between a component analysis of an array $\underline{\mathbf{X}}$ and using $\underline{\mathbf{X}}$ in a regression is that the component analysis tries to explain $\underline{\mathbf{X}}$ as well as possible and make the sum of squares of $\underline{\mathbf{E}}$ small while in a regression model the goal is primarily to make the residual $\mathbf{e}_y$ of $\mathbf{y}$ small and not necessarily to use all of $\underline{\mathbf{X}}$ for that purpose. Studies of the residual of $\underline{\mathbf{X}}$ in principal component regression or partial least squares regression should take that into account.

Influence

Influence is a term that is used for diagnostics that describe how influential individual samples (or variables) are for a given model. It is not desirable that, for example, one sample is primarily responsible for the outcome of a model. Usually, the model is sought to be descriptive of the general variation typical of *all* the samples. If the model is mainly due to one sample, it indicates that the sample is an outlier or that too few or too badly distributed samples have been used. Influence measures were first introduced in regression analysis [Belsley *et al.* 1980, Cook & Weisberg 1982]. The goal was to explain the role of influential samples and variables on the solution of multiple linear regression. Later on, the principles of influence were transferred to other regression methods [Næs 1989] and also to principal components [Critchley 1985]. Influence analysis has mainly been developed for regression models and most influence diagnostics are therefore related to regression analysis. However, it is also useful to evaluate the influence of samples, variables and/or individual elements in component models in order to identify outliers and important variables. There are many unresolved issues in two-way and especially multi-way influence analysis, but some important aspects are covered in the following.

Influence measures may be calculated for objects and variables in a two-way array and in this section, only the main influence measure, leverage, is treated. Leverage was strictly developed for multiple linear regression problems [Hoaglin & Welsch 1978]. The leverage for a matrix $\mathbf{X}$ ($I \times J$) in the regression problem $\mathbf{y} = \mathbf{Xb} + \mathbf{e}_y$ is defined as

$$\mathbf{h} = \mathrm{diag}[\mathbf{X}(\mathbf{X'X})^{-1}\mathbf{X'}] \tag{7.8}$$

where the vector $\mathbf{h}$ has I elements. The ith element is the leverage of the ith object and its value is between zero and one (between $1/I$ and one in case a column of ones is included in $\mathbf{X}$ for handling offsets). The leverage defines the influence of the of ith sample because the prediction through the regression vector $\mathbf{b}$ is given as

$$\hat{\mathbf{y}} = \mathbf{X}(\mathbf{X'X})^{-1}\mathbf{X'y} \tag{7.9}$$

If the ith leverage, i.e. ith diagonal element of $\mathbf{X}(\mathbf{X'X})^{-1}\mathbf{X'}$, is zero, then it can be shown that the whole ith row and column of $\mathbf{X}(\mathbf{X'X})^{-1}\mathbf{X'}$ are zero and it therefore follows that the response y_i has no impact on the regression model whatsoever [Cook & Weisberg 1980, Digby 1983]. Thus, the sample i is not influential at all. The opposite situation occurs if the ith element is one. In that case, all other elements in the ith row and column of $\mathbf{X}(\mathbf{X'X})^{-1}\mathbf{X'}$ are zero and hence, the prediction of the ith calibration sample is only governed by the response value of the ith sample. This is unfortunate because any error in the response is directly transported to the prediction. No averaging effect is obtained. It follows that monitoring the leverage of a sample provides a useful measure of influence.

It is noted that, although the leverage is a regression tool, the actual response value does not appear in the definition (for PLS though, the response value is implicitly present through its influence on the calculated score values). Viewed as a squared Mahalanobis distance, the leverage has sometimes been defined in a broader meaning and used even for component models [Martens & Næs 1989]. However, the term leverage, being defined for regression models only, has little meaning in a component model. Instead, only the more general term squared Mahalanobis distance will be used in the following. Unless specifically mentioned, it is always the Mahalanobis distance to the center of the data that is treated.

For a PCA model using R components the squared Mahalanobis distance is based on the model

$$\mathbf{X} = \mathbf{T}_R \mathbf{P}_R' + \mathbf{E}_R \tag{7.10}$$

and usually the influence within the modeled part of $\mathbf{X}$ is most interesting. Hence the squared Mahalanobis distance is then based on treating $\mathbf{T}_R$ as 'given' and calculate the squared Mahalanobis distances as

$$\mathbf{h}_T = \text{diag}[\mathbf{T}_R (\mathbf{T}_R' \mathbf{T}_R)^{-1} \mathbf{T}_R'] \tag{7.11}$$

$\mathbf{T}_R' \mathbf{T}_R$ is diagonal and contains the R largest eigenvalues of $\mathbf{X}' \mathbf{X}$ on its diagonal. It is readily seen that the squared Mahalanobis distance is closely related to Hotelling's T^2 statistic, the only difference being a scalar correction. This distance is used extensively for outlier detection, e.g. in multivariate statistical process control (see Chapter 10).

Calculating the squared Mahalanobis distances for the variable component matrix, the influence of individual variables can be explored:

$$\mathbf{h}_P = \text{diag}[\mathbf{P}_R (\mathbf{P}_R' \mathbf{P}_R)^{-1} \mathbf{P}_R'] = \text{diag}[\mathbf{P}_R \mathbf{P}_R'] \tag{7.12}$$

because the matrix $\mathbf{P}_R' \mathbf{P}_R$ is the identity matrix. The squared Mahalanobis distances in $\mathbf{h}$ are usually small for a one-component model and may grow to a value close to one for a model with many components. Objects or variables that have distances much larger than the remaining ones, identify objects or variables of high influence; and hence potential outliers. If their residuals are high as well, the objects are likely of a different constitution than the remaining ones. On the other hand, a small distance indicates a variable that is not relevant and may be only contributing noise (Figure 7.14).

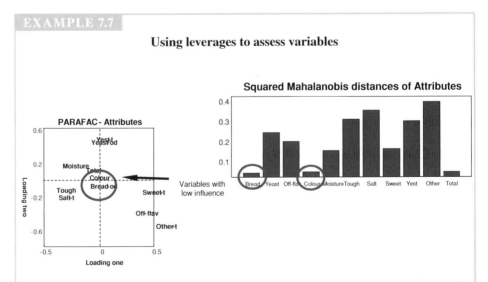

Figure 7.14. Squared Mahalanobis distances for a two-component PARAFAC model of the BREAD data centered across the bread mode.

A two-component PARAFAC model is fitted to a sensory data set of different breads. Ten breads are assessed by eight assessors on eleven different attributes. The left plot of Figure 7.14 shows a scatter plot of the attribute loadings (B) of the PARAFAC model. It is seen that two attributes have low loadings in both components. The right plot shows that this is also reflected by the small Mahalanobis distances, indicating that these attributes are not relevant for the problem or at least not consistent with the main variation in the data.

Once a tentative model rank is found, it is convenient to plot the leverages/distances as a bar plot to see if any objects or variables stand out. These may be candidates for outliers. An object with large squared Mahalanobis distance may not always be bad for the model if the object simply helps expanding the relevant domain. Whether this is the case can be monitored by comparing if the model changes significantly when the object is removed. However, a high-distance sample is indicative either of an erroneous outlier or of a poor sampling.

The Tucker3 model gives orthogonal loading matrices $\mathbf{A}$, $\mathbf{B}$ and $\mathbf{C}$ and also for these, the squared Mahalanobis distances can be defined

$$\mathbf{h}_A = \mathrm{diag}[\mathbf{A}(\mathbf{A}'\mathbf{A})^{-1}\mathbf{A}'] \tag{7.13}$$

$$\mathbf{h}_B = \mathrm{diag}[\mathbf{B}(\mathbf{B}'\mathbf{B})^{-1}\mathbf{B}'] \tag{7.14}$$

$$\mathbf{h}_C = \mathrm{diag}[\mathbf{C}(\mathbf{C}'\mathbf{C})^{-1}\mathbf{C}'] \tag{7.15}$$

Simplified equations result because $\mathbf{A}'\mathbf{A}$ equals the identity matrix $\mathbf{I}$, $\mathbf{B}'\mathbf{B}$ equals $\mathbf{I}$, and $\mathbf{C}'\mathbf{C}$ equals $\mathbf{I}$. The same equations are valid for PARAFAC models, but the middle cross-product is not identity. To summarize the properties of the squared Mahalanobis distances, the following can be said. Leverage can be defined for multiple linear regression as an influence measure. It is related to a specific Mahalanobis distance. The term leverage is sometimes also used for similar Mahalanobis distances in low-rank regression methods such as PCR and PLS. Then it becomes dependent on the rank of the model. The squared Mahalanobis distances can also be defined for PCA and multi-way models and can be calculated for both variables and objects.

7.6 Summary

In this chapter practical aspects of three-way modeling are discussed under the general framework of validation. There are many levels and objectives of validation and the most important one is to validate the model with respect to how well it explains, generalizes, or predicts the real-world problem it is aimed at describing. This aspect is not discussed much here, partly because it is problem dependent and mainly because it would only be a repetition of theory from standard multivariate analysis. Instead the focus is on data-analytical aspects specific to three-way modeling.

Specific cross-validation schemes for three-way data are given main emphasis. The choice of models and model hierarchy are explained. It is important to get a good fit and parsimony. The selection of appropriate model rank by the use scree plots, residual analysis and split-half analysis is introduced. Different ways of calculating residual statistics and leverages for three-way arrays are presented.

8
VISUALIZATION

8.1 Introduction

Decomposition of three-way arrays was shown in previous chapters. A three-way model has to be accepted, understood and presented to be of any use. Acceptance, understanding and presentation include outlier detection, detection of classes and patterns, validation, diagnostics, finding relations to known hard models of the data etc. The validation and diagnostics of Chapter 7 are an important part of this. The statistical tradition is to do the interpretation numerically, by producing a small number of statistics for possible use in hypothesis testing. However, much of the interpretation can be done more efficiently in the visual mode because humans are better at recognizing visual patterns than at dealing with numbers in tables. Visualization has a solid basis in two-way analysis and some of the established practices for two-way arrays will serve as an introduction to the topic for three-way visualization. Most of the readers will be familiar with the plots commonly used in fields such as principal component analysis, regression analysis and experimental design, and some of these need no thorough explanation.

Visualization has not been used much in multi-way analysis, but the same techniques as in two-way analysis can be used for almost any case in multi-way analysis. It is the interpretation of the plots that result from multi-way analysis that is sometimes special. The number of plotting techniques available is huge [Cleveland 1985] and any presentation of them is therefore limited but an attempt is made to give the most important techniques with comprehensive examples.

A selection of plotting and graphing techniques is shown in Table 8.1. Some graphing techniques that are mainly used in advertising and have not much scientific use are not mentioned, e.g. pie charts, 3D bar charts, Chernoff faces. On top of the graphical aspect, also greyvalues and colors can be used to represent aspects of data analysis. No color is used in this book.

Of the plots in Table 8.1 some are used regularly for three-way analysis, while others are less often used. All these plotting techniques have a high potential for helping the user interpret three-way analyses. What kind of plot is used is also dependent on the data and the

Multi-way Analysis With Applications in the Chemical Sciences.   A. Smilde, R. Bro and P. Geladi
© 2004 John Wiley & Sons, Ltd   ISBN: 0-471-98691-7

Table 8.1. An overview of plotting techniques

Line plots (spectra, chromatograms, concentration profiles)
Box plots
Quantile plots
Autocorrelation/spectral analysis plots
Two-dimensional plots (score and loading plots in PCA and for three-way
   analysis)
Three-dimensional plots
Biplots and joint plots in two dimensions
Biplots and joint plots in three dimensions
Response and other surfaces
Imaging or mapping of results and parameters
Movies of raw data, results and parameters
Correlation maps
Dendrograms
Advanced (interactive) visualization in more than 3 dimensions

---

problem to be solved. Different aspects of raw data, models and residuals can be plotted; a nonexhaustive summation is given in Table 8.2.

A *bar chart* or *bar plot* is a plot where the scalar values in a vector are shown as bars of different length. The positions of the bars indicate positions in the vector. Bar plots are useful for showing decomposition of the sum of squares in parts. A *line plot* is a plot of a vector where the horizontal axis is the position in the vector, often representing a continuous variable, and the vertical position of the points is given by the scalar values in the vector. For clarity, the points can be joined by straight or curved lines. These plots are useful for showing spectra, chromatograms, electrochemical curves, but also loadings based on them. A *scatter plot* is a plot of two vectors of the same size against each other, where the scalar values are used as coordinates for the points. Scatter plots can be used for two variables, two scores vectors, two loading vectors etc. They may clearly show outliers, groupings and gradients. Scatter plots can also be shown in three dimensions by using perspective or other three-dimensional viewing devices. *Three-dimensional scatter plots* can be rotated in space to give a better view of the data. Scatter plots can be put on top of each other as *biplots* or *joint plots*. These show how objects and variables or loadings from different modes are related to each other.

An *image* is a regular grid (composed of lines and columns) of squares or rectangles, where each rectangle is a pixel, representing a greyvalue. Together the pixels give a visual impression as a black and white picture. Also *color images* can be used. In this case, three greyvalues representing a basis for the color system of the human eye are given for each pixel. Television images, computer screen images, color photographs and color prints are all made up in this way. Images have obvious uses for all microscopic, medical imaging and remote sensing applications. A *movie* is a sequence of images. Movies are used when images evolve in time or to show a third dimensions or a change in variable. A *contour plot* is a surface given as a function of two variables. It can be compared to the altitude lines on a map. Most contour plots are based on a model function. The model function may be based on a limited amount of data points and use interpolation between them. Bilinear data from fluorescence spectroscopy and hyphenated techniques are often shown in this way.

Table 8.2. Items to be plotted

---

1. Raw data plotted as
   - Bar charts
   - 'Continuous' functions, line plots
   - Scatter plots
   - Images
   - Movies
   - Contour plots/3D surface plots
   - Histograms
   - Correlation spectroscopy/2D NMR
   - Autocorrelation/spectral analysis plots

2. Parameters and results (the model) plotted as
   - Scree plots
   - Loadings in line or bar plots
   - Loadings in scatter plots, 2D or 3D
   - Loadings as histograms
   - Loadings as autocorrelation/spectral analysis plots
   - Modeled data as contour/3D surface plots

3. Residuals plotted as
   - Line or bar plots
   - Images/movies
   - Scree plot analogs
   - Histograms
   - Autocorrelation/spectral analysis plots
   - Distance to model as line or bar plots
   - Diagnostics in line or bar plots
   - Diagnostics as images
   - Diagnostics as quantile plots

Contours can also be given in perspective as *3D surfaces*. *Correlation spectroscopy* [Noda 1993, Bax 1982] is a useful technique for the interpretation of spectral data employing many wavelengths. This technique makes it easy to interpret where things are covarying in spectral data.

A *histogram* is a bar plot where the bar position is based on regular intervals of a variable and the length of the bars are counts or frequencies. Histograms are used for showing properties of distributions or for comparing distributions. *Autocorrelation plots* are bar plots or line plots where the variable is replaced by lag (time) and its value by a correlation coefficient. *Spectral analysis plots* show frequency as bar position and a Fourier or Wavelet coefficient as the length of the bar. These plots can be used for showing properties of autocorrelated continuous variables. *Quantile plots* are scatter plots of the ordered values in a vector (horizontal axis) and the calculated probability for each value (vertical axis). Usually a linearizing transformation based on a chosen distribution is used for the probabilities. In this way the points fall on a straight line if they follow the distribution. The normal probability plot is an example of a quantile plot. In a *dendrogram*, the distances between objects (variables) on the horizontal axis are shown as vertical bars. A node connecting two bars indicates a cluster and distances between clusters or between clusters and objects (variables) can be shown in the same way. A dendrogram is a useful replacement for scatter plots when many scatter plots at the same time have to be used to visualize a situation. The dendrogram shows clustering of objects or variables, outliers etc., taking into account more than two or three variables at a time. It is less useful for showing gradients and for showing many (>100) objects.

The order of the topics treated is basically as plots would be used in an ongoing analysis: scree plots, line plots, scatter plots, special plots as an aid in understanding the model, residual plots, leverage plots.

ELABORATION **8.1**

Cartesian plots

Some ways of plotting were introduced in Tables 8.1 and 8.2. For the interval and ratio scales, plots are often made in squares or rectangles, with the scales and ticks just outside (or inside) the rectangular frame. This is the cartesian coordinate system [Cleveland 1985]. A cartesian plot describes a rectangular frame, horizontal and vertical scales defined by tick marks, a data region, data symbols and reference lines for indicating zero or some relevant threshold value.

For a geographical map, the horizontal and vertical coordinates are the same (distance) and distances and angles can readily be interpreted. In a map, every point has a physical meaning. Contour plots as in experimental design [Meyers & Montgomery 1995] follow the same principle, but here the horizontal and vertical scales do not necessarily mean the same thing and distances and angles do not always mean anything physical. Images are a special form of maps, with square pixels.

Square or rectangular plots can also be used as scatter plots. Scatter plots are usually made by plotting some measured result against some parameter in a cartesian coordinate system. Examples are pH in a lake against sampling time, yield of a reaction against temperature, current against voltage. The principle of these scatter plots is that the entries from two vectors of the same size are plotted pairwise in the cartesian coordinate system. In these plots, the area outside the plotted points is not used and may never become useful. Some regions in the plots may be physically impossible and therefore they always remain empty, which does not happen for maps. In plots like these, it is possible to construct a model (a continuous line or curve) from the plotted points and to show this model in the plot. In this way, data and model are shown together which increases the possibilities of interpretation.

Many plots made in the sciences show measured values in the units they were measured in. These may be SI units, practical empirical units or indices derived from these, e.g. by taking ratios, products, sums etc. If the horizontal and vertical units are different, an appropriate scaling to obtain a square or rectangular size of plot is chosen. This scale is often based on range scaling with some 5–10 % extra shrinking for keeping points from falling on the frame of the plot. This technique is so established that one usually looks at the plots without considering the basic principles that led to the result. A general name for this would be 'raw data plots'.

Latent variable methods that produce scores and loadings do not use physical units. The plots of these are therefore unit-free or use abstract units. Also, the items plotted against each other (scores, loadings) are based on the same measured data. This makes these plots special. In two-way data analysis by principal component analysis, the scores are orthogonal and the loadings are orthonormal, and this makes the coordinate systems (the loadings) used for the score plots orthonormal. With correct horizontal and vertical scaling, such plots can be used to show distances and angles in a meaningful way [Geladi *et al.* 2003]. A combined way of showing scores and loadings, with a meaningful interpretation of distances and angles is the biplot or joint plot, discussed in Section 8.11.

Three-way analysis does not automatically lead to orthogonal or orthonormal components. PARAFAC and Tucker models rarely have an orthonormal basis for the loadings produced by the model. This implies that distances between points are not directly related to distances in the raw data. Some cautionary words and remedies are given in Section 8.7

8.2 History of Plotting in Three-way Analysis

It is interesting to take a look at the historical development of three-way analysis and how different pioneers in psychometrics used plots for their data and results. The history of three-way analysis starts with the work of Cattell in 1952 and ends somewhere in 1980. Anything after 1980 can be called 'present time'. In the historical article of Cattell [1952], no plots are shown. The classical articles by Tucker [1964, 1966, 1967], also show no plots. The examples and the analyses are introduced as numbers. The same can be said about the article of Levin [1965].

Carroll & Chang [1970] make extensive use of plots to explain the results of multidimensional scaling. The three-way nature of their data comes from distance matrices where the ways are stimulus, distance and distance. They show loading plots for three factors and score plots for the same factors. They call the loadings 'stimulus space' and the scores 'subject space'. They also interpret directions in factor space. This is done for two examples. Even though the paper is about three-way analysis, the plotting techniques are borrowed from classical factor analysis. In a classical PARAFAC paper by Harshman [1970], vowel sound data are analyzed. The array is 4 formants × 8 vowels × 11 individuals. In the paper, a three-dimensional loading plot is shown for three-factor space. An interpretation is given for the three factors. In Tucker [1972], 87 subjects rate the similarity of 12 adjectives to give an (87 × 12 × 12) array. Scree plots are used to determine the dimensionality of the subject space and the adjective space and also scatter plots are used. The core matrix is (4 × 3 × 3).

The classical articles of Kruskal [1976, 1977] contain no figures at all, just algebra. In Lohmöller [1979] a three-dimensional plot is used to show an analysis of variance result. Sands & Young [1980] introduce a PARAFAC algorithm called ALSCOMP3. The analyzed data are ratings from 28 subjects of 4 political roles on 77 adjectives. The authors show a 2D plot with interpretation of the oblique factors. Kroonenberg & De Leeuw [1980] show loading plots for three factors for Tucker models. They also introduce the joint plot, which provides a way to jointly visualize loadings from several modes. The data used are 16 English consonants under 17 degrading conditions judged by five observers. An interesting article on displaying multidimensional scaling results that does not have any figures in it is provided by Gower [1984].

A general conclusion is that in the early years of three-way analysis, not much emphasis was put on displaying data or results. This is expected, given the algebraic and algorithmic difficulties encountered. Many authors preferred equations and tables over plots. The few figures that were published got their inspiration from traditional two-way factor analysis in psychology.

8.3 History of Plotting in Chemical Three-way Analysis

Three-way analysis outside psychology and sociology and more specifically in chemistry started around 1980. The goals were quite different from what the psychologists had been doing. Curve resolution was a major drive for chemists, who are used to plotting spectra and chromatograms and therefore are expected to use these plots also for evaluating and interpreting three-way models.

Early visualization in chemical three-way analysis was used in curve resolution methods, and especially GRAM. The extraction and visualization of unknown spectra (curve resolution) was considered very important. Tucker and PARAFAC analysis were used much in the exploratory sense and PARAFAC also for curve resolution. The first publication of Ho *et al.* [1978] has figures of eigenvalues expressed as a function of a parameter β, see Figure 8.1. The second publication of Ho *et al.* [1980] also contains contour plots for the excitation-emission spectra of polyaromatic compounds (see Figures 8.2 and 8.3). They also show excitation and emission spectra as line plots (see Figures 8.4 and 8.5). The plots are only used to show the raw data and show no results after analysis. An early publication of de Ligny *et al.* [1984] has all the results shown as numbers in tables.

The first article of Sanchez and Kowalski [1986] on generalized rank annihilation factor analysis is purely based on equations, but later articles of the same group on spectral curve resolution contain line plots of the estimated spectra after the rank annihilation analysis [Sanchez *et al.* 1987, Ramos *et al.* 1987]. Sanchez [*et al.* 1987] is mainly about simulated examples. See Figures 8.6 and 8.7.

Neither the psychometricians nor the chemometricians have used the full potential of plotting and visualizing their data as evidenced in the literature. Tables 8.1 and 8.2 can be a source of inspiration for future uses of graphics.

8.4 Scree Plots

The first plot used to check the results of a two-way principal component analysis is the scree plot. This was described in Chapter 7. The classical scree plot shows the component size (% of sum of squares, eigenvalue) as a function of ordered component number. The determination of pseudorank is done by selecting a spot where the eigenvalues become insignificant. This was shown in Figure 7.4, and is also seen in Figure 8.8. The scree plot is presented in different forms in the literature: line plots, a scatter plots or bar plots. What is shown is eigenvalues, λ (or singular values, $\lambda^{1/2}$) on the vertical axis and component number on the horizontal axis. A bar plot can be easier to interpret than a curve if not too many singular values are shown.

The scree plots for the eigenvalues can also be shown cumulatively because they add up to 100 % of the sum of squares (%SS). Alternatively, the percentage of the sum of

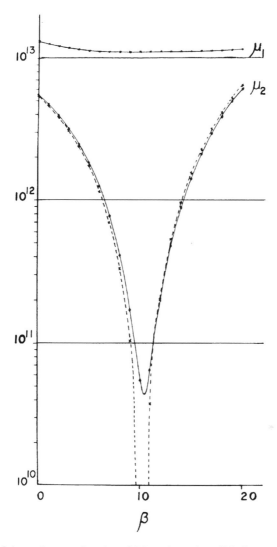

Figure 8.1. A plot of eigenvalues as a function of β from the rank annihilation paper of [Ho *et al.* 1978]. In the proper circumstances the minimum value of β expresses the desired relative concentration. Reprinted with permission from C. Ho *et al.* (1978), *Analytical Chemistry*, **50**, 1108–1113. Copyright (1978) American Chemical Society.

squares left in the residual can be used. See Figures 8.9 and 8.10. These two figures show complementary information. Also sums of squares resulting from test set validation or cross-validation can be used in this type of plots. This is useful for studying predictive ability of the models as also explained in Chapter 7.

Although the scree plot was defined for two-way analysis, it can also be used for showing component sizes for three-way models. The cumulative scree plot can show the variance explained for PARAFAC models of increasing rank. The components in the models change, though, when the effective three-way rank is increased. The first loading is different in all models of different rank. A scree plot for the peat example was shown in Chapter 7,

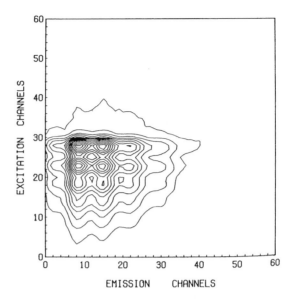

Figure 8.2. An excitation–emission contour plot from the fluorescence data used in [Ho *et al.* 1980]. The pure component is anthracene. This is an example of a contour plot. Reprinted with permission from C. Ho *et al.* (1980), *Analytical Chemistry*, **52**, 1071–1078. Copyright (1980) American Chemical Society.

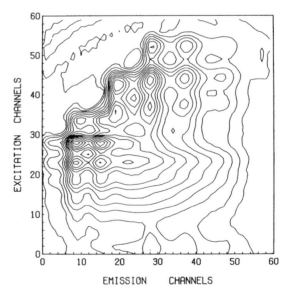

Figure 8.3. The excitation–emission contour plot of a mixture of six aromatic chemicals as used in [Ho *et al.* 1980]. The plot shows one of the planes of the $2 \times 60 \times 60$ array. Reprinted with permission from C. Ho *et al.* (1980), *Analytical Chemistry*, **52**, 1071–1078. Copyright (1980) American Chemical Society.

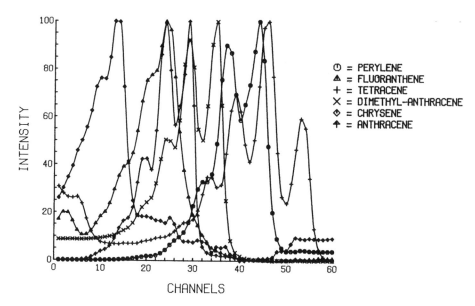

Figure 8.4. Line plots of the excitation spectra of the six chemicals used in [Ho *et al.* 1980]. Reprinted with permission from C. Ho *et al.* (1980), *Analytical Chemistry*, **52**, 1071–1078. Copyright (1980) American Chemical Society.

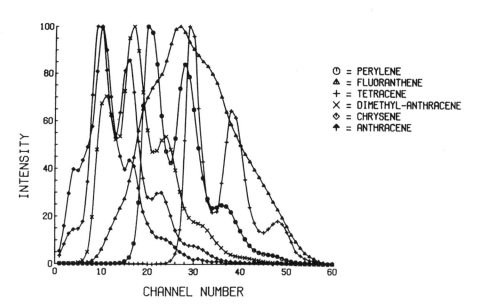

Figure 8.5. Line plots of the emission spectra of the six chemicals used in [Ho *et al.* 1980]. Reprinted with permission from C. Ho *et al.* (1980), *Analytical Chemistry*, **52**, 1071–1078. Copyright (1980) American Chemical Society.

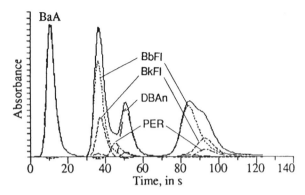

Figure 8.6. The line plot shows chromatograms resulting from a rank annihilation analysis of LC-UV data of a mixture of aromatic chemicals as used in [Ramos *et al.* 1987]. Reprinted from *Journal of Chromatography*, **385**, Ramos S, Sanchez E, Kowalski B, Generalized rank annihilation method. II. Analysis of bimodal chromatographic data, 165–180, Copyright (1987), with permission from Elsevier Science.

Figure 7.5. The figure indicates a pseudorank of three to five. Figure 7.6 indicates that a pseudo-rank of four should be chosen, even though the fourth component may not be very relevant.

The scree plots described here for component analysis models are also used in regression models of the type $\mathbf{y} = \mathbf{Xb} + \mathbf{e}_y$ where the total sum of squares of $\mathbf{y}$ is decomposed in a model $\mathbf{Xb}$ and a residual $\mathbf{e}_y$. Figures like Figures 7.4, 8.9 and 8.10 with percentage of the sum of squares for each component, percentage explained of the model or residuals as a function of the number of PCA or PLS components are used to study the regression model. With results from cross-validation or a test-set, a similar plot can be used to select a minimum: an effective rank for the regression model. This helps in rank determination for regression modeling. Note that $\mathbf{y}$ can be a vector or a matrix or any multi-way array and $\mathbf{X}$ can be a matrix or any multi-way array. The principle and the usefulness of the (cumulative) scree plot remain the same. Because these plots are not typical for three-way analysis, the reader is referred to the regression and calibration literature for more details [Brown 1993, Martens & Næs 1989, Beebe *et al.* 1998, Næs *et al.* 2002].

8.5   Line Plots

Line plots are used for showing continuous/homogeneous variables against variable number, where the variable number represents monotonically increasing or decreasing variables such as wavelength, wavenumber, mass number (m/e), retention time, process time, voltage, current etc. Line plots are useful for showing raw data, as spectra or chromatograms or any other results that are continuous and sometimes correlated. The power of simply looking at the raw data and asking: 'does this make sense from a spectroscopic, chromatographic, electrochemical, electrophoretical, flow injection analysis, time series, dynamics etc. point of view?' should not be forgotten. Line plots can be used to plot loading values. With coloring or other markings, a number of plots can be put in the same figure. See Figures 8.4–8.7 for some historical line plots. In this section two examples are introduced for highlighting

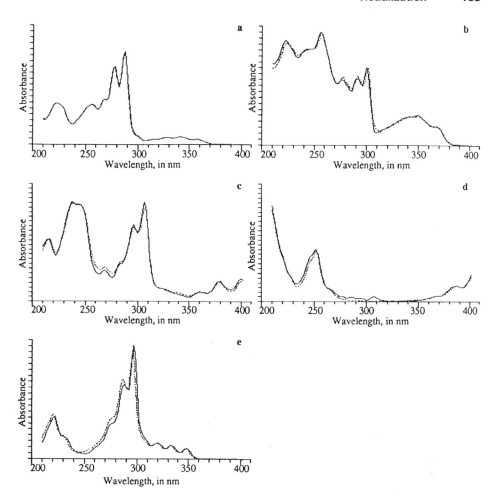

Figure 8.7. The line plots show spectra reconstructed from LC–UV data by rank annihilation analysis of the aromatic chemicals used in [Ramos *et al.* 1987]. Reprinted from *Journal of Chromatography*, **385**, Ramos S, Sanchez E, Kowalski B, Generalized rank annihilation method. II. Analysis of bimodal chromatographic data, 165–180, Copyright (1987), with permission from Elsevier Science.

plotting aspects. One is a batch synthesis monitored by near infrared spectroscopy, using many wavelengths and the other is a design with qualitative treatments for flocculating peat particles. They are described in Examples 8.1 and 8.2. The structure of the three-way data sets is visualized in Figure 8.11.

An important use of three-way analysis is in curve resolution. It is in many cases important to display the estimated 'pure' component spectra or chromatograms. Sometimes curve resolution by three-way techniques is the only way of getting the pure spectrum or chromatogram of an unstable species in for example a biological mixture or chemical synthesis. Here line plots of the spectra and/or chromatograms are useful. For the batch example (Example 8.1), the reduced wavelength mode has a high number of variables (701) and therefore line plots are excellent ways of studying loadings. An example is given in

Singular value

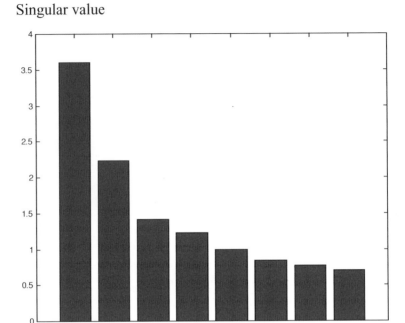

Component number

Figure 8.8. The scree plot of Figure 7.4, but now using singular values, here shown as a bar plot. Some authors recommend singular values for easier determination of pseudorank by the 'broken stick' method [Faber *et al.* 1995].

Figure 8.12. In this figure, the first PARAFAC loading is compared to a spectrum of a mixture of chemicals to show the similarity.

EXAMPLE 8.1

The batch synthesis example

The batch synthesis example is a three-way array of size ($14 \times 13 \times 1050$). Chapter 10 gives more details about this example from Geladi and Åberg [2001]. The batches are organic synthesis reactions for producing an ester from an acid and an alcohol. The reactions were carried out at reflux temperatures and using an acid catalyst. The near infrared measurements used the wavelength range of 400–2500 nm (every 2 nm) with a fiberoptic transflectance probe directly in the reaction mixture. The VIS–NIR measurements were performed every 10 min during 2 h, giving 13 sampling points. The array is shown schematically in Figure 8.11. The array is of the type batch number × time × near infrared wavelength. The PARAFAC analysis was carried out after Savitzky–Golay smoothing with first derivative as preprocessing and cropping of the wavelength range to 500–1900 nm (701 wavelengths). Three useful PARAFAC loadings could be

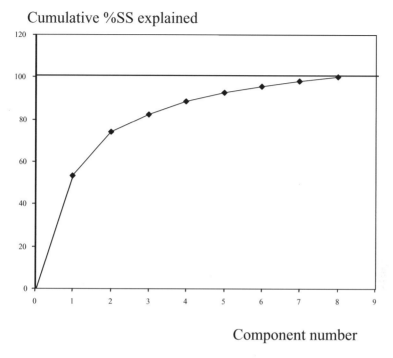

Figure 8.9. A plot of % of the sum of squares (%SS) in the PCA model against number of components. This is an analog of the scree plot and it is used for the same purpose. It is usually called 'cumulative scree plot'.

extracted and interpreted. The wavelength mode loadings (see Figure 8.12) could be related to specific chemicals or mixtures and the time mode loadings could be interpreted as changing concentrations.

EXAMPLE 8.2

The peat example

The example is a three-way array of size (5 × 7 × 21). For five major peat types, called S, I, M, A and N, seven treatments for better aggregation of colloidal particles are applied. The result of automated digital image analysis on the particles under a microscope gives a histogram of particle areas in 21 bins from 5–8 μm^2 for bin 1 to 0.14 mm^2 and higher for bin 21. The array is peat type × treatment × particle area. Figure 8.13 gives an example of the particle area histograms.

The seven treatments are: (UN) Untreated, (CH) Mild wet charring, (Z22) Zetag 22 (a floculation chemical), (A + M) AlCl$_3$ + Magnafloc (a floculation treatment), (PH) pH lowering to 2.5, (P + F) pH lowering combined with freezing and (FR) freezing only. The abbreviations are used in order not to clutter the plots with long names. Chapter 10 gives more details about this example from Geladi *et al.* [2000]. A schematic view of the

Residual %SS

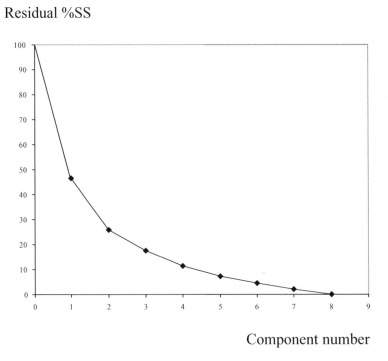

Component number

Figure 8.10. A plot of % of the sum of squares in the residual against number of components. This plot is the mirror image of Figure 8.9 and its purpose is the same: to find a cutoff where increasing the rank of the model becomes meaningless.

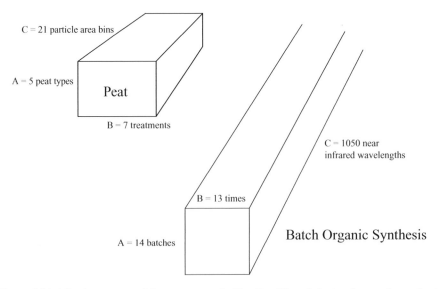

Figure 8.11. The data arrays of the peat example (5 × 7 × 21) and the batch organic synthesis example (14 × 13 × 1050) shown schematically.

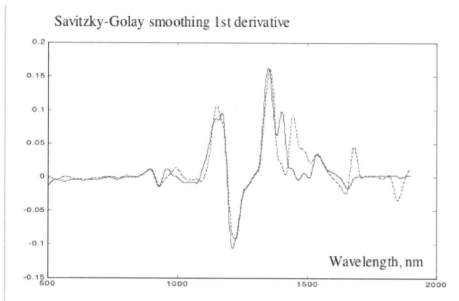

Figure 8.12. A PARAFAC loading from the batch organic synthesis example (full line) compared with a pure mixture spectrum (dashed line). The data are shown in Savitzky–Golay first derivative form. For a large number of variables, continuous curves are an excellent visualization form.

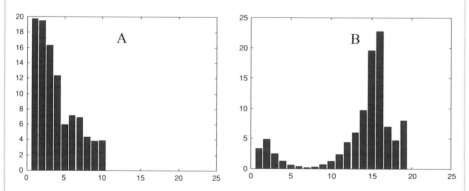

Figure 8.13. Some typical histograms of particle area for peat slurry filtration. These are bar plots, where the bar position gives particle area and the bar length total amount of that area in the sample. (A) A slurry with many small particles. (B) A slurry with many intermediate size particles. Bar plots are useful for showing an intermediate number of variables.

data array is given in Figure 8.11. For the peat example (four-factor PARAFAC model), C-loadings are shown as barplots in Figure 8.14. It may be observed that loading 1 is mainly a small particle loading. Loading 2 is a contrast between small and intermediate particles. These loadings may be compared to real particle size histograms as those in Figure 8.13.

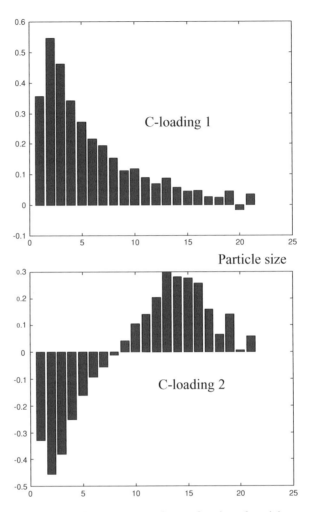

Figure 8.14. Loadings 1 and 2 of the peat example as a function of particle area are given as bar plots. Bar plots are excellent for showing an intermediate number of variables in a loading plot.

8.6 Scatter Plots

Scatter plots are used to plot a number of loadings (or scores for PCA) against each other. They can be made in 2D or 3D. When more than three loadings have to be shown, more plots can be arranged in one figure. Loadings (or scores) can be plotted very efficiently in this manner to give an overview of the reduced data space. There are many ways of making scatter plots. A different view of scatter plotting is also given under the heading 'joint plots' in Section 8.11.

Figure 8.15 shows the A-, B- and C-loadings for the first two components of the peat example as scatter plots. The first component is larger than the second one (sizes 31 000 and 7000). In the C-loadings, it can be seen that particle sizes are autocorrelated and that there is a gradient from small to large particles. The B and C loadings considered

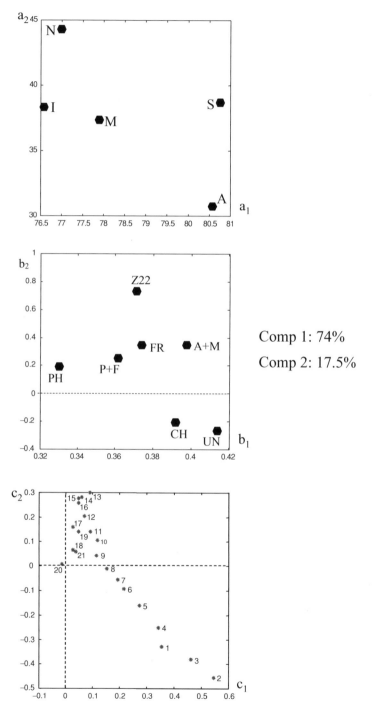

Figure 8.15. The A-, B- and C-loadings for component 1 and 2 of the peat example as scatter plots. The percentages of the total sum of squares for the components are also indicated. These are not additive as in PCA.

together show that two treatments, U and C, give mainly small particles. It was shown earlier that peat type is less important than treatment, so too detailed conclusions can not be given for the A-loadings. A conclusion is that the first component represents amount of small particles and that the second component is a contrast between small and intermediate particles.

Figure 8.16 gives the loading plots for components 3 and 4 for the peat example. These are small components of about equal size (3400 and 2800). The A-loadings show much more than those in Figure 8.15. The A-loading shows a grouping along the third component. These groups do not correspond to the botanical composition of the peat. The peat types S and A (high in the first loading) are of low humification and the peat types I, M and N (low in the first loading) are of high humification. The B-loadings show a grouping of treatments UN, CH,PH and Z22. Except for a sign reversal (mirroring), the group of treatments (UN, CH, PH, Z22) corresponds to small and intermediate particles. Other treatments give mainly large particles.

For showing three loadings at the same time, three-dimensional plots can be made. These mainly rely on using a parallelopiped (box) and some kind of perspective to give the illusion of three dimensions and depth on a flat screen or piece of paper. Examples of this are given in Figures 8.17 and 8.18. Many graphical and electronic techniques exist for showing perspective and depth. Three-dimensional scatter plots are sometimes better studied by interactive rotation than by looking at static plots.

A special scatter plot is used in regression modeling for plotting calculated or predicted values against measured values in a square plot. In these plots the points should fall on a diagonal of the square when calculated/predicted values match the measured values. Figure 8.19 shows some situations. In the normal case, all points fall close to a diagonal line. Nonlinearity is usually found for high values when predicted values are smaller than the measured ones. Bias means that all predicted values are systematically different from measured ones. An outlier can be an erroneous sample or still fit in the model by extrapolation. When the residual becomes larger for a range of values, heteroscedasticity is indicated. These plots are not unique for three-way regression and the reader is refered to the calibration/regression literature for more explanation [Brown 1993, Martens & Næs 1989, Beebe *et al.* 1998, Geladi *et al.* 1999, Geladi 2002].

8.7 Problems with Scatter Plots

(Non)orthonormality of the base

Scatter plots in PCA have special properties because the scores are plotted on the base **P**, and the columns of **P** are orthonormal vectors. Hence, the scores in PCA are plotted on an orthonormal base. This means that Euclidean distances in the space of the original variables, apart from the projection step, are kept intact going to the scores in PCA. Stated otherwise, distances between two points in a score plot can be understood in terms of Euclidian distances in the space of the original variables. This is not the case for score plots in PARAFAC and Tucker models, because they are usually not expressed on an orthonormal base. This issue was studied by Kiers [2000], together with problems of differences in horizontal and vertical scales. The basic conclusion is that careful consideration should be given to the interpretation of scatter plots. This is illustrated in Example 8.3.

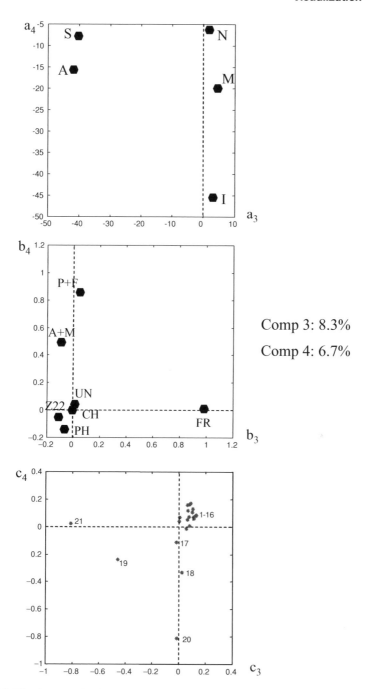

Comp 3: 8.3%

Comp 4: 6.7%

Figure 8.16. The A-, B- and C-loadings for components 3 and 4 of the peat example as scatter plots. The percentages of the total sum of squares for the components are also indicated.

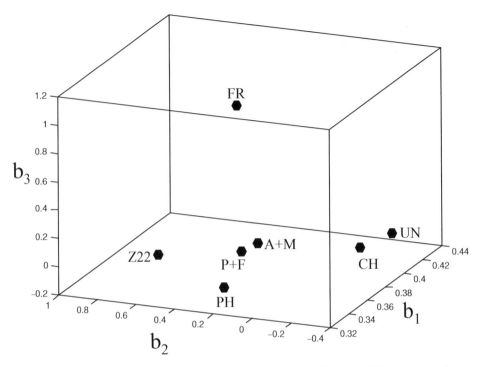

Figure 8.17. A 3D scatter plot for the first, second and third B-loadings of the peat example.

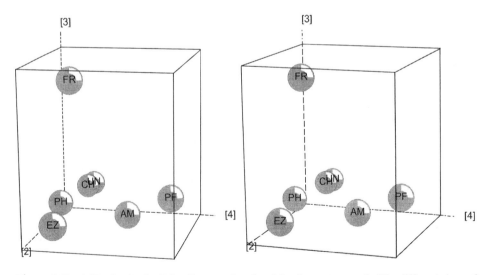

Figure 8.18. A 3D plot for the B-loadings number 2 to 4 for the peat example. The different sizes of the spheres improve perspective. The two figures can be viewed in 3D by using a 3D viewer. UN = untreated, CH = mild wet charring, EZ = Zetag 22, AM = AlCl$_3$ + Magnafloc, PH = pH lowering, FR = freezing and PF = pH lowering with freezing.

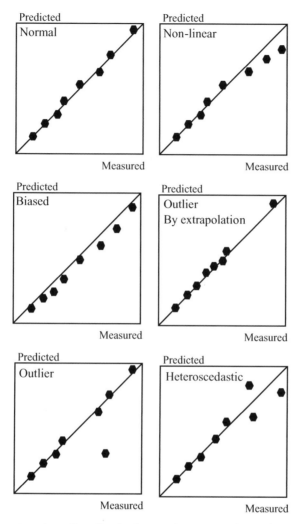

Figure 8.19. The scatter plots of predicted values against measured ones from regression models. Different situations are shown.

EXAMPLE 8.3

Scatter plotting illustration

In order to illustrate the importance of orthogonal and orthonormal bases, a small real example is given. Consider a mixture of two chemical compounds: ethanol and iso-propanol. A good way to show plotting principles is by making an orthogonal design in the concentrations. The columns of **C** are the concentrations of ethanol (column 1) and isopropanol (column 2).

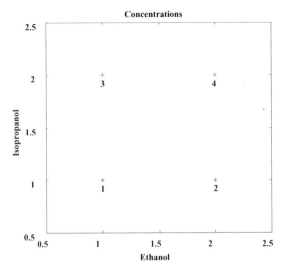

Figure 8.20. The orthogonal design in the concentrations of the alcohols (**C**).

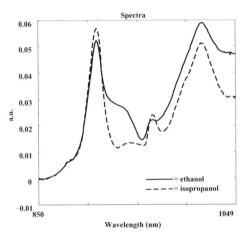

Figure 8.21. The spectra of the alcohols (**Z**) for unit concentration.

$$\mathbf{C} = \begin{bmatrix} 1 & 1 \\ 2 & 1 \\ 1 & 2 \\ 2 & 2 \end{bmatrix}$$

A scatter-plot of **C** is shown in Figure 8.20. The pure short-wave near-infrared spectra of the compounds are shown in Figure 8.21 and called **Z**, where each column of **Z** contains a pure spectrum for unit concentration. Now generate **X** according to $\mathbf{X} = \mathbf{CZ}'$, hence, the mixture obeys the Lambert–Beer law exactly. Figure 8.20 shows the orthogonality of the concentrations, because the concentrations are plotted as is (on 'concentration axes'). Note that the concentrations are the latent variables in the data set $\mathbf{X} = \mathbf{CZ}'$ The

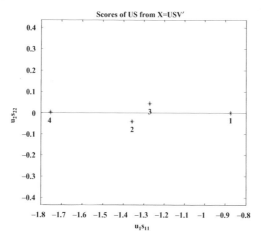

Figure 8.22. The score plot of **US**. The plot is made so that the points fit in a rectangle.

spectra are not orthogonal which can be seen in Figure 8.21 (they overlap severely; their angle is 11°)

A singular value decomposition $\mathbf{X} = \mathbf{USV}'$ can be performed on $\mathbf{X}$ and two components are sufficient to describe $\mathbf{X}$, because it has rank two. A plot of the rank two solution of the rows of $\mathbf{US}$–the scores–is shown in Figure 8.22. Apart from a rotation ($\mathbf{V}$), this is a plot of the rows of $\mathbf{X}$ expressed on an orthonormal base. Comparing the two score plots (Figures 8.20 and 8.22), it is clear that e.g. sample two and three are quite different in terms of concentrations but quite similar with respect to spectral properties.

In summary, Figure 8.20 expresses the data on the latent space and Figure 8.22 expresses the data in the raw measurement space. They are both valid representations of the same data.

The above has repercussions for component models. In component models, $\mathbf{X}$ is decomposed as $\mathbf{X} = \hat{\mathbf{X}} + \mathbf{E}$ and in order to study the model, score plots are used. Such score plots only reflect the rows of $\hat{\mathbf{X}}$ in the original domain (of $\mathbf{X}$) if an orthonormal basis is used to express the scores. For curve resolution types of studies, it is usually more insightful to express the scores (e.g. concentrations) on the basis of the estimated spectra ($\hat{\mathbf{Z}}$).

PARAFAC loadings (or scores) are usually not expressed on an orthonormal basis as are PCA scores. Scatter plots based on PARAFAC loadings should, therefore, be interpreted with care. The interpretation of distances in a score (or loading) plot of a PARAFAC model in terms of the raw data can be incorrect because the basis in which the scores (or loadings) are expressed is not orthonormal. Scatter plots are, however, directly interpretable in terms of the latent variable. For example, if the latent variables of a spectral data set are interpreted as relative concentrations, then two samples with similar scores have similar relative concentrations. The complication is merely that their spectra might not be similar. Of course, plots of loadings of the same component have a clear interpretation. In that case, these can be shown as line plots. Figure 8.23 gives such a plot, where the A-loadings are

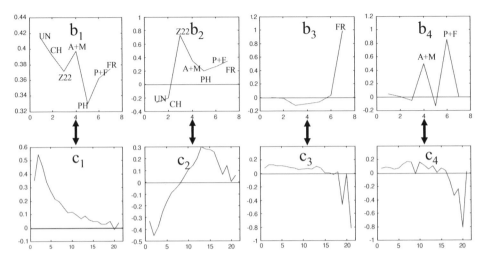

Figure 8.23. The B- and C-loadings of the peat example as line plots.

left out because they are less important. Interpretation should be made as the arrows show, between corresponding B- and C-loadings.

For a three-way model such as the PARAFAC the following is obtained after fitting the model:

$$\mathbf{X} = \mathbf{A}(\mathbf{C} \odot \mathbf{B})' + \mathbf{E} = \hat{\mathbf{X}} + \mathbf{E} \qquad (8.1)$$

which shows the problem of interpreting distances between rows of A (a score or loading plot) in terms of the original data, because the base ($\mathbf{C} \odot \mathbf{B}$) is not orthonormal. A simple way to overcome this problem is by performing an SVD on $\hat{\mathbf{X}}$

$$\hat{\mathbf{X}} = \mathbf{USV}' \qquad (8.2)$$

This decomposition of the properly matricized array $\hat{\mathbf{X}}$ ensures an orthonormal basis ($\mathbf{V}$) for a loading plot of the A-mode ($\mathbf{US}$). Similar reasoning leads to orthonormal bases for the **B** and **C** loadings and for the nonorthogonal loadings from Tucker3 models.

Figure 8.24 shows a scatter plot made according to Equation (8.2) for the peat example, matricized to 35 × 21, giving the particle area mode as **V**. The corresponding scatter plot for the PARAFAC C-loadings can be seen in Figure 8.15. The difference is not very big and the choice of using one plot or the other would often be a subjective one, but the plot in figure 8.24 has all the nice properties of a score plot from principal component analysis. In the case of the peat example, true spectral shape is not very important, but for the batch synthesis example, a true curve resolution type spectrum as in Figure 8.12 is needed for the interpretation. Loadings as in Equation (8.2) would not do as well.

A weak point of a decomposition as in Equation (8.2) is that the singular value decomposition of the matricized array is not the same for the A-, B- and C-loadings, so that, for example, a comparison of the first component for different modes as in Figure 8.23 becomes meaningless.

In Chapter 10, examples of the need for scatter plots of both types are given. Sometimes the underlying hard model is reflected very strongly in the PARAFAC or Tucker3 loadings

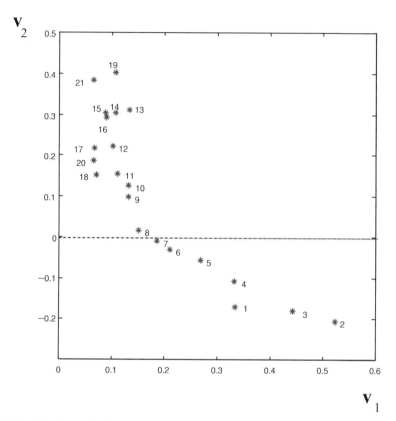

Figure 8.24. A loading plot for the peat example, made according to Equation (8.2). The plot is deliberately made square.

and then a latent variable interpretation of loading plots is needed. This would be typical for fluorescence spectroscopy data. For other situations, e.g. with heterogeneous variables for environmental data, the relationship between the underlying model and the PARAFAC model is rather weak and a recalculation of loadings as in Equation (8.2) may become useful.

Scale problems

The 'sizes' of PARAFAC components as in Figures 8.15 and 8.16 can be considered as some kind of component size. These sizes may be very different for 'large' and 'small' components. The scatter plots are always square or rectangular, but close to square (a 3:4 or 9:16 ratio). The human eye is most at ease with these kinds of plot proportions. This means that the horizontal and vertical axes do not use the same scale. Using the same scale will sometimes give rectangles of a ridiculous shape when component sizes are widely different. This problem also exists in two-way analysis by PCA. Also for the 3D plots in Figures 8.17 and 8.18 the proportions in the box may not reflect real component size. This matter is discussed further under the heading 'joint plots'. Sometimes square/rectangular plots are preferred for visual aid and sometimes true proportions are favored in order to preserve distances.

ELABORATION **8.2**

Psychology of plotting and proportions

Tufte [1983] suggested that rectangular plots with a width of 1.2–2 times the height are to be preferred for 2D plots. This is also the viewing proportion of movie and TV screens and computer terminals. This is preferred when plotting axes that are expressed in different units. When both axes are expressed in the same units, a square plot is to be preferred. An example is predicted against measured plots from calibration (see Figure 8.19), or score plots, where both axes are assumed to have the same importance and where diagonals are supposed to be at a 45° angle.

Sign indeterminacy

The sign indeterminacy can be explained by rewriting Equation (8.3) for the PARAFAC model in different ways. For the Tucker3 model, the problems are the same, though a bit more complicated due to the presence of the core array. The fit of the model is exactly the same in all cases:

$$x_{ijk} = a_{i1}b_{j1}c_{k1} + a_{i2}b_{j2}c_{k2} + \cdots + e_{ijk} \tag{8.3}$$

$$
\begin{aligned}
x_{ijk} &= (-a_{i1})(-b_{j1})c_{k1} + a_{i2}b_{j2}c_{k2} + \cdots + e_{ijk} \\
&= a_{i1}(-b_{j1})(-c_{k1}) + a_{i2}b_{j2}c_{k2} + \cdots + e_{ijk} \\
&= (-a_{i1})b_{j1}(-c_{k1}) + a_{i2}b_{j2}c_{k2} + \cdots + e_{ijk}
\end{aligned}
\tag{8.4}
$$

The same multiplication by –1 as in Equation (8.4) can also be done for a second, third component etc. A general equation is:

$$x_{ijk} = [(-1)^p a_{i1}][(-1)^q b_{j1}][(-1)^r c_{k1}] + [(-1)^s a_{i2}][(-1)^t b_{j2}][(-1)^u c_{k2}] + \cdots + e_{ijk} \tag{8.5}$$

with $p + q + r$ and $s + t + u$ even. The sign reversal causes a mirroring of the scatter plot. This problem exists in principal component loading and score plots, but there the score and loading plot are mirrored at the same time. This mirroring may be confusing for PCA, but the joint interpretation of scores and loadings remains the same. For PARAFAC models, the joint interpretation of A-, B- and C may be wrong with respect to the sign. By using background knowledge and sound judgment the problem can very often be circumvented. The problem disappears when nonnegativity constraints are applied because the signs are then fixed. In Figure 8.16 there is a typical example of sign reversal and mirroring (B- and C-loadings) that only background knowledge can handle. A sign reversal of e.g. the A- and B-loadings would give a more interpretable result for the C-loadings.

Including the origin

In PCA, because the data are often centered, scores are usually nicely centered around zero, meaning that the origin of the coordinate system is very often in the scatter plot. For a PARAFAC model mean-centering is not so often needed. Therefore, zero is not always included in the loading values. Scatter plots are usually made so that all plotted points fill the square or rectangle and then the origin may not be in the plot. This can be seen in Figure 8.15 for the A-loadings. These loadings are very correlated with the angle between the vectors being only 7°, but because the origin of the scatter plot is far outside the square, the data seem nicely spread and the unsuspecting user would guess that the loadings were nearly orthogonal, while the opposite actually holds.

The interpretation of zero in a scatter plot is that variables/objects lying close to zero do not influence the model very much for the components that were used for the plot. In Figures 8.15 and 8.16, the origin is included in the scatter plots where possible. For some of the components and some of the loadings, the origin is outside of the plot. The C-loadings in Figure 8.15 show that the largest particles do not contribute to the first two components. The B-loadings and C-loadings in Figure 8.16 show that small and intermediate sized particles and the treatments UN, CH, Z22 and PH do not contribute much to components 3 and 4.

A rule of thumb is to try to include the origin in a scatter plot and to indicate where it is. When the origin is far outside the scatter plot, care should be taken in interpreting the plot. The interpretation of distances, groupings, outliers etc. may be meaningless when the origin is not in the plot.

8.8 Image Analysis

Images were explained briefly in Section 8.1. They consist of rows, columns and square pixels having a greyvalue. Usually they have to be at least of the size 32×32 or 64×64 to be observed as images instead of just collections of grey squares. Multivariate image analysis deals with three-way and higher arrays of a special type, where two or three object indices are used to address pixels (square/rectangular elements of flat images) or voxels (cubic elements of volume images). Extra ways are created by measuring the images in more variables (e.g. wavelengths) and/or following them over time (see Figure 8.25). The analysis of multivariate images is very often simply based on eigenvalue/eigenvector analysis while temporarily ignoring the organization of the object ways, which means two-way analysis of the reorganized data [Geladi & Grahn 1996]. If an image is measured in many wavelengths and followed over time, a three-way array of the type wavelength $\times$ time $\times$ pixel can be constructed. Such an array can be modeled by a three-way method by ignoring the pixel ordering temporarily.

The interpretation of multivariate images and their analysis results, including residuals, makes extensive use of the properties of the images and the visual interpretation by the human eye. The interpretation can be through greyvalue or color images. Natural colors and synthetic colors are used. Also overlays of different image types are very common. Examples and details are given in [Geladi & Grahn 1996, Geladi & Grahn 2000].

A special example is the scatter plot of two (score) images. The images may be large (512×512 or larger) and the scatter plots have too many points in them and become cluttered. The way of showing these plots is by using density information in each location of the cartesian square. These densities are shown as greyvalues or color codes. They are also called three dimensional histograms.

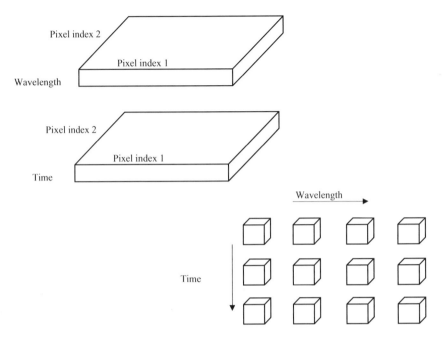

Figure 8.25. In image analysis, a flat image has two pixel indices and time or wavelength as a third way to form three-way arrays. By using volumes (three voxel indices) measured at different wavelengths and at different times five-way arrays can be produced. The visualization is special since the raw data are images and many model parameters and residuals can be visualized as images.

Figure 8.26 shows the peat 35×21 data array, the four-component PARAFAC model and the absolute value of the residual as images. Each data point is given by a pixel of the corresponding greyvalue. The image contains all the slices of the three-way array under each other. Transformations of intensity, contrast and linearity may aid in interpreting such images. Sometimes operations such as smoothing and sharpening may become useful, but these are preferably applied to larger images, from 128×128 and upwards in size. Figure 8.26 clearly shows how the three-way model is a smoothed version of the original data. The residual image shows how certain combinations of peat type and treatment form outliers in the large particle area bins.

8.9 Dendrograms

Scatter plots of loadings or scores are useful, but when the effective rank of the model becomes high, there are quite a number of plots to be studied. In these cases it may be more useful to use dendrograms [Beebe *et al.* 1998, Næs *et al.* 2002, Mardia *et al.* 1979]. With only few variables, the dendrograms rely on some distance based classification in the entire multivariate space, rather than using reduced space as with scatter plots. The dendrogram gives a good overview of the structure of clusters and outliers in the many dimensions used. When there are many variables, it is a good idea to make the dendrogram

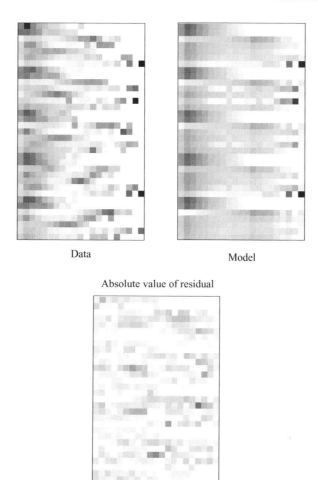

Data   Model

Absolute value of residual

Figure 8.26. The matricized arrays of the peat example (35 × 21) shown as images: the data, the model and the absolute value of the residual. The greyvalues represent the lengths of histogram bars in the 2D histogram. The horizontal axis represents the 21 particle area histogram bins, the vertical axis shows the five peat types × seven treatments.

on loadings (scores) explaining a large percentage of the total sum of squares instead of on the original variables. This saves calculation time and separates out the residual that is assumed to contain mainly noise. One complication with clustering is that there are many clustering methods and ways of preprocessing the data that may add more confusion than help.

For the four B-loadings of the peat example, the distance based classification was carried out after mean-centering and using the k-means method [Sharaf *et al.* 1986]. The results are shown in Figure 8.27. The figure shows clearly that FR is an outlier and that UN and CH are clustered close together. The loading plots of Figures 8.15 and 8.16 may make more sense

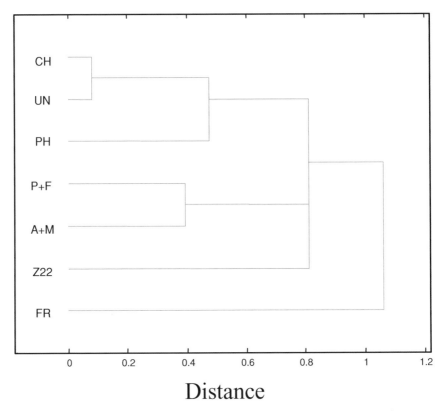

Distance

Figure 8.27. The dendrogram of the four B-loadings of the peat example.

when studied together with the dendrogram. The use of other clustering methods gave similar dendrograms, so it may be assumed that the dendrogram in Figure 8.27 has some robustness.

8.10 Visualizing the Tucker Core Array

Sometimes it is necessary to interpret the Tucker core array. This is not an easy task when the number of loadings in each mode becomes large and then visualization may be helpful. Figure 8.28 gives the $4 \times 4 \times 4$ core array for the peat example as an image (orthonormal Tucker loadings). There are four images in a row and each image has four rows and four columns. The sizes of the core elements are given as greyvalues for the pixels. White means a small core element and black means a large one. Shades of grey mean an intermediate size. Because images can only have positive intensities, a separate image is given for the negative core values, as absolute values. The intensities can also be given as squared values, which is helpful because they reflect the amount of variance explained. Interpretation of the intensities is not as accurate as that of numbers. The human eye can only distinguish between 16 and 32 greyvalues. The use of color may help distinguishing even more. The images give a useful overview of where the largest core elements can be found. Images like

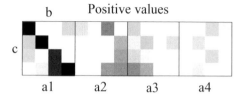

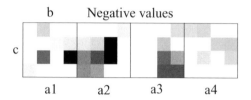

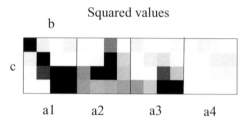

Figure 8.28. The image of the $4 \times 4 \times 4$ Tucker core array for the peat example. Positive, negative and squared values are given as three different images.

these are even more useful when larger core arrays are studied. Gemperline *et al.* [1992] visualize the sizes of the core elements as circles of different diameter.

8.11 Joint Plots

Introduction

Loading plots from different ways can be studied separately for A-loadings, B-loadings and C-loadings, but more information can sometimes be extracted by plotting the A, B and C data together in a joint plot or biplot. It may be tempting to imagine that scatter plots for different modes can be made on transparant sheets and that these can be superimposed and interpreted in that way. This is intuitively appealing and useful, but misinterpretations are possible in some situations.

It is useful to go back to two-way visualization in principal component analysis to find what really is seen in a plot. A score plot for a two-way PCA model has an orthonormal basis, because the loadings are orthonormal. This can be compared to projecting all points in multidimensional space on a movie screen using a strong light source at a large distance. What is seen in this projection is true Euclidean distance in the reduced space, if both components have the same eigenvalues and the plot is made square. Distances between points and clusters are interpretable in the projection. It is possible to do a probabilistic interpretation of the properties of clusters by superimposing binormal probability ellipses on them. If the two components have eigenvalues of very different size, a square score plot shows a distorted view of Euclidean space. A more realistic view is to use flat rectangles but this can lead to ridiculous proportions that are very unfit for human viewing. It should be pointed out here that many popular data analysis programs are not capable of making truly Euclidean plots. This is sometimes amended by giving the percentage of the variance explained on the axes.

The discussion above only uses geometrical arguments relating to the raw data. The scores in a score plot also have a latent variable interpretation. In a sensory example it may be found that two principal components are related to sweetness and crispness of a food product. It is possible to make a square score plot, even if the eigenvalues of sweetness and crispness are quite unequal, because the subjective interpretation of these properties is not really related to eigenvalue size. The sizes of some of the eigenvalues may also be the result of sampling and then they change drastically if samples are added or deleted. The above shows that there is a duality in interpreting scatter plots. In some cases, correct distances in the raw data space are of interest and in other cases the subjective interpretation of a latent variable is in focus. These two points of interest do not necessarily coincide.

When making biplots (overlayed scatterplots), a number of matters have to be considered (ignoring scaling of the variables before the analysis):

1. The relative scaling of the different components, e.g when making a scatterplot of loadings with a large and a small eigenvalue.
2. The relative scaling of the scores and loadings. Do variables and objects use the same scale or does each axis have two scales?
3. Compensation for an unbalance in number of objects and variables.
4. The physical scales used on the horizontal and vertical axes.

The biplot

A compromise was developed by Gabriel in 1971 and called the biplot [Gabriel 1971]. This is also described by Jackson [1991] and Brereton [1992]. It is useful to start with the simple case of two-way analysis. Principal component analysis of $\mathbf{X}$ is given as a singular value decomposition. A model with two principal components is (see Chapter 3):

$$\mathbf{X} = \mathbf{u}_1 \lambda_1^{1/2} \mathbf{v}_1' + \mathbf{u}_2 \lambda_2^{1/2} \mathbf{v}_2' + \mathbf{E} \tag{8.6}$$

A more general equation for R components is:

$$\mathbf{X} = \mathbf{U} \mathbf{L}^{1/2} \mathbf{V}' + \mathbf{E} = \mathbf{U} \mathbf{S} \mathbf{V}' + \mathbf{E} \tag{8.7}$$

$\mathbf{X}$: the pretreated data matrix (scaled, centered etc.)

$\mathbf{E}$: the matrix of residuals

$\mathbf{U}$: the orthogonal matrix of scores

V : the orthogonal matrix of loadings

S : the diagonal matrix of singular values

L : The diagonal matrix of eigenvalues of $\mathbf{X'X}$

$\mathbf{u}_1, \mathbf{u}_2$: the first two columns of $\mathbf{U}$

$\mathbf{v}_1, \mathbf{v}_2$: the first two columns of $\mathbf{V}$

λ_1, λ_2 : the first two eigenvalues in $\mathbf{L}$

The $\lambda^{1/2}$ can be distributed equally among the $\mathbf{u}$ and $\mathbf{v}$ parts (scores and loadings) for the purpose of forming new variables $\mathbf{h}$ and $\mathbf{g}$ to be plotted:

$$\mathbf{X} = (\mathbf{u}_1\lambda_1^{1/4})(\lambda_1^{1/4}\mathbf{v}_1') + (\mathbf{u}_2\lambda_2^{1/4})(\lambda_2^{1/4}\mathbf{v}_2') + \mathbf{E}$$
$$= \mathbf{h}_1\mathbf{g}_1' + \mathbf{h}_2\mathbf{g}_2' + \mathbf{E} \tag{8.8}$$

A more general formulation is:

$$\mathbf{X} = (\mathbf{u}_1\lambda_1^{c/2})(\lambda_1^{(1-c)/2}\mathbf{v}_1') + (\mathbf{u}_2\lambda_2^{c/2})(\lambda_2^{(1-c)/2}\mathbf{v}_2') + \mathbf{E} \tag{8.9}$$

with $0 \leq c \leq 1$. For $c = 0.5$ Equation (8.8) is obtained. The biplot is made by plotting $(\mathbf{u}_1\lambda_1^{c/2})$ against $(\mathbf{u}_2\lambda_2^{c/2})$ (the first and second scores) and $(\lambda_1^{(1-c)/2}\mathbf{v}_1')$ against $(\lambda_2^{(1-c)/2}\mathbf{v}_2')$ (the first and second loadings) and putting the plots on top of each other. The special cases of $c = 0, c = 0.5$ and $c = 1$ are described by Jackson, but also other values of c may be chosen for a subjective interpretation. What is shown in Equations (8.6)–(8.9) is also valid for further components. A general equation for Equation (8.9) for many components is for $c = 1/2$:

$$\mathbf{X} = [\mathbf{UL}^{1/4}][\mathbf{L}^{1/4}\mathbf{V}'] + \mathbf{E} \tag{8.10}$$

with $\mathbf{L}$ as the diagonal matrix of the eigenvalues. Setting $c = 1$ gives:

$$\mathbf{X} = [\mathbf{UL}^{1/2}][\mathbf{V}'] + \mathbf{E} \tag{8.11}$$

which is the row metric preserving version. It gives Euclidean distances between objects and Mahalanobis distances between variables. Setting $c = 0$ gives:

$$\mathbf{X} = [\mathbf{U}][\mathbf{L}^{1/2}\mathbf{V}'] + \mathbf{E} \tag{8.12}$$

which is the column metric preserving version. It gives Euclidean distances between variables and Mahalanobis distances between objects. It is also possible to have the sum of the exponents in Equations (8.10)–(8.12) different from 1/2. In that case the model fit is not correct, but the plots can still be made [Martens & Martens 2001].

What is missing in the above is a compensation for number of objects and number of variables. Equations (8.10)–(8.12) work well when the number of objects and variables is almost equal. With 1000 objects and 15 variables, all objects would end up close to the origin of the scatter plot and the variables would be more spread out in the plot. This is so because of the constraint of length one of $\mathbf{u}$ and $\mathbf{v}$ in the singular value decomposition. A compensation for number of objects (I) and variables (K) is made by introducing a fudge or zoom factor α:

$$\mathbf{X} = [(1/\alpha)\mathbf{UL}^{1/4}][\alpha\mathbf{L}^{1/4}\mathbf{V}'] + \mathbf{E} \tag{8.13}$$

A number of ways of calculating α can be imagined. The main point is to be able to see all variables and objects clearly in the plot.

EXAMPLE 8.4

The wine example

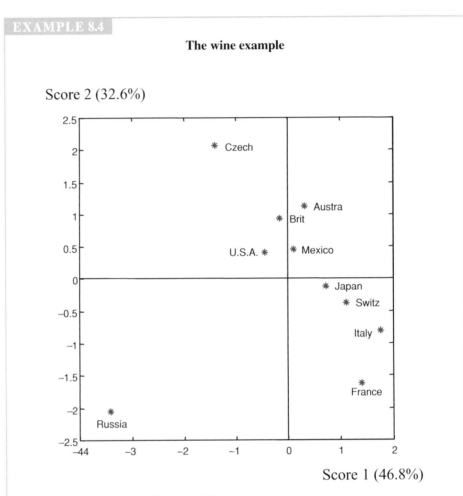

Figure 8.29. The plot of $\mathbf{u}_1\lambda_1^{1/2}$ and $\mathbf{u}_2\lambda_2^{1/2}$ for the wine example, given as a square plot (Austra = Australia). This plot would be called a score plot for PCA on the data matrix of Example 8.4. The percentages of the sum of squares explained by the components are given on the axes.

A simple example may serve to show how things work. The wine data set has ten objects and five variables [Wise & Gallagher 1998]. The objects are countries and the variables are the consumption of wine, beer and hard liquor, with life expectancy and heart disease rate as variables four and five. The table was published in *Time* magazine, in 1996. The interpretations are made for the sake of showing how the plots can be used. They are very dependent on the reliability of the original data and on the correctness of a two-component model. The data are analyzed by singular value decomposition after scaling of each variable by its standard deviation and mean-centering. The singular values for the components are: 4.55, 3.80, 2.29, 1.95 and 0.88. Figure 8.29 shows a square score plot for components 1 and 2 of the wine example. The plot shows Russia as an outlier. The other countries are clustered. The cluster has a gradient from the beer-drinking Czech republic to the wine-drinking mediterranean countries Italy and France. This interpretation is made by judging distances in an Euclidean manner. The

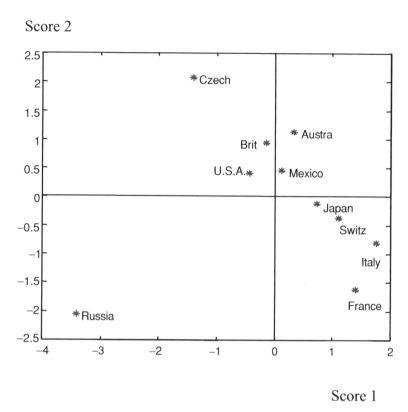

Figure 8.30. The same plot as in Figure 8.29, but now with equal scales for the horizontal and vertical axes, giving correct Euclidean distances to the origin and between objects.

vertical and horizontal scales are different, making the Euclidean interpretation slightly false. Luckily, the singular values 4.55 and 3.80 are of almost equal size, so there is no big problem.

Figure 8.30 shows the same plot as Figure 8.29, but now with identical vertical and horizontal scales. The plot becomes rectangular and the interpretation is now based on true Euclidean distances. A more dramatic plot is shown in Figure 8.31 for the first and third scores. The horizontal and vertical scales are the same, but the plot becomes a flatter rectangle because of the difference in singular values 4.55 and 2.29. The plot is truly Euclidean. Figure 8.32 is a loading plot for the first and second loadings. The plot is square, with identical horizontal and vertical scales. Loading vectors are normalized to length one and this is expressed in the plot. None of the variables have small loading values, so all variables contribute to the two-component model. By comparing Figures 8.30 and 8.32, conclusions about the importance of the variables for the structure in the score plot can be made.

The comparison in the previous paragraph between Figure 8.30 and Figure 8.32 immediately leads to the idea of plotting the two figures on top of each other as a biplot, but this is not as simple as just superimposing them. Figure 8.33 gives a biplot as in Equation (8.11). The horizontal and vertical scales are identical, so the position of the

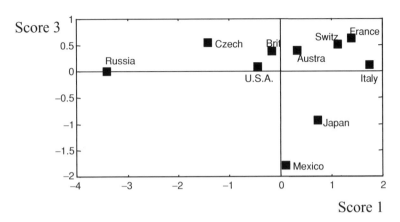

Figure 8.31. The plot of $\mathbf{u}_1\lambda_1^{1/2}$ and $\mathbf{u}_3\lambda_3^{1/2}$ for the wine example. Using Euclidean distances gives a flat rectangle. Note that the scales on the horizontal and vertical axes are equal.

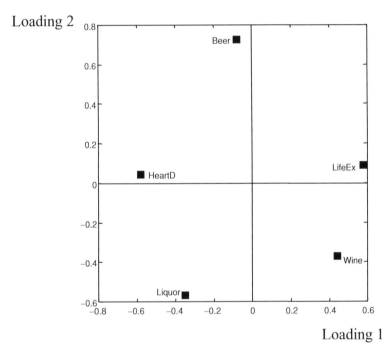

Figure 8.32. The loading plot for the wine example using $\mathbf{v}_1$ and $\mathbf{v}_2$. Horizontal and vertical axes have equal scales. The distances in this plot are Mahalanobis.

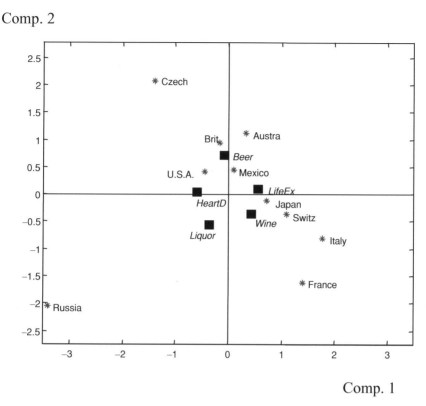

Figure 8.33. The biplot according to Equation (8.11) for the wine example. The horizontal and vertical axes are equal. The scales are identical for objects and variables. The objects have Euclidean distances. The center of the plot is cluttered because the variables have too small loading values.

objects is still subject to Euclidean interpretation, but now the variables are seen in the same plot. The center of the plot is too cluttered and therefore the variables need to be zoomed out by a fudge factor. This is seen in Figure 8.34. In this figure, there are arrows pointing to the variables. The length of an arrow indicates the importance of the variable in the model. Now the interpretation is easier. Beer is close to the Czech republic and wine is close to France and Italy. Loadings are actually directions of the principal components, so they should be interpreted as directions. The directions can be used for projection of the objects onto them. The orthogonal projections of France and Italy on the wine arrow show that France has a slightly larger wine consumption than Italy. The projections on the beer arrow show that the Czech republic has a large beer consumption, while the British and Australian consumption is less and almost equal. The lines pointing to the variables as in Figure 8.34 also go through the origin and become negative. This does not mean negative beer, wine or liquor consumption, just that the data were mean-centered and that the consumption is below the average for the data set.

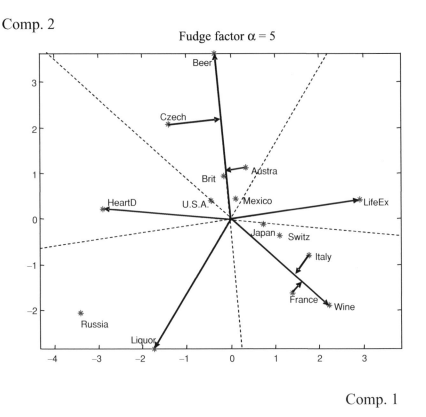

Figure 8.34. The same figure as Figure 8.33 but now with a fudge (zoom) factor of 5 for the variables. It is possible to make orthogonal projections from the objects onto the lines going from the origin through the variables. Dashed lines are negative parts of the arrows.

ELABORATION **8.3**

Biplots, some practical remarks

It may be useful to give a small recipe for making simple biplots. This is given for the first and second components, but it can easily be modified for other components.

1. Check the score $(\mathbf{u}_1\lambda_1^{1/2}, \mathbf{u}_2\lambda_2^{1/2})$ and loading $(\mathbf{v}_1$ and $\mathbf{v}_2)$ plots separately. It is important to have equal scales on horizontal and vertical axes. If that gives very narrow rectangles, a square plot can be made keeping in mind that components of differing eigenvalues are shown in the plot. This can then also be indicated on the axes. When the axes are stretched, some interpretations are not possible. For example, the angle between two variable vectors no longer reflects the correlation between them. Extreme outliers may have to be removed in both score and loading plots. Variables with small loading values may have to be marked and remembered as unimportant, possibly removed, especially in cases with many variables.

2. Use the vectors $[(\mathbf{u}_1\lambda_1^{1/2})'\ \mathbf{v}_1']$ and $[(\mathbf{u}_2\lambda_2^{1/2})'\ \mathbf{v}_2']$ (see Equation 8.11) and make the scatter plot with equal scales on the horizontal and vertical axes. In this plot the distances

between the objects are Euclidean and the angles between the variables and the axes are correct. It is possible to draw lines from the origin to the variables and to project the objects on these lines.

3. If the variables or the objects end up too close to the center of the plot, some compensation in the scale may be given using $[(\mathbf{u}_1 \lambda_1^{1/2})' \, \alpha \mathbf{v}_1']$ and $[(\mathbf{u}_2 \lambda_2^{1/2})' \, \alpha \mathbf{v}_2']$ for the plot, where α is an appropriate fudge or zoom factor. In this case, the scale for the objects can not be used for the variables.

4. It is possible to use Equations (8.10) and (8.12) as alternatives to Equation (8.11) in some special cases.

The joint plot

The joint plot for three-way analysis is related to the biplot. It was first used in the social sciences and psychology for Tucker components and it has some special properties. The uses in chemistry are not yet abundant. A good description is given in [Kroonenberg 1983] and [Kroonenberg 1994]. Figure 8.35 shows the Tucker 3 decomposition as described in Chapter 4. A component q of the B-loadings is selected and for this component the joint plots are going to be made. The corresponding plane of the core array is extracted and called $\mathbf{G}_q$. The core content of $\mathbf{G}_q$ is distributed between the A-loadings and the C-loadings as follows:

$$\mathbf{D}_q = \mathbf{A}\,\mathbf{G}_q\mathbf{C}' = \mathbf{A}[\mathbf{UL}^{1/2}\mathbf{V}']\,\mathbf{C}' = [\mathbf{AUL}^{1/4}][\mathbf{L}^{1/4}\mathbf{V}'\mathbf{C}'] = \mathbf{A}_q\mathbf{C}_q' \qquad (8.14)$$

$\mathbf{G}_q$ is decomposed by singular value decomposition and the eigenvalues are distributed equally between the A-loadings and the C-loadings as in Equation (8.14). See also Figure 8.36. The P columns of $\mathbf{A}_q$ and the R columns of $\mathbf{C}_q$ can be used for making biplots if $P = R$, otherwise the smallest of P and R limits the number of biplots. There are Q possible

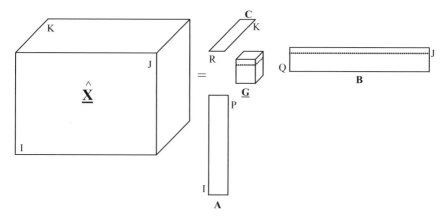

Figure 8.35. The Tucker3 decomposition. For the joint plot, one of the components for the B-loadings is selected and the corresponding plane $\mathbf{G}_q$ of the core is extracted (see dashed lines).

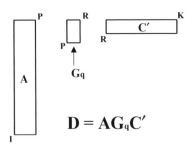

$$D = AG_qC'$$

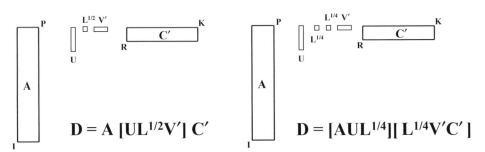

$$D = A [UL^{1/2}V'] C' \qquad D = [AUL^{1/4}][L^{1/4}V'C']$$

Figure 8.36. The selected plane G_q in the product $D = AG_qC'$ is decomposed by SVD and the eigenvalues are distributed equally to each side.

ways of defining G_q. One can also replace **B** or **A** by **C** in Figure 8.35. The matrix G_q in Equation (8.14) does not have to be square or of full rank. Sometimes a few eigenvalues in **L** suffice. With all the possible choices, $P \times Q \times R$ versions of Equation (8.14) can be made, and in each of these versions many plots can be made. If only a few Tucker components are used this is easy, but it is necessary to make a subjective selection when the number of Tucker components grows. If one of the modes only contains one Tucker component, the biplot is obtained. Just as for the biplot, the joint plot also contains a zoom factor. Rewriting Equation (8.14):

$$D_q = (1/\alpha)A_q \alpha C'_q \qquad (8.15)$$

Kroonenberg [1994] and Kroonenberg and de Leeuw [1977] use the following form of Equation (8.15):

$$D = (I/K)^{1/4}A_q C'_q (K/I)^{1/4} \qquad (8.16)$$

This equation is also given by Gemperline *et al.* [1992]. This article also contains a good application of Tucker3 analysis on trace element data and joint plots.

EXAMPLE 8.5

The crab disease example

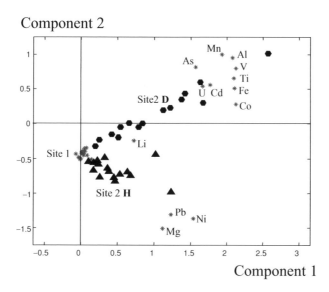

Figure 8.37. The joint plot of the blue crab disease data for the first plane of the core array. The plot shows individuals in three groups: site1 (asterisk), site2 healthy (triangle) and site2 diseased (hexagon) together with some selected elements (asterisk with an element symbol) that are related to the separation.

The study of Gemperline *et al.* [1992] used individual blue crabs from two river sites. One site had only healthy crabs and another more industrial site had healthy and diseased crabs. This gives three groups and for each group 16 crabs were analyzed. For each individual crab three organs were analysed: liver, muscle and gill. The ICP-AES analysis gave concentrations for 25 elements. This gives a (48 × 25 × 3) array: 48 individuals × 25 elements × 3 organs. The individuals and the sites are nested, so it is not possible to construct a four-way array. The goal of the analysis was to find the relationship between sites and shell disease through the elemental composition of the analysed organs. A Tucker3 analysis was carried out after variablewise scaling to give a (4,5,2) core array explaining about 70 % of the total sum of squares and corresponding loadings. The gills were shown to be the organs explaining the difference between healthy and diseased. Figure 8.37 shows a joint plot made according to Equation (8.17). The figure shows that the healthy crabs from site 1 are clustered very close together. The crabs for site 2 split into two groups: healthy and diseased, with the diseased ones having a very large within-group variation. The healthy crabs of site 2 overlap with the healthy crabs from site 1. The elemental composition giving this separation can be seen clearly in the figure. It is possible to give an environmental explanation for how the elements group and how these groups separate the crab classes. A more detailed description of all aspects of this analysis is found in the original article [Gemperline *et al.* 1992] and also in Chapter 10.

8.12   Residual Plots

It was shown in Chapter 7 how different subarrays of the residual $\underline{\mathbf{E}}$ can be made and how statistics can be calculated for them. The goals of the analysis often dictate how residuals should be viewed and treated. There are many possibilities and not all of them fit each specific three-way data array.

For some examples, it is useful to summarize this residual $\underline{\mathbf{E}}$ over the $\mathbf{E}_i$, $\mathbf{E}_j$ and $\mathbf{E}_k$. Over these matrices a bias and a sum of squares can be calculated and they may be shown as line plots or bar plots. For the peat example ($5 \times 7 \times 21$) the residual after four components was given as an image in Figure 8.26. Plots for the particle area way are given in Figure 8.38. The biases are very small. Figure 8.38 also gives SS_{res} values for the residual and for comparison also for the matching subarray of $\underline{\mathbf{X}}$, called SS_{tot}. Even though there may be seen some structure in the SS_{res} values, the SS_{res} are small compared to the SS_{tot} values. This is satisfactory. Such an interpretation is subjective. A priori information may allow a more objective interpretation if it is available. The data of the wavelength mode of the batch synthesis example ($14 \times 13 \times 701$) for a three-component PARAFAC model are given as line plots in Figure 8.39. Also here, the bias is very low. The residual sum of squares is reasonably low, but still has some structure. However, this structure has no meaningful three-way model to it.

In multi-way regression models there are the residuals $\underline{\mathbf{E}}_y$ and $\underline{\mathbf{E}}_x$ (see Chapters 4). These residuals can be used as described in Chapter 7 for calculating a number of different sums of squares. These can also be visualized as described here for component models.

8.13   Leverage Plots

Leverage is discussed in Chapter 7. It is a measure of influence in regression models, but the idea of calculating Squared Mahalanobis distances can also be used for PARAFAC or Tucker loadings. For the peat example ($5 \times 7 \times 21$) the Squared Mahalanobis distances $\mathbf{h}^A$, $\mathbf{h}^B$ and $\mathbf{h}^C$ as they are defined in Chapter 7, may be calculated for the peat types, for the treatments and for the particle sizes.

Figure 8.40 gives a bar plot of this for the treatments. It is also possible to plot the Squared Mahalanobis distances against the SS_{res} values as scatter plots. This leads to a combined interpretation of SS_{res} and Mahalanobis distance. This plot is very informative, see Figure 8.41. It can be seen that the treatments behave quite different from each other. Freezing has a large $\mathbf{h}^B$ and a small residual. It contributes well to the model with unique information. pH lowering has the largest residual and a low $\mathbf{h}^B$, so it is not modeled very well. The same may be concluded about the $AlCl_3+$ magnafloc treatment. The Zetag22 and pH lowering + freezing also have large $\mathbf{h}^B$ and intemediate residuals. They influence the model but do not fit as well as freezing. Untreated and wet charring have intermediate $\mathbf{h}^B$ and low residuals. They fit well to the model and do not influence it too much.

8.14   Visualization of Large Data Sets

The visualization of large data sets has special requirements. A typical example area is multivariate image analysis, especially in remote sensing, but hyperspectral imaging is also moving into other fields. Another is three-dimensional visualization of weather systems.

Bias

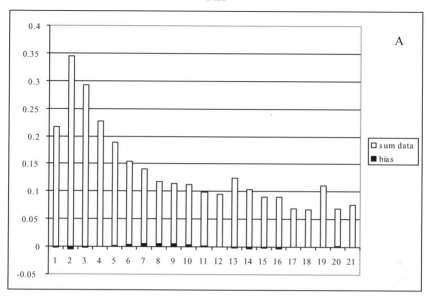

Sum of squares

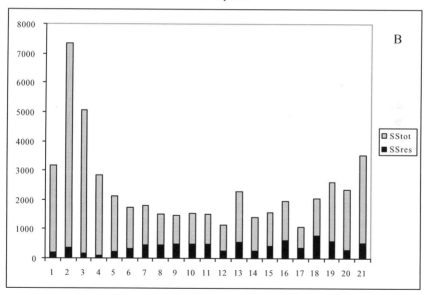

Figure 8.38. (A) The bias in the residual array (after four PARAFAC components) compared to the total sum of the raw data for the particle area way of the peat example. The bias is sufficiently small. (B) The residual sum of squares of the particle area compared to the total sum of squares.

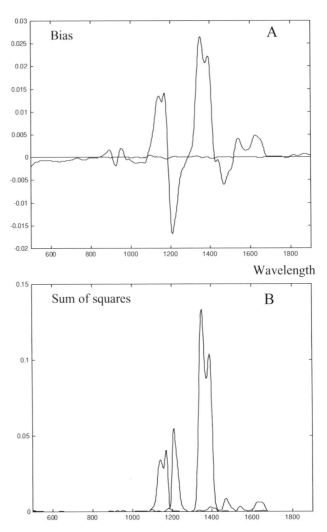

Figure 8.39. (A) The bias in the residual (after three PARAFAC components) compared to the total sum of the raw data for the wavelength mode of the batch synthesis example. The bias is sufficiently small. (B) The residual sum of squares as a function of wavelength compared to the total sum of squares. The lines closest to zero are for the residual.

But large data sets are also produced by simulations and other technical calculations of high spatial resolution. For large data sets, individual points in the plots can not be recognized because there are too many of them and other representation techniques have to be found.

Scientific visualization deals with visualizing large data sets. Both raw, derived and calculated data can be visualized. Large systems such as weather data (real data) and results of aerodynamic calculations (calculated data) are used frequently [Nielson *et al.* 1990, Jones 1996, Cleveland 1985, Cleveland & McGill 1988]. Scientific visualization uses images, color, volume rendering by interpolation, special symbols and animation. Sound is also included if appropriate. In this book, color is not used and animation is impossible with static figures, but they certainly have a future for visualization in three-way analysis.

Squared Mahalanobis distance

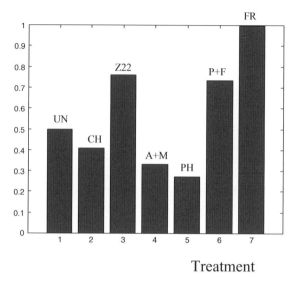

Treatment

Figure 8.40. The Squared Mahalanobis distance $\mathbf{h}^B$ for the treatments in the peat example.

Squared Mahalanobis distance

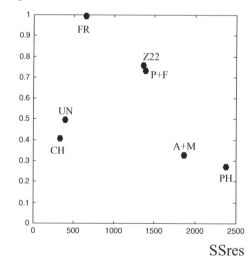

SSres

Figure 8.41. Squared Mahalanobis distance plotted against residual sum of squares for the treatments in the peat example.

8.15 Summary

The human eye is very good at subjectively interpreting complex patterns and this is used to an advantage by examining plots. There are quite a number of plots, all with their own special properties. These plots are used abundantly in two-way analysis. They can also be adapted for three-way analysis, even though sometimes small modifications in interpretation are needed. Three-way analysis requires checking of raw data, model parameters and residuals. The effective rank of the model has to be determined in scree plots and the loadings should be studied in line plots or scatter pots. When many scatter plots have to be studied, dendrograms may come in useful. Also biplots and joint plots can be made useful. Large amounts of data may need visualization as images. Residual plots and influence plots may be made and the influence and residual can be plotted against each other. There are peculiarities in plotting that one should be aware of for avoiding misinterpretation. These are: disproportional horizontal and vertical scales, sign indeterminacy, oblique loadings and not having the origin in the plot.

Especially scatter plots should be interpreted very carefully. Principal component analysis produces latent variables and at the same time orthogonal scores and orthonormal loadings. The latent variable and the Euclidean interpretation of scatter plots both come from the same model. For three-way models, the latent variables do not allow a direct Euclidean interpretation of loading plots. A recalculation can give this Euclidean interpretation, but then the original latent variable structure gets lost.

9
PREPROCESSING

9.1 Background

Mean-centering and scaling are usual operations in two-way analysis. In fact, they are so usual that many users of data analysis do not reflect on why and how they are used. There are good reasons for mean-centering two-way data. The averages are rarely of importance so the system of coordinates is shifted while the geometry of the multivariate space is conserved. Scaling in a set of nonhomogeneous variables is necessary to remove differences in measurement units and for homogeneous data it may also be appropriate to scale. These 'facts' can be misleading when centering and scaling are used in nontraditional ways. Three-way analysis is one such instance. Mean-centering and scaling become more complicated because they can be performed and combined in several ways. Although it is intuitively clear that certain operations should not be done on three-way arrays, this is not enough. A more theoretical basis for determining what is feasible and what is not should be presented.

The purpose and use of centering and scaling are discussed in this chapter. The background is explained using two-way bilinear data analysis for simplicity, and the results are then generalized to three-way data analysis. Other types of preprocessing are also relevant for three-way models, but these will only be lightly touched upon in Appendix 1 as the basic use of these is very similar to their use in standard two-way analysis.

It is shown that centering is most conveniently seen as a projection step, where the data are projected onto certain well-defined spaces within a given mode. This view of centering helps explaining why, for example, centering data with missing elements is likely to be suboptimal if there are many missing elements.

Building a model for data consists of two parts: postulating a structural model and using a method to estimate the parameters. Centering has to do with the first part: when centering, a model including offsets is postulated. Scaling has to do with the second part: when scaling another way of fitting the model is employed. It is shown that centering is simply a convenient technique for fitting models with certain offsets, but this does not work for all types of offsets. It is also shown that scaling is a way to fit models with a weighted least squares loss function and that sometimes this change in objective function cannot be performed by a simple scaling step.

Multi-way Analysis With Applications in the Chemical Sciences. A. Smilde, R. Bro and P. Geladi
© 2004 John Wiley & Sons, Ltd ISBN: 0-471-98691-7

Important notation, definitions and main results are given in Section 9.1. The derivation of the main results is based on a quite mathematical exposition in the following sections. For those less interested in these mathematical derivations, it is important to know at least the rules of centering and scaling given in Section 9.1.

Centering of two-way data is treated in Section 9.2 by first defining the purpose of centering, the way centering accomplishes these purposes, situations in which centering cannot fulfill the purposes and alternatives to centering in such situations. Scaling of two-way data is treated in Section 9.3 in the same fashion as for centering. In Section 9.4 the properties of simultaneous centering and scaling of two-way data are treated.

In Section 9.5 three-way centering and scaling is described based on the two-way results. The main overall results are described in Section 9.6.

Definitions

It is important to have a concise terminology for scaling and centering. The following convention is based on suggestions from the literature [Bro & Smilde 2003, Harshman & Lundy 1984, Kiers 2000, Kruskal 1989, ten Berge 1989].

The term 'an offset' – also sometimes called an intercept – is used for a part of the model that is constant across one or several modes. An R-component bilinear model of a data matrix, $\mathbf{X}$ $(I \times J)$ with elements x_{ij}, may be written in terms of scalars or in matrix notation as

$$\mathbf{X} = \mathbf{\Phi\Theta}' + \mathbf{E} \Leftrightarrow x_{ij} = \sum_{r=1}^{R} \phi_{ir}\theta_{jr} + \varepsilon_{ij} \tag{9.1}$$

where $\mathbf{\Phi}$ $(I \times R)$ and $\mathbf{\Theta}$ $(J \times R)$ hold the parameters, ϕ_{ir} and θ_{jr}, where Greek symbols are used to indicate population parameters. For bilinear data, $\mathbf{\Phi}$ would be the underlying scores and $\mathbf{\Theta}$ the underlying loadings. The matrix $\mathbf{E}$ holds the unknown errors. A simple example of such data is shown in Figure 9.1 (upper left). This is a one-component data set and the correlation to the concentration of the spectrally detected analyte is shown in the lower part of both the raw and the centered data.

Offsets may be constant *across* the first mode (rows). An example of offsets constant across the first mode (rows) is given in Figure 9.2. The model associated with such offsets is:

$$\mathbf{X} = \mathbf{\Phi\Theta}' + \mathbf{1\mu}' + \mathbf{E} \Leftrightarrow x_{ij} = \sum_{r=1}^{R} \phi_{ir}\theta_{jr} + \mu_j + \varepsilon_{ij} \tag{9.2}$$

where $\mathbf{\mu}$ $(J \times 1)$ holds the constant terms μ_j and where $\mathbf{1}$ is a one-vector of suitable size $(I \times 1$ in this case). Again, the Greek symbol $\mathbf{\mu}$ indicates a population value. An example using spectroscopy-like data is given in Figure 9.3. Note the different (and independent) offsets on the absorbance scale. These are exaggerated to stress that in assuming offsets across the first mode no assumptions are implied about the relations between these.

Offsets may also be constant across the second mode (columns) as shown in Figure 9.4. The underlying bilinear model with offsets across the second mode reads

$$\mathbf{X} = \mathbf{\Phi\Theta}' + \mathbf{\mu1}' + \mathbf{E} \Leftrightarrow x_{ij} = \sum_{r=1}^{R} \phi_{ir}\theta_{jr} + \mu_i + \varepsilon_{ij} \tag{9.3}$$

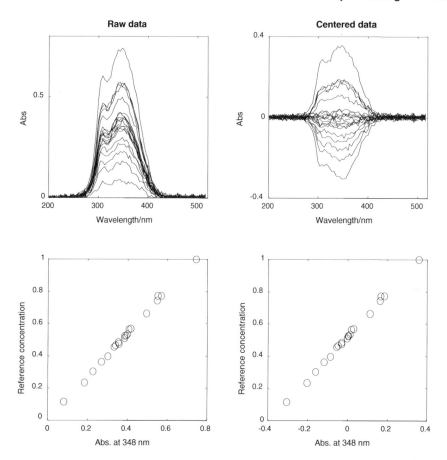

Figure 9.1. No offsets. Raw data (shown upper left) are simulated UV–Vis spectra of samples containing one analyte. To the right, the same spectral data are shown using 'ordinary' centering subtracting the column means (called centering across the first mode). As shown in the lower scatter plots of concentration versus wavelength 348 nm the quantitative relation to the concentration of the analyte is retained by centering.

1	6	8	9	3	5	6	7
1	6	8	9	3	5	6	7
1	6	8	9	3	5	6	7
1	6	8	9	3	5	6	7
1	6	8	9	3	5	6	7

Figure 9.2. Offsets constant across the first mode; $\mathbf{\mu} = (1\ 6\ 8\ 9\ 3\ 5\ 6\ 7)'$.

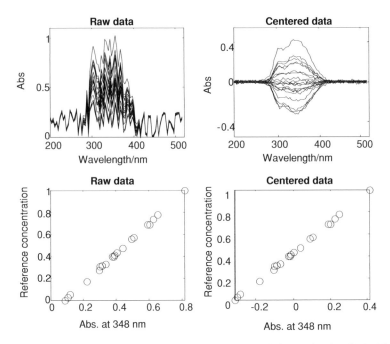

Figure 9.3. Offset across the first mode. Raw data (shown upper left) are simulated UV–Vis spectra of samples containing one analyte. To the right, the same spectral data are shown using 'ordinary' centering subtracting the column means (centering across the first mode). As shown in the lower plots the quantitative relation to the concentration of the analyte is retained by centering.

$$
\begin{array}{cccccccccccc}
1 & 1 & 1 & 1 & 1 & 1 & 1 & 1 & 1 & 1 & 1 & 1 \\
2 & 2 & 2 & 2 & 2 & 2 & 2 & 2 & 2 & 2 & 2 & 2 \\
5 & 5 & 5 & 5 & 5 & 5 & 5 & 5 & 5 & 5 & 5 & 5 \\
4 & 4 & 4 & 4 & 4 & 4 & 4 & 4 & 4 & 4 & 4 & 4 \\
8 & 8 & 8 & 8 & 8 & 8 & 8 & 8 & 8 & 8 & 8 & 8
\end{array}
$$

Figure 9.4. Offsets constant across the second mode; $\boldsymbol{\mu} = (1\ 2\ 5\ 4\ 8)'$.

where the vector $\boldsymbol{\mu}$ $(I \times 1)$ is now holding the offsets μ_i. An example is given in Figure 9.5. Offsets may also be constant across columns *and* across rows yielding

$$
\mathbf{X} = \boldsymbol{\Phi}\boldsymbol{\Theta}' + \mu\mathbf{1}\mathbf{1}' + \mathbf{E} \Leftrightarrow x_{ij} = \sum_{r=1}^{R} \phi_{ir}\theta_{jr} + \mu + \varepsilon_{ij} \tag{9.4}
$$

in which the single constant μ is the same for all elements of $\mathbf{X}$. Such a situation may arise, for example, in chromatography or capillary electrophoresis where a constant offset in the detector may appear, due to the way the detector's zero-level is determined. An example is shown in Figure 9.6. This type of offset *cannot* be eliminated by subtracting the grand mean from all elements. It will later be shown in detail what possibilities are available

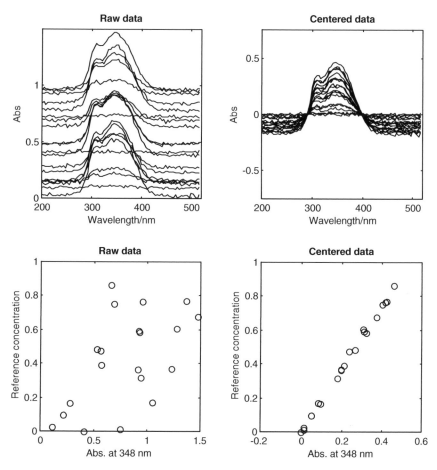

Figure 9.5. Offset across the second mode. Raw data (shown upper left) are simulated UV–Vis spectra of samples containing one analyte. To the right, the same spectral data are shown using centering across the second mode (i.e. row-wise – the opposite direction of Figure 9.4). As shown in the lower plots the linear relation to the concentration of the analyte is not present in the raw data but is obtained upon centering the data.

for handling this situation efficiently. Thus, for bilinear models there are two basic types of offsets: constants across one mode (columns or rows) or constants across both modes. Combinations of such offsets may also appear as is for example seen in analysis of variance settings.

Offsets are often handled by first centering the data and subsequently fitting the bilinear model to the centered data as also shown in the previous plots. If the data are centered by subtracting the column-average from every element in the column, this is referred to as *centering across the first mode*. Mathematically it can be expressed as

$$z_{ij} = y_{ij} - \frac{\sum_{i=1}^{I} x_{ij}}{I} \qquad (9.5)$$

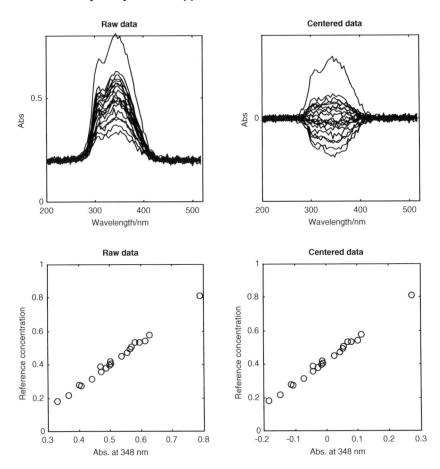

Figure 9.6. Common offset for all elements. Raw data shown upper left simulating UV–Vis spectra of samples containing one analyte. Note the offset on the absorbance scale. To the right, the same spectral data are shown using 'ordinary' centering subtracting the column means (centering across the first mode). As shown in the lower plots, the quantitative relation to the concentration of the analyte is retained by centering.

where z_{ij} is an element of the centered data matrix. If $\mathbf{m}$ ($J \times 1$) is a vector holding the jth column-average in its jth element, then centering across the first mode can also be expressed by

$$\mathbf{Z} = \mathbf{X} - \mathbf{1m'} \tag{9.6}$$

where $\mathbf{1}$ is an I-vector of ones and $\mathbf{Z}$ the matrix holding the centered data. Subtracting the row-average from each element in a row is referred to as *centering across the second mode* and can be expressed as

$$z_{ij} = x_{ij} - \frac{\displaystyle\sum_{j=1}^{J} x_{ij}}{J} \tag{9.7}$$

or using $\mathbf{m}$ ($I \times 1$) as a vector holding the ith row-average in its ith element

$$\mathbf{Z} = \mathbf{X} - \mathbf{m1}' \qquad (9.8)$$

In general, centering across one mode is also called *single-centering* and performing, for example, centering across the first mode and then a subsequent centering *of the outcome* across the second mode is called *double-centering*.

The term slab-centering, which is sometimes seen in the literature, refers to centering by subtracting, from each slab in a three-way array, the overall average of that slab. For two-way data, this simply corresponds to subtracting the average of all elements. Slab-centering is not generally useful as discussed in Section 9.5.

For scaling, the terminology used is the following. When a matrix is scaled such that each row is multiplied by a specific scalar, it is called *scaling within the first mode* ($z_{ij} = x_{ij}w_i$). If each column is multiplied by a certain scalar as in traditional autoscaling, it is referred to as *scaling within the second mode* ($z_{ij} = x_{ij}w_j$). In matrix notation scaling within the first mode can be written as

$$\mathbf{Z} = \mathbf{WX} \qquad (9.9)$$

where $\mathbf{W}$ is an $I \times I$ diagonal matrix holding the scalar w_i in its ith diagonal element.

Scaling within the second mode can be written as

$$\mathbf{Z} = \mathbf{XW} \qquad (9.10)$$

where $\mathbf{W}$ is now a $J \times J$ diagonal matrix holding the scalar w_j in its jth diagonal element.

ELABORATION **9.1**

Why centering across and scaling within?

It may seem strange why different words are used for scaling (within) and centering (across). However, scaling and centering are two quite different operations.

Centering is performed across a mode in the sense that *one* offset is subtracted from every element in a certain vector, i.e., the data are centered *across* the elements of one mode. For three-way data the same holds; from each element of a vector the average value is subtracted.

Scaling is performed by multiplying *all* elements of a certain variable (or object) by the same scalar. For two-way data scaling therefore also pertains to vectors, but e.g. for three-way data, this means that a whole slab (corresponding for example to the $I \times K$ matrix of the jth wavelength) has to be multiplied by the same scalar. Thus, scaling is performed *within the elements* of one mode.

Main rules for centering and scaling

Already at this point, it is useful to have an overview of the centering and scaling properties. A number of results are therefore listed here. These results are derived and substantiated in the following sections.

CENTERING

- Centering deals with the structural model because it implies that offsets are present additional to the bi- or tri-linear part of the model.
- Centering is a part of a two-stage procedure in which offsets are removed first and the multilinear terms are estimated in the second stage. This is only equivalent to the one-stage procedure of estimating all parameters simultaneously, if proper centering is used.
- Proper centering means that averages are subtracted from columns of the array. Such centering is called centering *across* a mode.
- For offsets that cannot be removed using proper centering, a one-stage procedure has to be used. This holds generally for data with missing elements.
- Centering can be performed across several modes sequentially.

SCALING

- Scaling only deals with the way the model is fitted and not the structure of it.
- Scaling provides a way to change the objective function by assuming certain weights. Some weight arrangements can be dealt with by scaling followed by ordinary least squares fitting. Only proper scaling is allowed.
- Proper scaling means multiplying all occurrences of, for example, a variable by the same scalar and is called scaling *within* a mode. For weighting schemes that cannot be dealt with by proper scaling, weighted least squares algorithms have to be used.
- Scaling can be performed across several modes but this has to be performed iteratively if scaling to a certain mean sum of squares within several modes is desired. Scaling to unit standard deviation within several modes is generally not possible.

CENTERING AND SCALING

- Incorrect centering or scaling introduces artificial variation. The amount of artificial variation introduced depends on the data and leads to models that are suboptimal to their 'correct' (least squares) counterparts. This is so, because the artificial variation has to be modeled additionally.
- Scaling does not affect centering across other modes, but centering affects scaling within all modes.

9.2 Two-way Centering

In order to understand when and how centering works, it is important first to consider the goals of centering and to realize how these goals are achieved in practice. These concepts are described in this section.

Reasons for centering

As stated by Harshman and Lundy [1984], quite subjective and qualitative reasons are often given for performing centering. Sometimes, centering is used to *estimate* offsets, but this aspect of centering is not considered here. Rather, the focus is on *removing* offsets.

It is possible to formulate rational reasons for centering on scientific grounds. Basically, centering should be performed only if there are common offsets in the data or if removing such offsets provides an approximately reasonable model. Centering should make a difference and this difference can manifest itself as follows:

REDUCED RANK OF THE MODEL (PARSIMONY)

If a component model of the raw data requires, say, $R + 1$ components to describe the data well, whereas a model of the centered data requires only R components, then centering is sensible because the model of the centered data only has $R(I + J) + J$ parameters. The J parameters pertain to the calculated averages, assuming that centering is performed across the first mode. The alternative of fitting the $(R + 1)$-component model to the raw data would lead to a model with $(R + 1)(I + J)$ parameters and thus would violate the parsimony principle [Judge *et al.* 1985, Seasholtz & Kowalski 1993, Weinberg 1964].

INCREASED FIT TO THE DATA

In some situations, the rank of the appropriate model is not reduced upon centering, but if the fit of the model of the centered data is significantly improved, then naturally introducing extra parameters is useful. It is possible to heuristically consider the offsets introduced by centering as one extra 'half' component of which either the scores (centering across first mode) or the loadings (centering across second mode) are known a priori to be equal to one. This holds in the sense that the fit of a model with R components is poorer than the fit of a model with R components *and* offsets, which again is poorer than a model with $R + 1$ components.

AVOIDANCE OF NUMERICAL PROBLEMS

In certain algorithms, it may be useful to center the data in order to minimize algorithmic problems. For fitting a bilinear model using principal component analysis, it is known that the ratio of the two largest eigenvalues is related to the convergence rate of the power method (and related techniques such as NIPALS). For PARAFAC it is also known that if some components are strongly correlated as evidenced through Tucker's congruence coefficient [Tucker 1951] then the fitting procedure may be complicated by so-called swamps. For both situations, it holds that centering across certain modes can be helpful in minimizing the cause of the problem because the resulting optimization problem is related to a different model with different (and hopefully better) properties with respect to numerical problems.

How centering works

If the vector **m** holds in its jth element the average of the jth column of **X**, then **m** can be expressed as

$$\mathbf{m} = I^{-1}\mathbf{X}'\mathbf{1} \Leftrightarrow \mathbf{m}' = I^{-1}\mathbf{1}'\mathbf{X} \qquad (9.11)$$

where **1** is an I-vector of ones and **X** has size $I \times J$. Then centering **X** across the first mode (column-centering) amounts to

$$\mathbf{Z} = \mathbf{X} - \mathbf{1m}' \qquad (9.12)$$

where $\mathbf{Z}$ $(I \times J)$ contains the centered data. As

$$\mathbf{1m}' = I^{-1}\mathbf{11}'\mathbf{X} \qquad (9.13)$$

the centered data can also be expressed

$$\mathbf{Z} = \mathbf{X} - \mathbf{1m}' = \mathbf{X} - I^{-1}\mathbf{11}'\mathbf{X} = (\mathbf{I} - I^{-1}\mathbf{11}')\mathbf{X} = (\mathbf{I} - \mathbf{P})\mathbf{X} = \mathbf{P}^{\perp}\mathbf{X} \qquad (9.14)$$

where $\mathbf{PX} = \mathbf{P}[\mathbf{x}_1, \ldots, \mathbf{x}_J] = [\mathbf{Px}_1, \ldots, \mathbf{Px}_J]$ and $\mathbf{x}_j$ is the jth column of $\mathbf{X}$. The matrix $I^{-1}\mathbf{11}' = \mathbf{P}$ is a symmetric and idempotent $(I \times I)$ matrix and is thus an (orthogonal) projection matrix (see Appendix 9.A). This shows that the column-averages are the orthogonal projections of the columns of $\mathbf{X}$ onto the direction of ones, i.e., the direction given by the vector $\mathbf{1}$. The centering matrix $(\mathbf{I} - I^{-1}\mathbf{11}') = \mathbf{P}^{\perp}$ is the projection matrix onto the nullspace of $\mathbf{1}'$ which equals range$(\mathbf{1})^{\perp}$ (where range$(\cdot)$ is the range of a matrix). Stated otherwise, centering may be interpreted as providing the residuals after regressing the columns of $\mathbf{X}$ onto $\mathbf{1}$. It follows, that centering may be viewed as the projection of the data onto a space with the common offset (given by the I-vector $\mathbf{1}$) removed.

Mathematically, centering is a projection onto the nullspace of $\mathbf{1}'$ and it is worthwhile to keep this in mind. Suppose that the true model of the data contains offsets across the first mode. Then the model can be written as

$$\mathbf{X} = \mathbf{\Phi}\mathbf{\Theta}' + \mathbf{1}\mathbf{\mu}' + \mathbf{E} \qquad (9.15)$$

Projecting these data onto the nullspace of $\mathbf{1}'$ leads to

$$\mathbf{P}^{\perp}\mathbf{X} = \mathbf{P}^{\perp}\mathbf{\Phi}\mathbf{\Theta}' + \mathbf{P}^{\perp}\mathbf{1}\mathbf{\mu}' + \mathbf{P}^{\perp}\mathbf{E} \qquad (9.16)$$

$$\Rightarrow \mathbf{P}^{\perp}\mathbf{X} = \mathbf{Z} = \mathbf{P}^{\perp}\mathbf{\Phi}\mathbf{\Theta}' + \mathbf{P}^{\perp}\mathbf{E} \qquad (9.17)$$

where $\mathbf{P}^{\perp}\mathbf{X} = \mathbf{Z}$ is the matrix holding the centered data and where the matrix $\mathbf{P}^{\perp}\mathbf{1}\mathbf{\mu}'$ vanishes, as $\mathbf{1}$ has no residuals when projected onto itself. The part $\mathbf{P}^{\perp}\mathbf{\Phi}\,\mathbf{\Theta}'$ is a bilinear model with scores $\mathbf{P}^{\perp}\mathbf{\Phi}$ and loadings $\mathbf{\Theta}$. Thus instead of fitting the bilinear model *and* the offsets to the original data, it is only necessary to fit the bilinear model to the centered data, $\mathbf{Z}$ with true structure, $\mathbf{P}^{\perp}\mathbf{\Phi}\mathbf{\Theta}' + \mathbf{P}^{\perp}\mathbf{E}$. Centering also leads to models with centered errors $(\mathbf{P}^{\perp}\mathbf{E})$ because these are also projected onto the nullspace of $\mathbf{1}'$, as is clear from Equation (9.17).

If $\mathbf{\Phi}$ and $\mathbf{\Theta}$ are nonnegative, then nonnegativity constraints can be imposed on the model for $\mathbf{X}$. When $\mathbf{X}$ is centered, e.g., across the first mode, such a constraint is not meaningful for $\mathbf{\Phi}$ anymore, because centering destroys nonnegativity of $\mathbf{\Phi}$ (but not of $\mathbf{\Theta}$). A geometrical view of centering is provided in Appendix 9.B.

As mentioned earlier, centering across a given mode is called single-centering. Single-centering an array across one mode that has previously been single-centered across another mode is called double-centering. Performing several single-centerings (for multi-way arrays as many can be performed as the number of ways) is unproblematic in the sense that centering across one mode, leaves the 'centeredness' intact in other modes [Harshman & Lundy 1984]. Further, the order of centering is immaterial. This means that if the data are first centered across the first mode, and subsequently the centered data are centered across the second mode, then the average of every column and every row will be zero. This follows, because centering across the first mode can be written as $\mathbf{P}_I^{\perp}\mathbf{X}$ where $\mathbf{P}_I^{\perp}$ is the centering operator for the first mode. Centering across the second mode can be written $\mathbf{X}\mathbf{P}_J^{\perp}$. Thus double-centering can be written $\mathbf{P}_I^{\perp}\mathbf{X}\mathbf{P}_J^{\perp}$, and hence: (i) the order of which centering is performed is immaterial and (ii) the double-centered array will have both column- and row-average zero, because $\mathbf{P}_I^{\perp}\mathbf{X}\mathbf{P}_J^{\perp}$ can be viewed as centering the matrix $\mathbf{X}\mathbf{P}_J^{\perp}$ across the first mode or centering the matrix $\mathbf{P}_I^{\perp}\mathbf{X}$ across the second mode.

Centering is a two-stage procedure for a least squares fitting problem

It is instructive to see that centering is merely a computational shortcut for fitting a specified model. Consider a two-way data set, which is generated as

$$\mathbf{X} = \mathbf{\Phi}\mathbf{\Theta}' + \mathbf{1}\mathbf{\mu}' + \mathbf{E} \tag{9.18}$$

where $\mathbf{\Phi}$ is $(I \times R)$, $\mathbf{\Theta}$ $(J \times R)$ and $\mathbf{\mu}$ $(J \times 1)$. It follows, that the data can be modeled by a bilinear model plus a common offset for each variable/column plus additional unmodeled variation held in the residual matrix $\mathbf{E}$. The least squares loss function for the above model is

$$\|\mathbf{X} - (\mathbf{A}\mathbf{B}' + \mathbf{1}\mathbf{n}')\|^2 \tag{9.19}$$

and this function is to be minimized directly over $\mathbf{A}$, $\mathbf{B}$, and $\mathbf{n}$, with $\mathbf{A}$, $\mathbf{B}$ and $\mathbf{n}$ being of the same dimensions as $\mathbf{\Phi}$, $\mathbf{\Theta}$ and $\mathbf{\mu}$ respectively. The matrices $\mathbf{A}$, $\mathbf{B}$, and $\mathbf{n}$ contain estimates of $\mathbf{\Phi}$, $\mathbf{\Theta}$, and $\mathbf{\mu}$ respectively, but it is intrinsic to the problem, that these estimates do not uniquely recover the underlying parameters. For example, $\mathbf{\Phi}$ and $\mathbf{\Theta}$ can, at most, be found up to a rotation.

As shown above the parameters can be estimated in two steps. Centering the data across the first mode will remove the offsets $\mathbf{\mu}$, and the bilinear model is subsequently fitted to the centered data $\mathbf{Z}$, thus minimizing the loss function

$$\|\mathbf{Z} - \mathbf{C}\mathbf{D}'\|^2 \tag{9.20}$$

only over $\mathbf{C}$ and $\mathbf{D}$ that are of the same dimensions as $\mathbf{A}$ and $\mathbf{B}$ above. It holds that

$$\min\|\mathbf{X} - (\mathbf{A}\mathbf{B}' + \mathbf{1}\mathbf{n}')\|^2 = \min\|\mathbf{Z} - \mathbf{C}\mathbf{D}'\|^2 \tag{9.21}$$

i.e. the fit of the model fitted directly and the bilinear model fitted to the centered data will be exactly the same (see proof in Appendix 9.C) even though the actual parameters will usually differ. This is an important result because it guarantees the optimality of the model even if the offsets are calculated separately from the bilinear parameters.

The solution for minimizing Equation (9.19) directly is not unique for $\mathbf{n}$. That is, the two-stage solution of centering first and then fitting the bilinear part gives one solution of many to the problem in Equation (9.19). Therefore, centering removes $\mathbf{\mu}$, but the averages are not necessarily estimates of the elements of $\mathbf{\mu}$. This nonuniqueness is explained in short. Centering involves subtracting from each column its column-average. The matrix

$$\mathbf{P}\mathbf{X} = \mathbf{P}\mathbf{\Phi}\mathbf{\Theta}' + \mathbf{P}\mathbf{1}\mathbf{\mu}' + \mathbf{P}\mathbf{E} \tag{9.22}$$

holds in each row the vector $\mathbf{m}'$ containing the average value of each column of $\mathbf{X}$ ($\mathbf{P}$ is

$11'/I$, as above). If the part $\mathbf{P\Phi\Theta'}$ is a matrix of zeros, then $\mathbf{m}$ will be an estimate of the true offsets μ (with error $\mathbf{PE}$), because

$$\mathbf{P1}\mu' = \mathbf{1}\mu' \tag{9.23}$$

by construction. However, assuming that $\mathbf{\Phi\Theta'}$ is nonzero and $\mathbf{\Theta'}$ is full rank, $\mathbf{P\Phi\Theta'}$ will only be a zero-matrix if $\mathbf{\Phi}$ is orthogonal to $\mathbf{P}$. That is, $\mathbf{P\Phi} = \mathbf{0}$. Thus, the offsets will only equal the true offsets if the column-space of $\mathbf{\Phi}$ is orthogonal to $\mathbf{1}$, usually meaning that $\mathbf{\Phi}$ is centered already (see also Appendix 9.D on rank reduction and centering).

When centering does not work

Viewing centering as a projection step rather than as a simple subtraction of averages has more than theoretical importance. In practice, situations often occur where subtraction of averages does not work and may, in fact, lead to models that fit the original data more poorly than if the data had not been preprocessed. This will be shown for two different problems: handling missing data and modeling a single common offset.

When data are missing, centering by subtracting averages from columns or rows does not lead to elimination of offsets. Rather, the offsets have to be eliminated simultaneously with the fitting of the bilinear part [Bro & Smilde 2003, Grung & Manne 1998]. This is so because the equivalence between subtracting average values and projecting onto the nullspace of vectors of ones no longer holds as the projection cannot be calculated with missing elements (see example in Appendix 9.E). Centering is an extension of the bi- (or multi-) linear model where offsets are assumed to be present in the model of the data. Data with missing elements constitute one situation in which such a model cannot be fitted in a least squares sense using centering. An alternative to eliminating offsets by preprocessing is given in Appendix 9.G.

The traditional centering across the first mode easily leads to the belief that subtracting averages with the same structure as the offsets will generally eliminate these offsets. This holds for offsets constant across one mode but it does not hold in general as elaborated on by Bro & Smilde [2003] and also described in Appendix 9.F. Some possible solutions can be seen in Appendix 9.G.

9.3 Two-way scaling

Unlike centering, scaling does not change the structure of the model. Scaling is used to change the importance attached to different parts of the data in fitting the model.

Because a least squares model is based on describing variation, it holds that variables with large variation are implicitly assumed to be important. However, if the variation is primarily due to noise or due to the use of a different scale (e.g. 1000 mg instead of 1 g) then it is necessary to adjust the algorithm such that this irrelevantly large variation does not influence the model more than necessary. Scaling provides a simple preprocessing step for adjusting for such matters (see Example 9.1). As will be described later, the impression of scaling as a preprocessing step is a little misleading. It is an algorithmic issue pertaining to how the model is fitted.

EXAMPLE 9.1

Considerations in scaling

Table 9.1. Rainwater data, some examples of the raw data. The concentrations are in ppb (mg/l). Of the 22 samples, only four are given

Volume (ml)	57	51	66	82
pH	4.35	4.75	4.00	4.75
SO_4^{2-}	3700	3000	2800	1500
NH^{4+}	315	240	20	110
Cl	2680	180	1680	900
NO	460	355	900	300
Na	2000	280	1800	230
K	140	170	210	240
Ca	550	200	250	105
Mg	190	55	160	50
Zn	28	45	9	6
Cd	0.45	0.28	0.25	0.67
As	1.80	0.90	3.80	0.60
Sr	0.93	0.08	0.46	0.08
Cu	29	15	40	13
Pb	58	20	10	3
Mn	5.7	7.1	4.0	4.7

An example using analysis of a simple two-way array will show some of the issues of concern in linear scaling. The rainwater data is a classical example given by Knudson *et al.* [1977]. The rainwater samples were collected during a thunderstorm at 22 locations in Seattle. During the measurement, a copper smelter was in use, so part of the study is done in order to follow the trace metals from the smelter to the sampling sites. The variables measured are: volume, pH, SO_4^{2-}, NH^{4+}, Cl^-, NO^{3-}, Na, K, Ca, Mg, Zn, Cd, As, Sr, Cu, Pb, Mn. The concentrations are given in ppb (mg/l). The data are shown in part in Table 9.1. The goal is to identify the sources of the ions, to find the stoichiometry of the ions related to the sources and to find contributions of the sources to the samples.

A first observation is that the volume is different in nature from the other variables. Some questions raised are: Should this volume be considered at all? Should it be used to compensate for dilution effects? These questions are not answered here but the answer depends on the goal of the study. A second observation is that the pH is not a concentration. Should it be made into a H^+ concentration or not? It is easily done but whether it is appropriate is not always obvious. A third problem is typical for chemistry or rather stoichiometry. The concentrations in ppb are not in stoichiometric form. To get real chemical meaning they may have to be expressed in molarity, molality or normality values. In this way the balance between cations and anions may be studied. Again, the choice is not always easy. For dilute samples, molarity and molality are close to each other, but normality may be determined by the type of reaction under consideration.

After having taken all these questions into consideration, the analysis to be done determines further treatment of the data. For principal component analysis e.g., there might be a problem regarding scaling. The data expressed in ppb or molarity are very unbalanced. The balance can be restored by using the standard deviations for scaling each variable. Note, that the stoichiometry information is not lost upon such a scaling. As discussed later, even though the scaled data are expressed in a new domain (scaled variables), this is really just an intermediate domain used for fitting the model. The model parameters are subsequently transformed back to the original domain in which the stoichiometry holds.

A two-way analysis shows that without centering across samples and scaling within variables (auto-scaling) there is no contribution to the model from the trace metals with their low concentrations, and without a stoichiometric scaling, leaving out centering, the ion balances of certain important sources are not really found. Thus, different ways of looking at the data may be needed for different purposes.

Scaling is usually done by some objectively calculated scaling factor: a standard deviation, an interquartile distance, a range or a robustified range. It is also possible to give subjective weights to certain variables. If, for instance, 800 spectral variables are used in conjunction with one pressure variable, it is likely that the pressure variable will not have a chance of influencing the model. Because the number of spectral variables is so high, and each contributes with variance to the data, the model will focus on explaining the spectral variables. However, by block-scaling the spectral variables so that their total variation is similar to the total variation of the single pressure variable, both blocks of data are given an equal chance of entering the model.

Reasons for scaling

Scaling is used for several reasons. Some important ones are

(i)   To adjust scale differences;
(ii) To accommodate for heteroscedasticity;
(iii) To allow for different sizes of subsets of data (block-scaling).

Regarding (i): it is quite common to use, for instance, autoscaling (centering across the first mode and scaling to unit standard deviation within the second mode) to let the variance of each variable be identical initially. Thereby, all variables have the same variance, and as the subsequent fitting of a model is performed so as to describe as much systematic variation as possible, every variable has the same initial opportunity of entering the model. This type of scaling is especially useful when the variables are measured in different measurement units (e.g. Pascals, Degrees, Celsius, . . .). Regarding (ii): the ordinary least squares fitting of a model is statistically optimal in a maximum likelihood sense, if the errors are homoscedastic, independent and Gaussian. If the variances of the distributions are not the same, though the same within, e.g., a specific variable, it is possible to accommodate the fitting procedure accordingly by initially scaling the data within the variable mode. By scaling each variable with the inverse of the standard deviation of the residual variance, the fitted model will be optimal in a maximum likelihood sense. Regarding (iii): when the data are made up of several subsets of very different sizes, it may sometimes be advantageous to scale each block separately in order to ensure that all the different blocks are allowed to influence the model as explained earlier. Thus, in this case scaling is used from an information point of view to ensure that important information can enter the model, irrespective of the variance of the different sources of information.

All the above-mentioned reasons can be put under the same heading by the term weighted least squares fitting, which is a general and broader approach to fitting models than merely the use of scaling. This will be elaborated on in the next section.

How scaling works

Scaling is a subject often treated in conjunction with centering. However, the purpose of scaling is very different from the purpose of centering. Scaling is a way of introducing another loss function than the least squares loss function normally used. As for centering, scaling has to be performed in a specific way in order not to introduce artificial structure that needs to be modeled. This becomes even more apparent when going to three-way models.

Scaling is usually performed by multiplying each column or each row in the data matrix by a scalar. There are two types of scaling that are relevant for two-way matrices. One is scaling within the first mode where every row is multiplied by a specific number

$$\mathbf{Z} = \mathbf{WX} \tag{9.24}$$

where $\mathbf{W}$ is an $I \times I$ diagonal matrix with the scaling parameter for the ith row on its ith diagonal element. This is the type of scaling used, for example, in standard normal variate correction [Barnes *et al.* 1989, Guo *et al.* 1999, Helland *et al.* 1995] where the norm or area of each row of $\mathbf{X}$ is scaled to the same scalar value by using an appropriate $\mathbf{W}$.

The other main type of scaling is within the second mode where every column is multiplied by a specific number

$$\mathbf{Z} = \mathbf{XW} \tag{9.25}$$

where $\mathbf{W}$ is here a $J \times J$ diagonal matrix with the scaling parameter for the jth column on its jth diagonal element.

An example of a scaling according to Equation (9.25) is shown in Figure 9.7. One-component bilinear data with huge random residual variation in the last half of the variables (upper left) is generated. The resulting $\mathbf{X}$ matrix has the spectra in its rows. The first principal component is seen to correlate well with the reference score generating the data (lower left). However, when the data are initially weighted as in Equation (9.25) by the inverse of the residual standard deviation (upper right), the right part of the data is downweighted substantially. The correlation between the true known score ('concentration') and the estimated score of the scaled data is even higher (lower right) than for the raw data (left). Even in this extreme case, the influence of the noise is not at all as drastic as might be expected. This illustrates that scaling is not as critical as centering as long as the scaling is reasonable and the variables relevant. Note that the scaling in this example is not the usual autoscaling using the inverse standard deviation of the data. Rather, the inverse standard deviation of the residual variation, for instance, assessed by replicates, is used (see Elaboration 9.2).

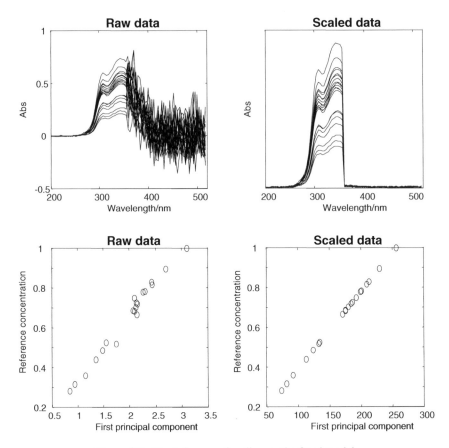

Figure 9.7. The influence of scaling on the fitted model.

ELABORATION **9.2**

Scaling and weighted least squares fitting

Given a data matrix $\mathbf{X}$ ($I \times J$) and a model $\hat{\mathbf{X}}$ ($I \times J$) which may, for example, be a bilinear model ($\hat{\mathbf{X}} = \mathbf{AB}'$), a standard approach for determining the model and its parameters is to fit the model in a least squares sense by minimizing the loss function $\|\mathbf{X} - \hat{\mathbf{X}}\|^2$ which can also be expressed $\sum_{i=1}^{I} \sum_{j=1}^{J} (x_{ij} - \hat{x}_{ij})^2$. When the different elements of the data have different uncertainties or relevances, it is possible to fit the model using a weighted least squares loss function. In one of its simplest form this can be expressed $\|(\mathbf{X} - \hat{\mathbf{X}}) * \tilde{\mathbf{W}}\|^2$ where $\tilde{\mathbf{W}}$ ($I \times J$) holds in its ijth element the weight of the ijth element of $\mathbf{X}$, and $*$ is the Hadamard (elementwise) product. Often the weight of an element is set equal to the inverse of the standard deviation of the residual variation. It is also possible to use more elaborate weights, if there are certain correlations between the residuals [Bro *et al.* 2002, Martens *et al.* 2003, Paatero & Tapper 1994, Wentzell & Lohnes 1999]. The above weighted loss function can also be expressed in scalar notation as $\sum_{i=1}^{I} \sum_{j=1}^{J} (x_{ij} - \hat{x}_{ij})^2 \tilde{w}_{ij}^2$.

In a maximum-likelihood sense, this loss function is optimal if the weights reflect the uncertainty of the individual elements, if there is no correlation between the residuals of different elements, and if the residuals are normally distributed. If the uncertainty of a given variable is the same over all objects, the above model turns into $\|(\mathbf{X} - \hat{\mathbf{X}})^* \tilde{\mathbf{W}}\|^2 = \|(\mathbf{X} - \hat{\mathbf{X}})\mathbf{W}\|^2$ where the weight matrix, $\mathbf{W}$, is now a diagonal matrix which holds the column-specific weights in the diagonal. The loss function may be transformed as follows $\|(\mathbf{X} - \hat{\mathbf{X}})\mathbf{W}\|^2 = \|\mathbf{XW} - \hat{\mathbf{X}}\mathbf{W}\|^2$.

In case of a bilinear model the above reads $\|(\mathbf{X} - \mathbf{AB}')\mathbf{W}\|^2 = \|\mathbf{XW} - \mathbf{AB}'\mathbf{W}\|^2 = \|\mathbf{XW} - \mathbf{AH}'\|^2$. Thus, by fitting the bilinear model $\mathbf{AH}'$ to the data scaled within the second mode, $\mathbf{XW}$, *in an ordinary least squares sense*, the weighted loss function is automatically optimized. This is the basic mathematical rationale behind scaling.

If the sought model, $\hat{\mathbf{X}}$, has the structure $\mathbf{AB}'$ then fitting a bilinear model to the scaled data provides the model in the form $\mathbf{AH}'$. Thus, the score matrix (or a rotated version of it) is directly provided in $\mathbf{A}$ whereas the loadings of the original problem are found by premultiplying the found loadings, $\mathbf{H}$, with $\mathbf{W}^{-1}$, as $\mathbf{B}'\mathbf{W} = \mathbf{H}' \Rightarrow \mathbf{B}' = \mathbf{H}'\mathbf{W}^{-1} \Rightarrow \mathbf{B} = \mathbf{W}^{-1}\mathbf{H}$.

Sometimes, only the scaled data are considered and model parameters are not transformed back to the original domain. The appropriateness of this approach is still, however, governed by the fact that scaling as outlined above maintains the bilinear structure assumed reasonable for the raw data.

There is a direct connection between the minimizing solution to $\|(\mathbf{X} - \mathbf{AB}')\mathbf{W}\|^2$ and $\|\mathbf{XW} - \mathbf{AH}'\|^2$. However, there is no direct connection between the solution to $\|(\mathbf{X} - \mathbf{AB}')\mathbf{W}\|^2$ and $\|(\mathbf{X} - \mathbf{TP}')\|^2$ unless the model has perfect fit. That is, fitting the bilinear model to scaled and unscaled data are two different problems with no direct relation.

When scaling does not work

When scaling within several modes is desired, the situation is complicated, because scaling one mode affects the scale of the other mode. For example, scaling to a standard deviation of one within both the first and second mode will generally not be possible, not even using iterative scaling [Harshman & Lundy 1984]. If scaling to a mean square of one is desired within both modes, this has to be done iteratively, until convergence [Harshman & Lundy 1984, ten Berge 1989]. Using mean-squares rather than, for instance, standard deviations for scaling has the attractive property that iterative scaling is guaranteed to converge, in case no centering is included in the iterative scheme [Harshman & Lundy 1984].

Iterative preprocessing may seem unsatisfactory, because it tends to complicate the subsequent evaluation and validation of the preprocessing, since more than one set of scaling parameters for each mode has to be used. These several matrices holding the scaling parameters from each iteration may be combined, though, into one single matrix for each mode (see [Harshman & Lundy 1984] and Equation (9.37)).

Alternatives to scaling

As shown in the preceding paragraphs, scaling can be considered to be a special case of using a weighted least squares loss function. When more complicated weights are needed, it is not always possible to fit the model indirectly by fitting the least squares model to the scaled data. In such situations an alternative to scaling is to use algorithms that directly handle a weighted least squares optimization criterion [Bro 1998, Kiers 1997, Paatero 1997]. This can be relevant, for example, when the residual variation is correlated across both rows and columns [Andrews & Wentzell 1997, Bro *et al.* 2002, Wentzell *et al.* 1997].

9.4 Simultaneous Two-way Centering and Scaling

A complicating issue in preprocessing is the interdependence of centering and scaling [ten Berge & Kiers 1989]. Not all combinations of centering and scaling will work as anticipated (see e.g. [Harshman & Lundy 1984, Kruskal 1983]). Generally, only centering across both modes is straightforward or scaling within one mode combined with centering across the other mode, which is exactly what e.g. autoscaling amounts to.

Scaling disturbs centering across the same mode, but not other modes

Scaling within one mode disturbs prior centering across the same mode, but not across other modes [ten Berge 1989]. This holds for two-way arrays as well as higher order arrays. The reason for this is illustrated in Figure 9.8. The vertical arrow shows a typical column vector and the horizontal line a typical row-vector. When scaling within the first mode, the elements of any column are multiplied by different numbers and hence prior centering across the first mode is destroyed.

Consider a two-way array $\mathbf{X}$ ($I \times J$). If the array is scaled within the first mode, this can be expressed

$$\mathbf{Z} = \mathbf{W}\mathbf{X} \tag{9.26}$$

where $\mathbf{Z}$ is the scaled array, $\mathbf{W}$ is an $I \times I$ diagonal matrix holding the scaling parameters and $\mathbf{X}$ is the original array. As can be seen, scaling within the first mode amounts to a multiplication of every row by a scalar. This does not affect any centering of the vectors across the second mode, because every element in a row-vector is multiplied by the same number. The average of any row will be the original average of the row scaled down accordingly, and therefore, if the average is zero, it will stay zero. In the first mode, however, each element in a column is multiplied by a different scalar. If centering is performed across the first mode, these column-vectors will not necessarily preserve their zero average after subsequent scaling within the first mode. Mathematically, the centered matrix $\mathbf{P}^{\perp}\mathbf{X}$ becomes

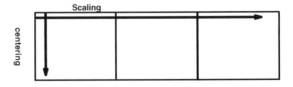

Figure 9.8. A two-way array showing the relation between centering and scaling.

$\mathbf{WP}^\perp\mathbf{X}$ upon scaling. As $\mathbf{1}'\mathbf{WP}^\perp\mathbf{X} \neq \mathbf{0}$, the preprocessed matrix is no longer guaranteed to be centered. Offsets constant across the first mode, however, *will* still be removed, because

$$\mathbf{P}^\perp(\mathbf{\Phi\Theta}' + \mathbf{1}_{\mathrm{I}}\mu') = \mathbf{P}^\perp\mathbf{\Phi\Theta}' \Rightarrow \mathbf{WP}^\perp(\mathbf{\Phi\Theta}' + \mathbf{1}_{\mathrm{I}}\mu') = \mathbf{WP}^\perp\mathbf{\Phi\Theta}' \qquad (9.27)$$

Note also the interesting fact that if scaling is performed before centering, the result will be different. In that case, the original offsets will not be removed, but the data will be centered (yielding centered scores and residuals), because

$$\mathbf{P}^\perp\mathbf{W}(\mathbf{\Phi\Theta}' + \mathbf{1}_{\mathrm{I}}\mu') = \mathbf{P}^\perp\mathbf{W\Phi\Theta}' + \mathbf{P}^\perp\mathbf{W1}_{\mathrm{I}}\mu' \neq \mathbf{P}^\perp\mathbf{W\Phi\Theta}' \qquad (9.28)$$

This holds, because unlike for $\mathbf{P}^\perp\mathbf{1}$, it does not hold that $\mathbf{P}^\perp\mathbf{W1}$ is a zero vector in general.

Thus, the order of preprocessing has to be chosen depending on the purpose of the analysis. For example, for decomposition models, removal of offsets is typically important in which case centering should be performed before scaling. In regression it may be more important that the scores are centered and hence, centering should be performed after scaling.

Centering across one mode disturbs scaling within all modes

Centering across one mode disturbs scaling within all modes. This holds for two- as well as multi-way arrays, but there are certain cases for which it does not hold. One of these special cases is the situation in which two-way data are scaled within the second mode to a standard deviation of one and subsequently centered across the first mode (autoscaling). This subsequent centering will not disturb the scaling within the second mode (though it would disturb scaling within the first mode, had that been performed). The reason is that the scaling is specifically performed relative to the center of the columns (standard deviations are based on centered data). Hence, any change in offset is immaterial for the standard deviation. Scaling by other means than standard deviations will not have this property. In multi-way analysis it is common to use mean squares for scaling instead of standard deviations because such scaling more often converges when implemented in an iterative scheme and because scaling by standard deviations implicitly assumes an offset, which may or may not be present depending on the structural part of the model.

9.5 Three-way preprocessing

The basic properties of preprocessing are unchanged for three-way arrays. Centering has to be performed across a specific mode and scaling has to be performed by a transformation within a specific mode. Most difficulties in preprocessing three-way arrays arise because of the problems outlined so far, which all generalize to multi-way arrays. The problems are sometimes enhanced, because three-way data are often rearranged (matricized) to two-way arrays before preprocessing. This is unfortunate, because it introduces a column-mode that is a combination of two of the original modes. Transformation within or across this combined mode should be avoided if multiway models are to be fitted, because the mode is not a 'real' mode, but merely a computational construct.

Centering

The observations on centering of two-way data are helpful in discussing centering of three- and higher-way arrays. If the basis of two-way centering is understood, then three-way centering is quite simple. In the following, it will be assumed that the underlying structure of the data follows a PARAFAC model plus possible offsets, but the conclusions hold for any multilinear model.

Consider a three-way array. Conceptually, offsets may occur in three different ways. Constant across all modes (Figure 9.9), constant across two modes (Figure 9.10) or constant across one mode only, as shown in Figure 9.11. In the figures, the first-mode loadings are held in the $I \times R$ matrix Φ, the second-mode loadings are held in the $J \times R$ matrix Θ and the third-mode loadings are held in the $K \times R$ matrix Ω. The matrix $\mathbf{D}_k$ is an $R \times R$ diagonal matrix holding the kth row of Ω in its diagonal.

Regardless of the structure of the offsets, the basic principle of centering is that the data must be preprocessed, so that they are projected onto the nullspace of vectors of ones in a

All elements have the same offset

$$X_k = \Phi D_k \Theta' + \lambda\, 1_I 1_J'$$

$$X_{(I \times JK)} = \Phi(\Omega \circ \Theta)' + \lambda\, 1_I (1_K \circ 1_J)'$$

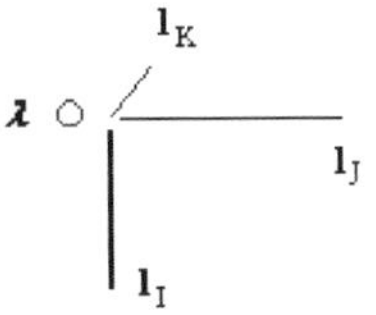

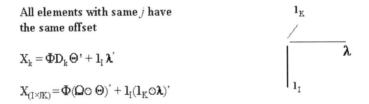

Figure 9.9. Structure of offsets in three-way (trilinear) data when all elements have the same offset held in the scalar λ. Two alternative but equivalent ways of writing the PARAFAC model are shown: the slab notation using submatrices X_k and the matricized notation.

All elements with same j have the same offset

$$X_k = \Phi D_k \Theta' + 1_I \lambda'$$

$$X_{(I \times JK)} = \Phi(\Omega \circ \Theta)' + 1_I (1_K \circ \lambda)'$$

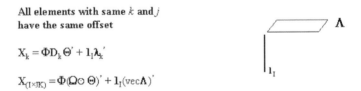

Figure 9.10. Structure of offsets in three-way (trilinear) data when all elements in each vertical slab have the same offset. The offsets are held in the J-vector λ.

All elements with same k and j have the same offset

$$X_k = \Phi D_k \Theta' + 1_I \lambda_k'$$

$$X_{(I \times JK)} = \Phi(\Omega \circ \Theta)' + 1_I (\mathrm{vec}\Lambda)'$$

Figure 9.11. Structure of offsets in three-way (trilinear) data when all elements in each vector have the same offset (case three). The offsets are held in a matrix Λ ($J \times K$) whose jkth element holds the offset of the vertical column with second and third mode index j and k. The vector λ_k is the kth row of the matrix Λ.

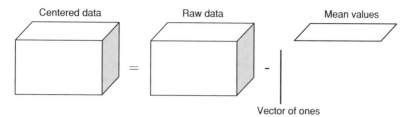

Figure 9.12. Centering across the first mode. For each column of the raw data, a mean value is calculated and subtracted from each element in the column. Thus, a two-way matrix of mean values is obtained. This centering is identical to matricizing the data to a two-way structure and centering these two-way data across the first mode.

particular mode. For the first mode, projecting a data array $\underline{\mathbf{X}}$ onto the nullspace of $\mathbf{1}'$, i.e., centering across the first mode amounts to

$$\mathbf{Z}_{(I \times JK)} = \mathbf{P}^{\perp} \mathbf{X}_{(I \times JK)}, \tag{9.29}$$

where $\mathbf{P}^{\perp} = \mathbf{I} - (\mathbf{11}'/I)$. For the second and third modes, the centering can be performed similarly. As mentioned earlier, such centering is referred to as *single-centering*. Centering, for example, across the first mode of an array can thus be done by matricizing the array to an $I \times JK$ matrix, and then centering this matrix across the first mode as in ordinary two-way analysis:

$$z_{ijk} = x_{ijk} - \frac{\sum\limits_{i=1}^{I} x_{ijk}}{I} \tag{9.30}$$

The column-mean is subtracted from every element, as depicted graphically in Figure 9.12. As can be seen, single-centering is similar in structure to the type of offsets shown in Figure 9.11. Such single-centerings performed successively across two modes, are referred to as double-centering. That is, double-centering is performed by first centering across one mode, and then centering the outcome across another mode. The order of centerings is immaterial, but it is essential that they are performed sequentially. For all three situations depicted in Figures 9.9–9.11 centering across the first mode will remove the shown offsets, because the offsets are constant across the first mode.

As for the two-way case, *only* single-centering (possibly several such) leads to the properties sought in centering (removal of offsets). Other types of centering, such as subtracting the overall mean, will introduce artifacts that have to be modeled additionally to the inherent systematic variation. Similar incorrect centerings are often used in three-way analysis. For example, matricized data are centered across a combined mode. For example, if an array of structure (variables × time × samples) is centered by subtracting the average calculated across samples *and* time, then in line with the above example, artificial offsets are introduced and the subsequent model will have to fit this additional variation as well. This can obscure validation and exploration of the model and will lead to models that do not provide overall least squares solutions.

Scaling

As explained for the two-way case, scaling is a transformation of a particular variable (or object) space. Instead of fitting the model to the original data, the model is fitted to the data transformed by a (usually) diagonal scaling matrix in the mode whose variables are to be scaled. This means that whole matrices instead of columns have to be scaled by the same value in three-way analysis. For a four-way array, three-way slabs would have to be scaled by the same scalar. Mathematically, scaling *within the first mode* can be described as

$$z_{ijk} = w_i x_{ijk}, \quad i = 1, \ldots, I; j = 1, \ldots, J; k = 1, \ldots, K \tag{9.31}$$

where $\underline{Z}$ with elements z_{ijk} is the scaled array and, for instance, setting

$$w_i = \frac{1}{\sqrt{\sum_{j=1}^{J} \sum_{k=1}^{K} \frac{x_{ijk}^2}{JK}}} \tag{9.32}$$

will scale to a unit mean square within the sample mode. Using matricized arrays, scaling may be expressed as

$$\mathbf{Z}_{(I \times JK)} = \mathbf{W} \mathbf{X}_{(I \times JK)} \tag{9.33}$$

where $\mathbf{W}$ is an $I \times I$ diagonal matrix holding the scaling values in its diagonal. As for the two-way case, even non-diagonal weight matrices can be applied, but this is outside the scope here. If the assumed structural model of the data is a multilinear model, e.g. a PARAFAC model $\hat{\mathbf{X}} = \mathbf{A}(\mathbf{C} \odot \mathbf{B})'$, then with the above scaling, a similar structural model, $\hat{\mathbf{Z}} = \mathbf{P}(\mathbf{R} \odot \mathbf{Q})'$, will hold for the transformed data., because

$$\|\mathbf{Z} - \mathbf{P}(\mathbf{R} \odot \mathbf{Q})'\|^2 = \|\mathbf{W}\mathbf{X} - \mathbf{P}(\mathbf{R} \odot \mathbf{Q})'\|^2 = \|\mathbf{W}(\mathbf{X} - \mathbf{W}^{-1}\mathbf{P}(\mathbf{R} \odot \mathbf{Q}'))\|^2 \tag{9.34}$$

The loss function minimized when fitting a model with a weighted criterion is

$$\|\mathbf{W}(\mathbf{X} - \mathbf{A}(\mathbf{C} \odot \mathbf{B})')\|^2 \tag{9.35}$$

and hence it holds that the parameters of this model can be found by fitting the scaled data and setting

$$\mathbf{A} = \mathbf{W}^{-1}\mathbf{P}, \mathbf{B} = \mathbf{Q}, \text{ and } \mathbf{C} = \mathbf{R} \tag{9.36}$$

Thus, fitting the scaled data in a least squares sense provides a solution not only to the problem posed as fitting a model to $\mathbf{Z}$, but to the problem of fitting a model to $\mathbf{X}$ where the first mode loadings obtained are transformed by the scaling matrix $\mathbf{W}$. As for two-way scaling, it is emphasized that the model parameters found have no direct relations to the parameters found when fitting the model to the raw data in a least squares sense although $\mathbf{B}$ and $\mathbf{C}$ will be estimates of the same parameters in the two situations.

Scaling has to be applied by transforming the data within a given mode. It is not appropriate to scale an array within two combined modes which can happen, for example, when autoscaling a matricized array. Such an inappropriate scaling will lead to the inclusion of artificial components in the data (Example 9.2).

Incorrect scaling of three-way arrays

To show the influence of incorrect scaling, consider a synthetic data set with PARAFAC structure $\mathbf{X}_{(I \times JK)} = \mathbf{A}(\mathbf{C} \odot \mathbf{B})'$, where $\mathbf{A}$ is a 4×2 matrix of random numbers and $\mathbf{B}$ and $\mathbf{C}$ are defined likewise. It is not important how these matrices are generated, as long as they have full column rank. Consider the following alternative two-component PARAFAC models:

(1) Using $\underline{\mathbf{X}}$;
(2) Using $\underline{\mathbf{X}}$ centered across the first mode and scaled within the combined second and third mode (autoscaled as a two-way matrix, $\mathbf{X}_{(I \times JK)}$);
(3) Using $\underline{\mathbf{X}}$ scaled within, e.g., mode two.

The fit values of these three models always using two components are given below in percentages of the sum-of-squares of the preprocessed data:

(1) 100.00%;
(2) 98.80%;
(3) 100.00%.

As can be readily seen, a two-component model is appropriate and should be so, even after scaling as in case (3). However, using ordinary two-way scaling methods as in case (2) destroys the multilinear structure of the data and deteriorates the model.

Simultaneous centering and scaling

The exact same rules for interdependence of preprocessing steps apply for multi-way data as for two-way data (Section 9.4), also with respect to treating missing data etc. Any preprocessed array may be written in matrix notation using the matricized $I \times JK$ data array $\mathbf{X}$ and the preprocessed array $\mathbf{Z}$

$$\mathbf{Z} = \mathbf{M}_I \mathbf{X} (\mathbf{M}_K \otimes \mathbf{M}_J)' \qquad (9.37)$$

where, for example, $\mathbf{M}_I$ is an $I \times I$ array holding either the centering or scaling transformation matrix for the first mode or even a combination of such. The exact content of these transformation matrices depends on the type of preprocessing chosen and in case of iterative preprocessing $\mathbf{M}$ may be a product of several matrices [Harshman & Lundy 1984].

Combined centering and scaling in one operator $\mathbf{M}$ is generally not going to retain all the properties of the two individual operations (see Section 9.4). For example, scaling within the second and third mode together with centering across the first mode is problematic. If the preprocessing is performed so that the data are centered after the weights are determined (iteratively) and applied

$$\mathbf{Z} = \mathbf{P}^{\perp}(\mathbf{X}(\mathbf{W}_K \otimes \mathbf{W}_J)') \qquad (9.38)$$

then the centering operation will destroy the property of e.g. suitable mean-squared error in the second and third mode (the brackets indicate the proper order of the preprocessing

Centering across the second mode

$- m_{ik}$

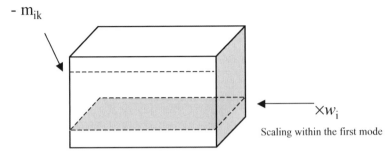

$\times w_i$

Scaling within the first mode

Figure 9.13. Three-way array. Proper centering must be done across a mode, exemplified here by proper centering across the second mode. From all elements of a specific row (fixed i and k) the same scalar m_{ik} is subtracted. Proper scaling, e.g., within the first mode, is performed such that all elements of a specific horizontal slab are multiplied by the same scalar w_i.

steps). If, on the other hand, the preprocessing is performed as

$$\mathbf{Z} = (\mathbf{P}^{\perp}\mathbf{X})(\mathbf{W}_K \otimes \mathbf{W}_J)' \tag{9.39}$$

this is not the case. In this case, the data are first centered and then the weights are determined from the centered array rather than from the raw data. Hence, the weights in Equation (9.39) are usually preferred.

9.6 Summary

A number of important features of the common preprocessing steps centering and scaling have been discussed.

- Centering deals with the structural model; scaling with the way this model is fitted.
- Centering is a part of a two-stage procedure in which offsets are removed first and the multilinear terms are estimated in the second stage. This is only equivalent to the one-stage procedure of estimating all parameters simultaneously, if proper centering, as defined here, is used. Proper centering is shown in Figure 9.13 (always across one mode at a time).
- Additionally to removing offsets, centering also leads to centered components (and residuals), but this property has been shown not to be linked intrinsically to removal of offsets.
- Centering, e.g. across the first mode, will also remove offsets that are constant over all modes.
- For offsets that cannot be removed by using proper centering, a one-stage procedure has to be used. This holds generally for data with missing elements.
- Scaling provides a way to change the objective function by assuming certain weights. Some weight arrangements can be dealt with by scaling followed by ordinary least squares fitting. Only proper scaling is allowed. Proper scaling is shown in Figure 9.13. For weighting schemes that cannot be dealt with by scaling, weighted least squares algorithms have to be used.
- Scaling can be viewed as applying a certain weighted least squares loss function or alternatively as standardizing variables or objects to a certain well-defined variability

level (typically unit mean square or unit standard deviation). These two views have been shown not to coincide in all situations.

- Incorrect centering or scaling introduces artificial variation. The amount of artificial variation introduced depends on the data and leads to models that are suboptimal to their 'correct' (least squares) counterparts. This is so, because the artificial variation has to be modeled additionally.

Two-way results

- Proper centering can always be written as $\mathbf{P}^{\perp}\mathbf{X}$.
- Several centerings can be performed sequentially.
- Proper scaling can always be expressed as $\mathbf{WX}$.
- Several scalings can be performed sequentially, but will generally need iterations to establish the scaling constants and this may not converge.
- Unproblematic combinations of centering and scaling can be expressed as $(\mathbf{P}^{\perp}\mathbf{X})\mathbf{W}$.

Similar results hold for transposed matrices.

Three-way results

- Proper centering can always be written $\mathbf{P}^{\perp}\mathbf{X}$ where $\mathbf{X}$ is the three-way array matricized, so that the mode to be centered across is the first mode.
- Several such single-centerings may be performed sequentially across several modes.
- Proper scaling can always be expressed $\mathbf{WX}$ for a matricized array as above.
- Several scalings can be performed sequentially, but will generally need iterations and may not converge.
- Proper combinations of centering and scaling can be expressed similarly to the two-way case. That is, scaling does not affect centering across other modes, but centering affects scaling within all modes.

The appropriate centering and scaling procedures can most easily be summarized as in Figure 9.13. Centering must be done by subtracting scalars from individual vectors of the array, while scaling must be performed by multiplying individual slabs.

Not much attention has been paid to regression models in this chapter. However, the rules and guidelines for centering and scaling in a multi-way regression problem are basically the same as for decomposition models. The only important difference is that centering across the first mode plays an especially significant role in regression. Such a centering is usually performed to remove offsets in the *relation* between the scores of the dependent and independent data. Thus, apart from the guidelines given above, it is necessary to consider that centering the first mode to eliminate offsets should be given first priority in analogy to two-way regression.

Appendix 9.A: Other Types of Preprocessing

This section describes transformations of the data that are different from scaling and mean-centering. They are often referred to as nonlinear transformations. There are not many practical applications of these transformations in the three-way literature yet, but their use is expected to grow in the future and they are a major element in two-way analysis. The transformations are very diverse in nature and it can be difficult to obtain an overview of the possibilities. However, the transformations are usually applied to solve some problem-specific issues and hence, a good experience with the data-type can help guide the selection. Furthermore, in many cases, the quality of the transformation can be assessed by means of assessing the quality of the model with respect to its basic purpose.

With instrumental measurements, all available data have undergone some type of signal processing. This processing is usually hidden in the instrument construction or algorithms. Many of the transformations explained below can also be considered as signal processing. The signal is corrected for known problem sources and the corrected data are used in the three-way analysis.

In environmental and clinical studies, it is often popular to take logarithms of the data in order to avoid very skew distributions of the modeled variables:

$$z_{ijk} = \ln(x_{ijk}) \tag{9.40}$$

A more general form of the logarithmic transform is to take an exponent α:

$$z_{ijk} = (x_{ijk})^{\alpha} \tag{9.41}$$

Convenient starting values for α are: $-1, 1/2, 2$. The exponent can be different for different variables. The difference made by a transformation as in Equation (9.41) is that the shape of the distribution, expressed as a histogram, is changed. Centering only moves the center of the distribution to zero, and scaling only changes the magnitude of the spread. A more advanced form of Equation (9.41) is the Box–Cox transformation, well known in experimental design [Box *et al.* 1978, Meyers & Montgomery 1995]. Xie *et al.* [1999] give examples where the lognormal distribution is assumed to be valid and a logarithmic transformation therefore needed. The authors use adaptive algorithms, so that the scaling is not separated from the calculation. The authors also present their resulting three-way components on a logarithmic scale. The strict use of the Box–Cox and similar transformations is for obtaining more normally distributed residuals, but in many cases the experience of the experimenter indicates that a more symmetrical distribution of the raw data by taking logarithms or exponents gives better results.

Transformations such as those described above are applied to each variable separately, without looking at neighboring variables, but for homogeneous variables, also transformations that look at a neighborhood of the value to be transformed are useful. For simplicity, the term spectra is used to signify an ensemble of homogeneous variables, even though the homogeneous data may be chromatographic, electrophoretic, electrochemical or time series.

In two-way analysis, it is often found useful to transform whole spectra when these are available. This is, for example, often done for spectral data of nonhomogeneous media: suspensions, colloids, emulsions and mixed solids. All organic and living material and many clinical samples belong to this category. A general term that has emerged for this type of applications is diffuse spectroscopy. A field where diffuse spectroscopy has received much attention is near infrared spectroscopy in the region 780–2500 nm. In this spectral region, organic and biological samples interact with the radiation, but the samples are not always clear and transparent. This leads to multiple-path interactions between the radiation and the material under study, making a direct interpretation of the spectral information hard. It has been shown in numerous cases that spectral transformations can simplify the data analysis and subsequent interpretation appreciably, in the analysis, classification and regression of two-way data. It is obvious that these transformations can also help when three-way analysis is intended.

The Multiplicative Scatter Correction (MSC) and Standard Normal Variate (SNV) transformations of whole spectra are used to remove baselines and sloping baselines that are often the result of differing powder size of the samples and have nothing to do with chemical composition [Barnes *et al.* 1989, Geladi *et al.* 1985, Isaksson & Kowalski 1993]. Derivation of spectra against wavelength is done in spectral windows. Figure 8.12 shows an example of the use of first derivative NIR spectra. A very popular derivation is the Savitzky–Golay [Savitzky & Golay 1964] smoothing derivation, but other forms have also been used [Hopkins 2001]. It was shown by Geladi & Åberg [2001] (see also Chapter 8) and Geladi & Forsström [2002] that first and second derivatives of NIR spectra may allow a PARAFAC decomposition when the model of the raw data does not give useful results. In Geladi & Forsström [2002], MSC and a Savitzky–Golay first derivative transformation was used in combination on the spectra in a $6 \times 40 \times 776$ data set. A four-component PARAFAC model explaining 99.2 % of the original sum of squares, made spectral interpretation of the PARAFAC loadings possible.

For regression and calibration work, it has been shown that transformation of the data is sometimes necessary and the above-mentioned transformations have been used frequently to build and improve calibration models using two-way data.

Other transformations of data with homogeneous variables (such as spectra) are the Fourier and wavelet transforms [Daubechies 1998, Qin & Shen 2000, Trygg *et al.* 2001]. There are two main ways of using these transforms. One is to simply use the transform to minimize the presence of noise in selected frequency regions by reconstruction of spectra from selected coefficients. Another use is to replace complete spectra by Fourier or wavelet coefficients and to build the data analysis or regression model on selected coefficients. Good overviews of data pretreatment methods are given in the literature [Beebe *et al.* 1998, Geladi 2002, Katsumoto *et al.* 2001].

Data can also be transformed by adding transformed variables to the untransformed ones. An example would be adding squared values of the variables in an $I \times K$ matrix to make it an $I \times 2K$ matrix. For three-way data this has not been used often but could be as beneficial as for two-way analysis.

Appendix 9.B: Geometric View of Centering

Let

$$\mathbf{X} = \begin{bmatrix} 3 & 1 \\ 1 & 3 \\ 5 & 2 \end{bmatrix} \tag{9.42}$$

The average of the first and second variable is three and two respectively and the centered data, $\mathbf{Z}$, are therefore given as

$$\mathbf{Z} = \begin{bmatrix} 0 & -1 \\ -2 & 1 \\ 2 & 0 \end{bmatrix} \tag{9.43}$$

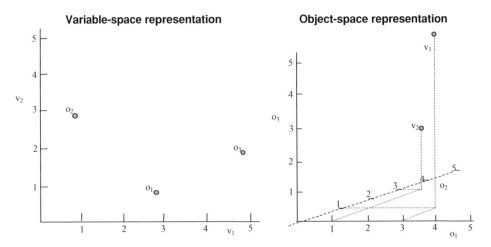

Figure 9.14. Variable- and object-space representation of the raw data given in Equation (9.42). o = objects and v = variables.

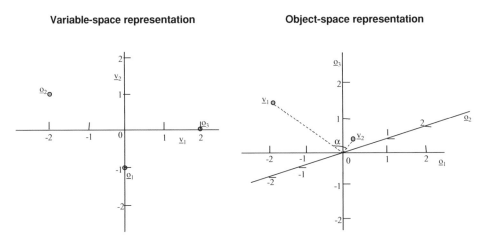

Figure 9.15. Variable- and object-space representation of the data given in Equation (9.43). Variables and objects are underlined to indicate the difference with the original ones

leading to a data set where the average of each column is zero. There are two ways of plotting the data. The objects may be plotted in a space where the variables constitute the axes or conversely the variables may be plotted in a space where the objects constitute the axes (Figure 9.14). For the variable-space representation of $\mathbf{X}$, the objects (rows of $\mathbf{X}$) are seen as points in the space. The Euclidean distances between different objects are reflected in the plot. For the object-space representation of $\mathbf{X}$ the variables are columns of $\mathbf{X}$ and are points in the space. The Euclidean distances between variables are also reflected in the plot.

When the data have been centered across the first mode the geometrical representation is given in Figure 9.15. In the variable-space representation of $\mathbf{Z}$ the coordinate system is translated, so the origin is now in the center of the data, but the distances between objects (rows) is kept intact. In the object-space representation of $\mathbf{Y}$ the variables are projected perpendicular to the line through $(1,1,1)'$ in Figure 9.14 (right) and the residuals of that projection gives $\underline{v}_1$ and $\underline{v}_2$. In this plot the distances between variables are changed, and now the cosine of the angle between the centered variables is also the correlation between variables because the common offset has been removed. Another way of saying it: the vector $(3,1,5)'$ is regressed on $(1,1,1)'$, and the residuals of that regression are the column-centered values. Hence, the common offset $(1,1,1)'$ is removed.

Appendix 9.C: Fitting Bilinear Model Plus Offsets Across One Mode Equals Fitting a Bilinear Model to Centered Data

Theorem

Given $\mathbf{X}$ of size $I \times J$ and the column-dimension R of a sought bilinear model. Then

$$\min\|\mathbf{X} - (\mathbf{AB}' + \mathbf{1m}')\|^2 = \min\|\mathbf{Z} - \mathbf{CD}'\|^2 \tag{9.44}$$

where $\mathbf{Z}$ is the original data, $\mathbf{X}$, with the column-averages subtracted.

Proof

The proof has been given by Kruskal [1977] and Gabriel [1978]. Understanding that centering is a projection, it is simple to prove the above theorem. Let the loss function be

$$\|\mathbf{X} - \mathbf{AB}' - \mathbf{1m}'\|^2 \tag{9.45}$$

and partition it into two orthogonal parts

$$\|\mathbf{P}^{\perp}(\mathbf{X} - \mathbf{AB}' - \mathbf{1m}')\|^2 + \|\mathbf{P}(\mathbf{X} - \mathbf{AB}' - \mathbf{1m}')\|^2 \tag{9.46}$$

using the Pythagorean fact that the squares of two orthogonal parts equal the square of the total. This equation can be further developed as

$$\|\mathbf{P}^{\perp}\mathbf{X} - \mathbf{P}^{\perp}\mathbf{AB}'\|^2 + \|\mathbf{PX} - \mathbf{P}(\mathbf{AB}' + \mathbf{1m}')\|^2 \tag{9.47}$$

And at the minimum this equals

$$\|\mathbf{P}^{\perp}\mathbf{X} - \mathbf{P}^{\perp}\mathbf{AB}'\|^2 \tag{9.48}$$

because $\|\mathbf{PX} - \mathbf{P}(\mathbf{AB}' + \mathbf{1m}')\|^2$ will be zero by setting

$$\mathbf{m}' = (\mathbf{1}'/I)(\mathbf{X} - \mathbf{AB}') \text{ since } \mathbf{1}(\mathbf{1}'/I) = \mathbf{P} \tag{9.49}$$

Setting $\mathbf{C} = \mathbf{P}^{\perp}\mathbf{A}$ and $\mathbf{D} = \mathbf{B}$ will therefore provide a solution with exactly the same fit as would be obtained by minimizing the original loss function. The solution may be computed using any bilinear algorithm for fitting a principal component analysis model of $\mathbf{Z}$. The scores will automatically be centered, because they are linear combinations of the columns of **X**. If the columns are centered, so are their linear combinations.

Appendix 9.D: Rank Reduction and Centering

In some cases centering reduces rank and in some cases not. Column centering of **X** $(I \times J)$ reduces the rank of **X** if and only if $1 \in$ range(**X**), where range(**X**) is the range of **X** (see Amrhein [1998] p. 156, and Amrhein *et al.* [1996]). Intuitively this is understandable. Centering is a projection. If the axis on which the projection takes place is a part of the range of **X**, then the residuals of this projection do not have this direction available anymore. Hence, the rank of the matrix of residuals lowers by one. This simple fact has several repercussions for centering across the first mode (column-centering).

The following can be said about the noiseless case. Suppose that **X** $(I \times J)$ is noiseless and can be modeled as

$$\mathbf{X} = \mathbf{\Phi}\mathbf{\Theta}' + \mathbf{1}\boldsymbol{\mu}' = [\mathbf{\Phi} \ \mathbf{1}] \begin{bmatrix} \mathbf{\Theta}' \\ \boldsymbol{\mu}' \end{bmatrix}$$

where $\mathbf{\Phi}$ is $(I \times R)$ of full rank; $\mathbf{\Theta}$ $(J \times R)$. Assume that

$$\begin{bmatrix} \mathbf{\Theta}' \\ \boldsymbol{\mu}' \end{bmatrix}$$

has full rank $R + 1$ which will be fulfilled in general. For **Z**, the column-centered **X**, two cases can be distinguished:

1. $\mathbf{1} \in$ range($\mathbf{\Phi}$) $\Rightarrow$ rank(**X**) $= R \Rightarrow$ rank(**Z**) $= R - 1$
2. $\mathbf{1} \notin$ range($\mathbf{\Phi}$) $\Rightarrow$ rank(**X**) $= R + 1 \Rightarrow$ rank(**Z**) $= R$

Hence, in both cases the rank of **X** is reduced by one. The reverse also holds: if for the noiseless case no rank reduction of **X** is obtained upon column-centering, then the above model is not valid. To summarize, in the noiseless case there is a simple relation between the validity of above model and rank reduction upon column-centering.

The situation with noise added is less simple. Suppose that **X** $(I \times J)$ also contains noise and the model for **X** is

$$\mathbf{X} = \mathbf{\Phi}\mathbf{\Theta}' + \mathbf{1}\boldsymbol{\mu}' + \mathbf{E} = [\mathbf{\Phi} \ \mathbf{1}] \begin{bmatrix} \mathbf{\Theta}' \\ \boldsymbol{\mu}' \end{bmatrix} + \mathbf{E}$$

If $I > J$, then in general $1 \notin$ range(**X**). Although $1 \in$ range('noiseless' **X**), this property is destroyed upon adding noise to **X**. In the case $I \leq J$ and if **X** has full rank I, then **1** is by definition in the range of **X**, whether or not there exists an offset term $\mathbf{1}\boldsymbol{\mu}'$. Hence, in the case with noise there is no simple relationship anymore between the validity of the above model and mathematical rank reduction upon column-centering.

Appendix 9.E: Centering Data with Missing Values

Consider the matrix $\mathbf{X}^{(1)}$ shown below:

$$\mathbf{X}^{(1)} = \begin{bmatrix} 10 & 20 \\ 7.5 & 15 \\ 1 & 2 \\ 1.5 & 3 \\ 0.5 & 1 \end{bmatrix}, \mathbf{Z}^{(1)} = \begin{bmatrix} 5.9 & 11.8 \\ 3.4 & 7.8 \\ -3.1 & -6.2 \\ -2.6 & -5.2 \\ -3.6 & -7.2 \end{bmatrix}, \mathbf{X}^{(2)} = \begin{bmatrix} ? & 20 \\ ? & 15 \\ 1 & 2 \\ 1.5 & 3 \\ 0.5 & 1 \end{bmatrix}, \mathbf{Z}^{(2)} = \begin{bmatrix} ? & 11.8 \\ ? & 7.8 \\ 0 & -6.2 \\ 0.5 & -5.2 \\ -0.5 & -7.2 \end{bmatrix}$$

This is a rank one matrix and will remain so, even after centering across the first mode. The averages of the two columns are 4.1 and 8.2 respectively and the centered matrix reads as $\mathbf{Z}^{(1)}$ which is also a rank one matrix.

Consider now a situation in which the first two elements in the first column are missing. The data then reads as $\mathbf{X}^{(2)}$. This data set is naturally still perfectly modeled by a rank one bilinear model, as no new information has been added. The average of each column is now 1 and 8.2 and subtracting these values leads to centered matrix $\mathbf{Z}^{(2)}$. This matrix *cannot* be described by a rank one model. This is easily realized by only looking at the last three rows. This is a rank two submatrix, and the addition of the first two rows cannot change this. Thus, by subtracting averages from the data with missing elements, the structure of the data has been destroyed and meaningful results cannot be expected. Prior centering no longer leads to elimination of the true offsets as centering ordinarily does.

Appendix 9.F: Incorrect Centering Introduces Artificial Variation

Consider a data set structured as a bilinear part plus a constant identical for all elements; that is, all elements have the same common offset, as also shown in Equation (9.4). It might seem natural to remove this offset by initially subtracting the grand mean m from the data. However, this will not simplify the subsequent modeling of the data, and in fact, it obscures interpretation of the model, because such a centering leads to artificial mathematical components that also need to be modeled.

To explain this, assume that $\mathbf{X}$ is perfectly described by a bilinear part plus a common offset

$$\mathbf{X} = \mathbf{\Phi}\mathbf{\Theta}' + \mathbf{1}_I \mathbf{1}_J' \mu \qquad (9.50)$$

Centering by removing the overall mean of all elements of $\mathbf{X}$ can be written

$$\mathbf{Z} = \mathbf{X} - \mathbf{P}_I \mathbf{X} \mathbf{P}_J \qquad (9.51)$$

where $\mathbf{P}_J$ is the projection matrix of $\mathbf{1}_J (= \mathbf{1}_J \mathbf{1}_J' / J)$ and $\mathbf{P}_I$ of $\mathbf{1}_I (= \mathbf{1}_I \mathbf{1}_I' / I)$. Then $\mathbf{P}_I \mathbf{X} \mathbf{P}_J$ is a matrix of the same size as $\mathbf{X}$ holding the overall average of $\mathbf{X}$ in all its elements. Inserting

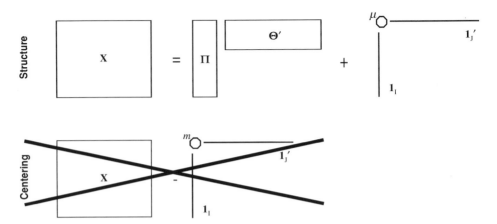

Figure 9.16. Hypothesized structure (top) of a data matrix, **X**, as a bilinear model plus offsets constant over all elements. Centering by subtracting the grand mean of **X** (bottom) will not remove the offsets from the data. The scalar m holds the grand average of **X**.

the true model of Equation (9.50) in Equation (9.51) leads to

$$
\begin{aligned}
\mathbf{Z} &= \mathbf{\Phi\Theta}' + \mathbf{1}_I\mathbf{1}_J'\mu - \mathbf{P}_I(\mathbf{\Phi\Theta}' + \mathbf{1}_I\mathbf{1}_J'\mu)\mathbf{P}_J \\
&= \mathbf{\Phi\Theta}' + \mathbf{1}_I\mathbf{1}_J'\mu - \mathbf{P}_I\mathbf{\Phi\Theta}'\mathbf{P}_J - \mathbf{P}_I\mathbf{1}_I\mathbf{1}_J'\mu\mathbf{P}_J \\
&= \mathbf{\Phi\Theta}' + \mathbf{1}_I\mathbf{1}_J'\mu - \mathbf{P}_I\mathbf{\Phi\Theta}'\mathbf{P}_J - \mathbf{1}_I\mathbf{1}_J'\mu \\
&= \mathbf{\Phi\Theta}' - \mathbf{P}_I\mathbf{\Phi\Theta}'\mathbf{P}_J \\
&= \mathbf{\Phi\Theta}' - \mathbf{1}_I\mathbf{1}_J's
\end{aligned}
\tag{9.52}
$$

The scalar s is the overall average of the true bilinear part $\mathbf{\Phi\Theta}'$. Even though the overall mean, μ, has been removed, a new common offset, s, has been introduced into the preprocessed data, and hence the same number of components is still necessary for modeling the data. Depending on the true parameters in the underlying model ($\mathbf{\Phi\Theta}'$), the model fitted to the preprocessed data may therefore explain less or more of the original data than if the data had not been preprocessed! Clearly, preprocessing the data by subtracting the overall mean is generally *not* useful.

As subtracting the overall level does not remove the offset, another approach must apparently be adopted for handling situations with one common offset. There are at least two different ways of treating the problem. The best way is to optimize the loss function of the problem directly in a one-step procedure, rather than trying to use a two-step procedure where the offsets are first removed.

Another simpler way of dealing with a constant offset follows from the observation that the model $\mathbf{X} = \mathbf{\Phi\Theta}' + \mathbf{1}_I\mathbf{1}_J'\mu$ may equivalently be written $\mathbf{X} = \mathbf{\Phi\Theta}' + \mathbf{1}_I\mu'$ where $\mu = \mathbf{1}_J\mu$. Posed in this way it is evident that a model with one global offset is a special case of the situation treated earlier where each variable (or sample) has a specific offset. Therefore, the constant offset may be eliminated by using ordinary centering across the first mode. As the offset is constant across rows, this centering removes the constant. An alternative procedure is to center across columns instead of rows, because the offset is also constant across columns.

EXAMPLE 9.3

Modeling data with grand mean offset

Table 9.2. Percentage of variation explained for different models of the raw (left) and corrected data (right)

#LV	Raw Data (%)	Overall Average Subtracted (%)
1	99.19	98.05
2	99.58	99.08
3	99.95	99.67
4	100.00	100.00
5	100.00	100.00

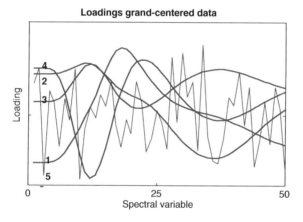

Loadings grand-centered data

Figure 9.17. Loading plot from principal component model of the data matrix where the overall average is subtracted.

An example is given for illustration. Consider a spectral data set. The data have been synthesized according to a bilinear three-component part plus a scalar offset of one. Thus, the model is $X = \Phi\Theta' + \mathbf{1}\mathbf{1}'$ and no noise is added. Using PCA to model the raw data, four components are needed for describing the variation as expected (Table 9.2 left). Modeling the data centered by subtracting the overall mean leads to a situation where four components still have to be used for describing all systematic variation (Table 9.2 right). In fact, the three-component model explains less of the original data after preprocessing in this case. If the true offset was known, subtracting this would lead to a simpler model as expected, but this true offset cannot be found from any average of the raw data. Even though only three systematic components should be present, the loading plot (Figure 9.17) clearly shows that the first four components are 'spectral'-like. With proper preprocessing only three loading vectors will be systematic.

Single-centering involves fitting several parameters (I or J respectively). When there is only one constant parameter in the true model, as is the case here, a risk of overfitting is introduced with this approach. It is advisable, therefore, to center across the mode of largest dimension such that as few offsets as possible need to be estimated.

Centering across one mode removes offsets constant across that mode *as well as* offsets constant across both modes [Harshman & Lundy 1984]. This important rule also extends to multiway data of arbitrary order. Thus, centering across one mode removes offsets constant across that mode *as well as* offsets constant across *several* modes involving that mode. This generalization follows from realizing that multiway models can always be considered as a constrained version of a bilinear model. Hence, offsets constant across an arbitrary number of modes can always be considered a constrained version of a model with offsets constant across one mode. Centering across one of these modes will therefore eliminate the offsets because of the projection properties of centering.

Appendix 9.G: Alternatives to Centering

Instead of modeling the data in two steps – removing offsets by centering and fitting a model to the residuals – it is possible to fit the model in one step, alleviating the need for projecting the offsets away. Two examples are given.

The example of missing data (Appendix 9.E) can be fitted directly in the following way. Assume that the offsets are, for instance, constant across the first mode and that a principal component analysis model is sought including offsets across the first mode. Such a PCA model of the data held in the matrix $\mathbf{X}$ including offsets reads

$$\mathbf{X} = \mathbf{\Phi\Theta}' + \mathbf{1\mu}' + \mathbf{E} \tag{9.53}$$

where $\mathbf{\mu}$ is a J-vector. A very simple way to fit this model to a data matrix with missing elements in a least squares sense, is by the use of an alternating least squares approach where the missing elements are continuously exchanged with their model estimates. Such an algorithm may proceed as follows.

1. Initialize missing values with reasonable values. Then the data-set is complete and can be modeled by ordinary techniques.
2. Fit the model including offsets to the (now complete) data set. For PCA, this amounts to centering the data and fitting the PCA model.
3. Exchange missing values in the data matrix with the current model estimates. These estimates will improve the current estimates and thus provide a data set where the estimated missing elements are closer to the correct ones according to the (yet unknown) true least squares model.
4. Proceed from step 2 until convergence.

This approach can be shown to converge, because it is an alternating least squares algorithm and hence has a nonincreasing loss function. Upon convergence to the global minimum, the imputed missing data will have no residuals and hence no influence on the model. The model parameters computed from the complete data are exactly those that would have been obtained, had the model only been fitted to the nonmissing data directly. This approach can be viewed as a simple special case of expectation maximization [Little & Rubin 1987].

For specific models or specific offsets, other approaches can also be feasible, but the above approach is general and easily implemented.

The problem with one common offset (Appendix 9.F) can be dealt with in the following way. The loss function for a bilinear model with constant overall offset is expressed as

$$\|\mathbf{X} - \mathbf{AB}' - \mathbf{1}_I \mathbf{1}'_J m\|^2 \tag{9.54}$$

Instead of fitting the over-parameterized model

$$\mathbf{X} = \mathbf{AB}' + \mathbf{1m}' + \mathbf{E} \tag{9.55}$$

in a two-step procedure, it is possible to fit a 'correct' structural model

$$\mathbf{X} = \mathbf{AB}' + m\mathbf{11}' + \mathbf{E} \tag{9.56}$$

directly. Instead of J parameters in $\mathbf{m}$, only one parameter, m, has to be estimated. The loss function may be optimized in several ways leading to a least squares model [Cornelius *et al.* 1992, van Eeuwijk 1996]. A simple algorithm is based on alternating least squares where first the offset is set to an initial value. Then the structural model is fitted to the corrected data $\mathbf{X} - m\mathbf{11}'$ using an ordinary PCA algorithm in this case. This provides an update of the bilinear parameters. Subtracting the new interim PCA model from the data leads to

$$\mathbf{X} - \mathbf{AB}' = m\mathbf{11}' + \mathbf{E} \tag{9.57}$$

Therefore, m may be determined conditional on $\mathbf{A}$ and $\mathbf{B}$ as the overall average of $\mathbf{X} - \mathbf{AB}'$. By alternating between updating the loading parameters and the offset, the model parameters will be estimated upon convergence.

10
APPLICATIONS

10.1 Introduction

The proof of the pudding is in the eating. The previous nine chapters gave an overview of the detailed aspects of multi-way analysis as known at the present time, but the more practically oriented user demands an integrated overview of how multi-way analysis is used. The last chapter of this book is a collection of explanatory examples from the literature where it is shown how active research scientists solve a problem from its definition to the final conclusions and how they use three-way analysis to do this. An attempt has been made to include examples of different important areas of chemical research where multi-way analysis is used. Each example uses selected aspects of what was explained in the previous chapters.

One of the oldest fields of use of multi-way analysis in chemistry is chromatography. Different chemicals, different solvents and different stationary phases may form the modes of a three-way array and the actual measurements to be modeled can be retention times (Figure 10.1).

Another development, also with a history in chemistry, is second-order calibration, where rank annihilation was developed for analyzing data from typically hyphenated instruments. This includes excitation–emission fluorescence spectra of different samples, liquid chromatography with ultraviolet (UV) detection for different samples and gas chromatography with mass spectrometric detection for different samples, giving an array. An illustration is given in Figure 10.2.

Process chemometrics is a field where multi-way methods were introduced more recently. A batch process gives rise to an array of batches followed over time and measured by multiple process sensors such as temperature and pressure sensors or perhaps spectroscopy, as in Figure 10.3. Multi-way component models are used for the construction of control charts and multi-way regression is used for constructing predictive calibration models.

Multi-way analysis has also been used in environmental studies where many chemicals are measured in different locations over time, giving an array of the type (chemical × location × time), as in Figure 10.4. This has been used in the calculation of emission profiles and source contributions.

Multi-way Analysis With Applications in the Chemical Sciences.   A. Smilde, R. Bro and P. Geladi
© 2004 John Wiley & Sons, Ltd   ISBN: 0-471-98691-7

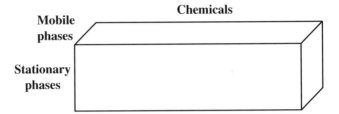

Figure 10.1. Example of three-way array of retention times as a function of different stationary phases, mobile phases and different chemicals.

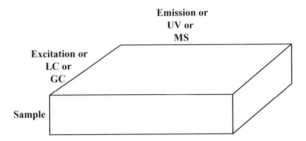

Figure 10.2. Example of typical structure of data for second-order calibration. LC, liquid chromatography; GC, gas chromatography; UV, ultraviolet; MS, mass spectrometry.

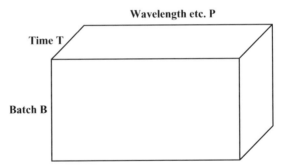

Figure 10.3. Typical batch process three-way data. For every batch, B, and at every time point, T, a set of variables (e.g. a spectrum) is measured.

There are also situations where two of the modes of the three-way array can be expressed as qualitative variables in a two-way ANOVA and where the response measured is some continuous (spectral, chromatogram, histogram or otherwise) variable. Calling the ANOVA factors treatment A and treatment B a (Treatment × Treatment × Variable) array is made, see Figure 10.5. An example for particle size distribution in peat slurries was explained in Chapters 7 and 8. The two-way ANOVA can become a three-way ANOVA in treatments A, B and C, leading to a four-way array etc. Having three factors and only one response per experiment will lead to a three-way structure and examples will be given on how these can be modeled efficiently even when the factors are continuous.

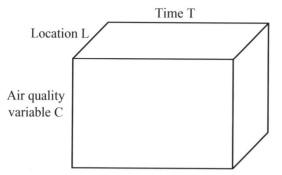

Figure 10.4. Typical organization of environmental data. For example, concentrations of air quality variables as a function of location and time.

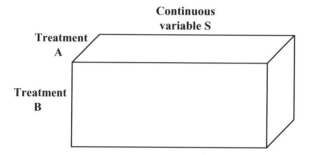

Figure 10.5. Example of the structure of data from an experimental design with two factors.

The common setup for the examples will be: (i) a short introduction into the field, (ii) the background of the example, (iii) building a three-way model, (iv) using the model, and (v) summary and some comments including a short overview of the application of multi-way analysis in the particular field.

10.2 Curve Resolution of Fluorescence Data

Using fluorescence spectroscopy

Fluorescence spectroscopy is often used in analytical chemistry, food analysis, environmental analysis etc. It is a very sensitive spectroscopic technique which can be performed nondestructively and provides qualitative and quantitative information of diverse types of chemical analytes [Andersson & Arndal 1999, Aubourg *et al.* 1998, Beltran *et al.* 1998a, Bright 1995, Bro 1999, Ferreira *et al.* 1995, Guiteras *et al.* 1998, Jensen *et al.* 1989, Jiji *et al.* 2000, Ross *et al.* 1991, Wolfbeis & Leiner 1985]. This application explains an example of estimating relative concentrations and pure analyte spectra from fluorescence measurements of chemical analytes in mixtures. Similar problems also arise frequently in other types of spectroscopy, in chromatography and other areas. Several names are used for this problem: unmixing, curve resolution, source separation etc. Specifically, the application illustrates the following issues

- Preparing data for three-way analysis;
- Selecting the number of components;
- Detecting and handling irrelevant variables;
- Detecting and handling outliers;
- Validating the number of components in the model;
- Characterizing the quality of the model;
- Using the model on new data.

Curve resolution is the goal and as fluorescence data are known to approximately follow a PARAFAC model (see next), only PARAFAC models will be investigated. Hence, in this application, there will not be any investigation as to which model structure is preferred.

PARAFAC AND FLUORESCENCE SPECTROSCOPY

In this particular problem, a set of 27 samples containing different amounts of L-phenylalanine, L-3,4-dihydroxyphenylalanine (DOPA), 1,4-dihydroxybenzene and L-tryptophan have been measured by fluorescence spectroscopy [Baunsgaard 1999, Riu & Bro 2002]. The data are important as a model system for many biological systems, e.g. in food and environmental analysis. The goal here is to develop a PARAFAC model of the measured data because a PARAFAC model will ideally resolve the pure spectra as well as the relative concentrations of the analytes [Bro 1997, Bro 1998, Bro 1999, Leurgans *et al.* 1993, Leurgans & Ross 1992, Ross & Leurgans 1995].

In fluorescence excitation–emission spectroscopy, each sample is measured by exciting the sample at several wavelengths and measuring the emitted light at several wavelengths. The result of such a measurement is an excitation–emission matrix (EEM). A typical example is given in Figure 10.6. The fluorescence intensity reflecting the number of photons emitted is plotted versus the excitation and emission wavelengths and the resulting landscape is a function of the amounts and types of fluorophores in the sample. Theoretically, the systematic chemical part of the EEM can be described in a simplified way as follows. For a measurement of one sample with F fluorescing analytes measured at emission wavelength j and excitation wavelength k the intensity x_{jk} is

$$x_{jk} = \sum_{f=1}^{F} a_f b_{jf} c_{kf}, \quad j = 1, \ldots, J; k = 1, \ldots, K; \tag{10.1}$$

The parameter a_f is linear in the concentration of analyte f and includes the quantum yield of the analyte. The parameter b_{jf} is specific to analyte f and expresses the fraction of photons emitted at wavelength j. Finally, the parameter c_{kf} holds the intensity of the incident light at excitation wavelength k multiplied by the absorption at wavelength k.

This model of fluorescence data is based on the assumption that the total absorbance is small and that there is no energy transfer between analytes [Ewing 1985]. If the model is extended to several (I) samples, it is identical to a PARAFAC model

$$x_{ijk} = \sum_{f=1}^{F} a_{if} b_{jf} c_{kf}, \quad i = 1, \ldots, I; j = 1, \ldots, J; k = 1, \ldots, K; \tag{10.2}$$

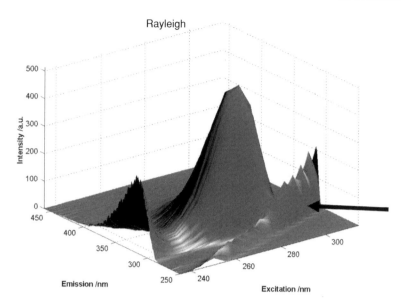

Figure 10.6. A typical excitation–emission landscape. Note the diagonal Rayleigh scatter ridge to the right at the diagonal where excitation and emission wavelengths are equal. (a.u., arbitrary units).

where x_{ijk} is now the intensity of sample i at emission wavelength j and excitation wavelength k and where a_{if} is the relative concentration of analyte f in sample i.

Background of the example

PREPARING THE DATA FOR ANALYSIS

Samples were prepared from stock solutions according to the design in Table 10.1. The design is made to serve specific purposes (pure samples, high concentrations etc.) and does not follow a classical design. Fluorescence excitation-emission landscapes were obtained of the 27 samples using a Perkin-Elmer LS50 B fluorescence spectrometer with excitation wavelengths between 200 and 350 nm and emission wavelength range between 200 and 450 nm. The samples 5, 10 and 15 were arbitrarily chosen to be validation samples whose concentrations will be predicted using a PARAFAC model built on the remaining samples.

A typical sample is shown in Figure 10.7. The measurements between the Rayleigh scatter ridge and its reflection contain the important chemical information used in the analysis. The data have parts that do not reflect the fluorescence of the analytes. Emission below excitation does not exhibit any fluorescence and the intensity is simply zero. If not removed, this part of the data can bias the estimated parameters towards zero [Andersen & Bro 2002]. There are several ways to deal with these nonchemical areas. In this case, these elements are simply set to missing [Bro *et al.* 2001, Bro 1997, Bro & Heimdal 1996, Heimdal *et al.* 1997, Jiji *et al.* 2000, Wentzell *et al.* 2001]. In addition to first order Rayleigh scatter other sources of scatter also occur. For these data a minor banded ridge of solvent related Raman scatter is observed for some of the samples. This type of scattering is less

Table 10.1. Concentrations of four fluorophores in 27 samples (10^{-6} M). Samples 5, 10 and 15 are not part of the initial model

Sample Number	Dihydroxybenzene	Tryptophan	Phenylalanine	DOPA
1	0.0	0.0	0.0	55.0
2	0.0	0.0	0.0	220.0
3	275.0	0.0	0.0	0.0
4	0.0	25.0	0.0	0.0
5	46.0	4.0	2800.0	18.0
6	17.0	2.0	4700.0	28.0
7	20.0	1.0	3200.0	8.0
8	10.0	4.0	3200.0	16.0
9	6.0	2.0	2800.0	28.0
10	0.0	0.0	5600.0	0.0
11	0.0	8.0	0.0	0.0
12	56.0	0.0	0.0	0.0
13	28.0	0.0	0.0	0.0
14	0.0	0.0	0.0	5.0
15	0.0	0.0	700.0	0.0
16	0.0	16.0	0.0	0.0
17	3.5	1.0	350.0	20.0
18	3.5	0.5	175.0	20.0
19	3.5	0.3	700.0	10.0
20	1.8	4.0	1400.0	5.0
21	0.9	2.0	700.0	2.5
22	28.0	8.0	700.0	40.0
23	28.0	8.0	350.0	20.0
24	14.0	8.0	175.0	20.0
25	0.9	8.0	1400.0	2.5
26	1.8	8.0	700.0	5.0
27	3.5	2.0	700.0	80.0

problematic in the sense that it is usually of minor magnitude and can be almost completely removed by subtracting measurements of pure solvent.

With these considerations, it is now possible to perform initial analysis on the data. All excitation spectra up to emission at 250 nm are removed to avoid low-information emission variables with too many missing data. This leaves a calibration set of size $24 \times 210 \times 24$.

Building a three-way model

PREPROCESSING

There are no expected offsets in the data (zero signal means zero analyte) hence centering is not expected to be useful. As all measurements are of similar type and quality, scaling is also not warranted at least not initially. More thorough analysis may point to some use of scaling but this is not pursued here [Appellof & Davidson 1981, Bro *et al.* 2001].

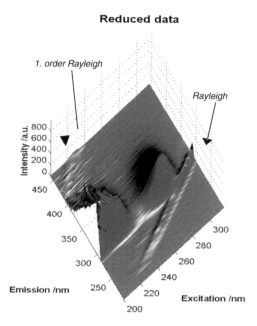

Figure 10.7. A full excitation–emission landscape as obtained from the fluorescence instrument. The Rayleigh ridge and its first order reflection are clearly seen as diagonal ridges. The Rayleigh scatter ridge to the right is at the diagonal where excitation and emission wavelengths are equal.

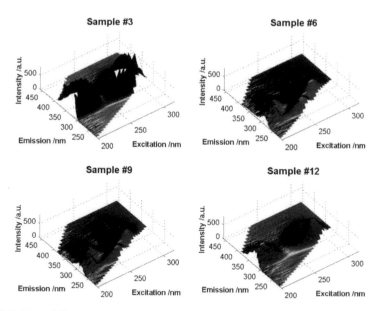

Figure 10.8. Four different samples after removal of nonchemical information. Notice the out-of-range signals in sample three which are treated as missing data in this analysis.

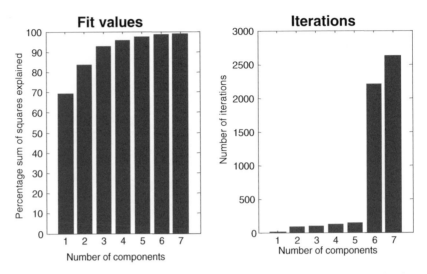

Figure 10.9. Fit values (left) as a function of the number of components. To the right, the corresponding number of iterations of the ALS algorithm is shown.

SELECTION OF NUMBER OF COMPONENTS

Although it is known from the chemistry of the problem that four PARAFAC components are likely to be appropriate, the following analysis will proceed as if the number of components were not known beforehand for didactical reasons. Initially, PARAFAC models with one to seven components are fitted to the data. From these models it is hoped that it will be possible to determine the approximate number of components to use as well as to assess whether all variables are appropriate to keep in the model.

In Figure 10.9, the fit of the different models is shown together with the number of iterations used by the algorithm. A large increase in the number of iterations is often indicative for over-factoring. This is only a rule-of-thumb because other aspects (e.g. highly correlated components) can lead to many iterations and because over-factored models can also sometimes be easy to fit within few iterations (e.g. if minor nontrilinear structure in the residuals can be well approximated by a PARAFAC component). Still, ad hoc diagnostics such as the number of iterations used are usually instructive to include when evaluating different models. In this case, the development in the number of iterations seems to indicate that up to five components are feasible. The fit-values are also pointing towards a four- a five- or even a six-component model as it is known that close to 100 % variation must be explained given the noise level of these fluorescence data.

From these initial results, a four- and a five-component PARAFAC model was examined in more detail. The initial results from these indicated that there were some potential non-extreme outliers and that the low-excitation wavelengths were very noisy. This can be seen, for example, from the estimated five-component excitation loadings as well as from plots of the residuals (Figure 10.10). Apparently, the low excitations are not well modeled leading to noisy line shapes in that area and to relatively high residuals compared to the remaining part of the data. The reason for this noise in the data is likely to be found in the instrument.

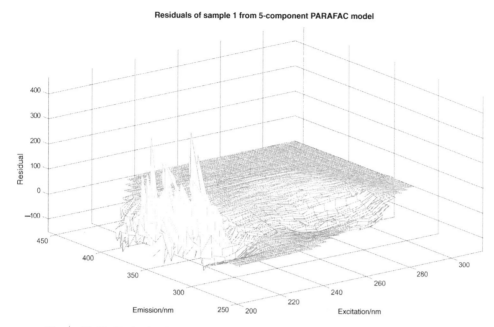

Figure 10.10. Typical residual of one sample from a five-component PARAFAC model.

The ultraviolet part of the data is highly influenced by the condition of the xenon lamp as well as by the physical environment [Baunsgaard 1999]. It was hence decided to leave out excitations from 200 to 230 nm. Correspondingly, emission below 260 nm was excluded because it mainly contained missing elements.

EXCLUDING LOW EXCITATION WAVELENGTHS

In Figure 10.11, a plot similar to Figure 10.9 is shown for the data where the low excitations have been removed. It is quite clear now that there are no more than four components in the data. After four components, the number of iterations increases considerably and the fit of the four-component model is close to 100 % as expected.

There is a significant difference between Figure 10.9 and Figure 10.11 illustrating that different conclusions can be reached at different steps in the data analysis. Before removing the noisy variables, it was difficult to build a reasonable model, but after proper selection of variables, the indications are much clearer.

At this point, it seems that a reasonable model complexity has been determined and it is useful to look more carefully for outliers and other details of the model. In the parameters of the model (Figure 10.12), it is seen that emission (Figure 10.12 (C)) and excitation (Figure 10.12 (D)) mode loadings seem to have a reasonable shape resembling pure spectra. In the Figure 10.12 (A), some samples have high relative scores, but looking at the corresponding concentrations in Table 10.1, the high scores are apparently in agreement with the nature of the data. Hence, there are no apparent gross outliers. This is also evidenced in the residual plot (Figure 10.12 (B)). The percentages explained variation in Figure 10.12 (C) are for a four-component model specifically. These percentages will change for other components.

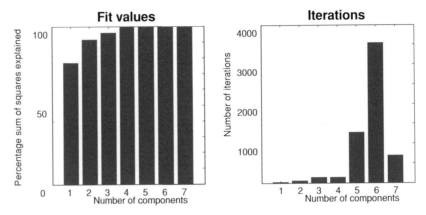

Figure 10.11. Similar to Figure 10.9 but with low excitations removed. Fit values (left) as a function of the number of components. To the right, the corresponding number of iterations of the ALS algorithm is shown.

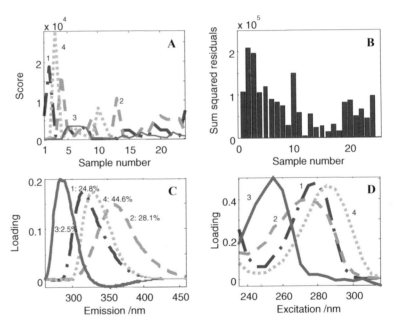

Figure 10.12. A four-component PARAFAC model on all calibration samples with low excitations removed. (A) Scores in sample mode. (B) Squared residuals summed across the first (sample) mode. (C) Emission mode loadings. (D) Excitation mode loadings. The percentage variation explained by each component is given in (C).

ASSESSING OUTLYING SAMPLES AND VARIABLES

At this point a more thorough residual and influence analysis is needed, e.g. to verify if more subtle outliers are present. In Figure 10.13, it is seen that for the four-component model, no samples have significant errors (relatively), but a few samples (2,3,4) have fairly high leverages. Given the high concentrations of these samples, it is likely due to slightly

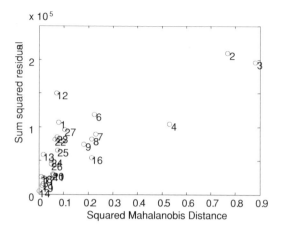

Figure 10.13. Influence plot of four-component PARAFAC model. The sum-squared residuals summed within the sample mode are plotted against the Squared Mahalanobis distance of the first mode components in A. Each sample is labeled by number.

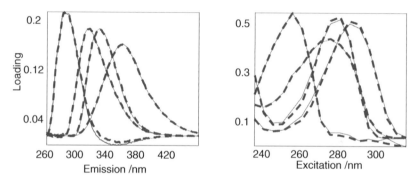

Figure 10.14. Emission and excitation loading plots from a four-component model with (thin solid lines) and without (thick dashed lines) samples 2, 3, and 4.

changed line-shapes of the spectra at the high concentrations (due to inner-filter effects and similar).

In order to check whether these high-leverage samples influence the model significantly, a new model is intermediately fitted without these samples. In Figure 10.14 the excitation and emission loadings obtained are compared to the model of all samples. The loadings with and without the potential outliers are almost completely overlapped and therefore, it can be concluded that the high-leverage samples merely have high leverages due to their high concentrations and not because they behave fundamentally different from the remaining samples. These samples are hence maintained in the model. Further scrutinizing parameters and residuals, it was substantiated that the model was sound and reliable.

VERIFYING THE NUMBER OF COMPONENTS

In order to assure that the chosen number of components is reasonable, several additional tests are performed. A split-half test was conducted in the following manner. The sample

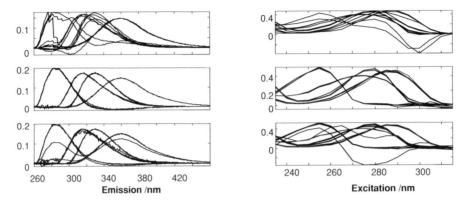

Figure 10.15. Results from split-half analysis. The top row represents three-component models and lower one five-component models. Left, emission mode loadings are given and right, excitation mode loadings. The loadings from all four models are superimposed.

set was divided into two separate sets consisting of the first 12 and the last 12 samples respectively. This split is reasonable because of the nature of the data (designed) as each of the sets will have representative information on all four chemical analytes. For each number of components one to six, a PARAFAC model was fitted to these two data sets separately. In order to substantiate the results, the same approach was used but now having sample set one equal to sample 1, 3, 7, 9, 11 etc. and sample set two 2, 4, 6, 8, etc. Thus, again the two samples sets are independent. For each number of components, four different models are thus obtained (pair-wise independent).

It is clear from the graphical representation in Figure 10.15 that at four components, all four models provide similar results whereas this is not the case for other numbers of components. The only other case is the one-component model (not shown) which is not relevant for other reasons (low variance as evidenced in Figure 10.11).

In order to quantify the difference, the following numerical display was produced. For each of the two splits, the similarity between (optimally matched) components in the emission and excitation mode was calculated as the Tucker congruence coefficient (Equation 5.22), thus giving one similarity measure for each component in each mode and split. The congruence of the emission and excitation mode was multiplied. As for correlations, the closer the congruence is to one, the more similar are the components. For each set and each number of components, these similarities are given in Table 10.2. Again, these numbers verify the conclusion that four components are adequate because only here will at least one of the splits (both in this case) provide the same result (high congruence) for all components.

CROSS-VALIDATION

A cross-validation of the PARAFAC model was also investigated. The expectation maximization approach was used by setting several elements to missing in each cross-validation segment. For an S-segmented cross-validation, the missing elements in the sth segment were chosen as the elements $s, s + S, s + 2S, \ldots$, in the vectorized array. A 17-segmented cross-validation was performed using one to seven components. Other numbers of segments were also investigated but yielded similar results and a 17-segmented cross-validation provided

Table 10.2. Similarity between models with different number of components. All similarities higher than 0.98 are bold

		1	2	3	4	5	6
1 component	split 1	**0.99**					
	split 2	**0.99**					
2 components	split 1	0.97	0.94				
	split 2	**0.99**	0.91				
3 components	split 1	0.56	0.64	0.84			
	split 2	0.46	0.88	**1.00**			
4 components	split 1	**1.00**	**0.99**	**1.00**	**1.00**		
	split 2	**1.00**	**1.00**	**1.00**	**1.00**		
5 components	split 1	0.54	−0.20	0.61	**0.99**	1.0	
	split 2	**1.00**	**1.00**	0.90	0.94	0.79	
6 components	split 1	−0.57	0.89	0.51	0.69	0.87	**1.00**
	split 2	0.57	0.71	0.88	0.85	0.83	0.98
Component no.		1	2	3	4	5	6

a reasonable compromise between computational load and the desire to leave out relatively few data elements in each segment. As can be seen from the results in Figure 10.16, cross-validation suggests either two components (first local maximum) or five components (first number of components with satisfactory explained variation). This contradicts the earlier findings that clearly suggested four components. Looking into the parameters of the two- and the five-component models (not shown here) substantiated that four components is a more adequate number of components to use. The cross-validation results are interesting because they illustrate that mathematical and statistical validation diagnostics are not providing final evidence of the correct choices to make. The diagnostics are merely useful tools

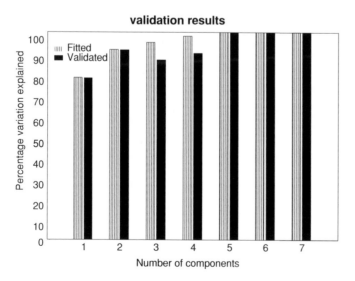

Figure 10.16. Cross-validation results using PARAFAC models with one to seven components. For each model a cross-validation was performed using 17 segments.

that help in making the right choices, but the responsibility for the final choice is left for the user and is preferably based on several different properties of the model: mathematical and statistical as well as chemical.

As a final check, the four-component PARAFAC model was fitted several times from different random starting points in order to assess whether there could be local minima problems. Ten realizations of the model were estimated and the fit values in terms of sum-of-squared residuals of these are (rounded): 181 226; 181 226; 181 227; 181 224; 181 227; 181 227; 181 223; 181 227; 181 227; 181 227. As can be seen, none of them deviate more than to the sixth digit and given a convergence criterion of the relative change in fit being smaller than 10^{-6}, this is expected. Hence, no problems with local minima are apparently present in this case.

Using the model on new data

FINDING SCORES OF NEW DATA

Three samples were not included in building the above PARAFAC model. Assuming that these are new samples, their corresponding score values can be found by fitting each test sample EEM to the loadings of the found model. Hence, for the data of each sample, $\mathbf{X}_i$ ($J \times K$), the scores are found from solving the problem

$$\min_{\mathbf{D}_i} \|\mathbf{X}_i - \mathbf{B}\mathbf{D}_i\mathbf{C}^\mathrm{T}\| \tag{10.3}$$

which, as shown on p. 129, is a simple regression problem. From solving this problem the scores *and* residuals of the new samples are obtained.

ASSESSING THE VALIDITY OF NEW SAMPLES

By comparing the size of the residuals of the test data with the residuals from the calibration samples, possible outlying behavior can be detected. For the three left-out samples, the sum-of-squared residuals are shown together with the sum-squared residuals obtained in the calibration model. For the calibration samples, the sum-of-squared residuals are given both as the fitted residuals and those obtained from leave-one-sample-out cross-validation. The fitted residuals are usually small because the model has been found specifically to minimize these. This makes them difficult to compare with residuals from new samples. By using residuals from leave-one-sample-out cross-validation, this overfit problem is minimized because the residuals from the calibration samples are obtained under similar conditions to the new samples.

In Figure 10.17, the resulting sum-of-squared residuals are shown. Most samples have residuals of similar magnitude and the difference between calibration and cross-validation is small. Two calibration samples stand out as having significantly high cross-validation errors: samples number two and three. As shown above though, these samples are not so influential that they have to be removed. For the three test samples (5, 10, 15), sample five stands out as having a high residual variation. The immediate conclusion is to exclude this sample from the present modeling and do further analysis on the sample to verify the reason for the outlying behavior. A formal test of outlying behavior can be based on either assessing

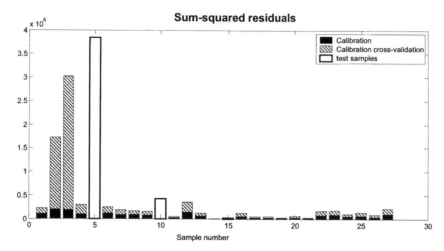

Figure 10.17. Sum-squared residuals of individual samples obtained during calibration, during cross-validation and from the test set.

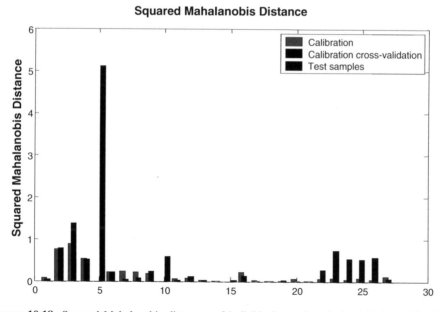

Figure 10.18. Squared Mahalanobis distances of individual samples obtained during calibration, during cross-validation and from the test set.

the sum-of-squared residuals with the distribution of the cross-validated calibration sample sum-of-squared residuals or by assessing the influence of the sample when taken into the calibration set. This can be done visually as in Figure 10.17 or quantitatively as discussed by Riu and Bro in the context of jackknifing [Riu & Bro 2002].

The Squared Mahalanobis distances can be calculated from the scores as discussed in Chapter 7. These are shown in Figure 10.18 and lead to exactly the same conclusion as

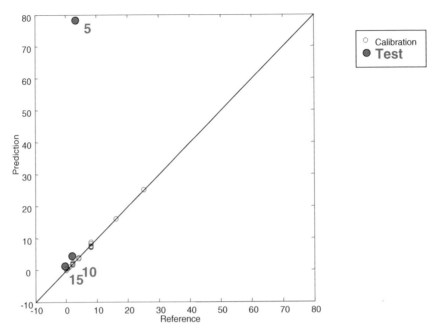

Figure 10.19. Predicted versus reference concentration of tryptophan. Test set samples are shown with solid circles and calibration samples with open circles.

the sum-of-squared residuals. The outlying test sample five has a very high leverage which clearly suggests that the sample is extreme. All other samples have leverages below or close to one as expected.

USING THE TEST SAMPLE SCORES IN A CALIBRATION MODEL

From the scores of the calibration data, regression models can be made for predicting the concentrations of the four analytes in new samples. A regression model for tryptophan was made using the score vector of component four of the PARAFAC model as independent variable and tryptophan as dependent variable in a univariate regression model without offset. Component four was chosen because the estimated emission and excitation spectra (e.g. Figure 10.14) of that component resembled the pure spectra of tryptophan. The scores of this component were also the ones correlating the most with the known concentrations of tryptophan.

Using the score of component four, a univariate regression model is built predicting concentration from scores of the calibration samples (without an offset). As can be seen in Figure 10.19 the model of calibration data is excellent. The unknown scores in the three test samples can subsequently be found simply by multiplying their score of component four with the regression coefficient. As expected, sample number five is not well predicted, whereas the two remaining ones are well predicted.

A similar exercise was performed on phenylalanine using component one of the PARAFAC model. The predictions are shown in Figure 10.20 and again, sample five is not well predicted. One calibration sample is apparently also incorrectly predicted, but

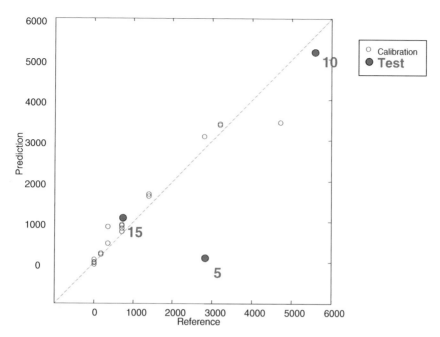

Figure 10.20. Predicted versus reference concentration of phenylalanine. Test set samples are shown with solid circles and calibration samples with open circles.

as the fluorescence data show no outlying behavior of this sample, it is possible that the apparent incorrect prediction is caused by an incorrect reference value.

Summary

In this example, a description has been provided on how to develop an appropriate PARAFAC model of a fluorescence data set. It has been shown that it is essential to know how to use the mathematical diagnostics as well as to know the nature of the data in order to reach to an adequate description of the data. Evaluating different diagnostics is important, but as shown in the example, the value of the diagnostics depends on the adequacy of the data. Outliers or irrelevant variables can blur the picture and such samples or variables have to be identified and properly handled. Even then, diagnostics may be misleading as was the case for the cross-validation results in this case. It is therefore essential that the judgments are based on the overall picture rather than on one specific diagnostic. It was also shown that outlier detection is important when using a model on new data. Otherwise, the results cannot be trusted.

Some other applications in the field

A vast literature exists on the use of multi-way models for curve resolution. A brief overview of this literature will be offered in the following in order to guide the reader to other important sources of information on multi-way curve-resolution.

MASS SPECTROMETRY

Gardner *et al.* [2001] explored the use of direct sampling mass spectrometry for the analysis of sample mixtures containing analytes with similar mass spectra. Water samples containing varying mixtures of toluene, ethyl benzene, and cumene were analyzed by purge-and-trap/direct sampling mass spectrometry using data sets of size $I \times 200 \times 105$ where the number of samples I varied. Multivariate calibration models were built using PLS, trilinear PLS (tri-PLS), and PARAFAC, using the temporal profiles as the third mode. For prediction samples that contained unmodeled, interfering compounds, PARAFAC outperformed the other analysis methods. The uniqueness of the PARAFAC model allowed for estimation of the mass spectra of the interfering compounds, which could be subsequently identified via visual inspection or a library search. Moreda-Pineiro *et al.* [2001] showed how PARAFAC can be used for studying different types of errors in multi-element determination in tea leaves by ICP-AES and ICP-MS. The results differed markedly from more traditional approaches and were shown to be robust and simple to interpret.

UV–VIS SPECTROMETRY

Durell *et al.* [1990] used PARAFAC for analyzing UV absorption spectra of plastocyanin obtained in different oxidation states and at different pH levels. Treating the pH and oxidation state as independent modes a four-way array was obtained (sample × wavelength × oxidation state × pH) which was then successfully decomposed using a four-way PARAFAC model. Trilinear and bilinear models were also estimated by unfolding the four-way array accordingly. It was shown that unique two-way decomposition was only possible using external knowledge about target spectra, while for the three- and four-way analysis no such knowledge was required. The unique results enabled detailed exploring of different plastocyanins verifying earlier proposed hypotheses and elucidating new aspects. Gurden *et al.* [2001] showed how additional information about the chemical system can be incorporated into the model using a so-called grey model. The principle was demonstrated using data from a first order chemical batch reaction monitored by UV–vis spectroscopy (27 experiments × 201 wavelengths × 130 times). The data were modeled using a Tucker3 structure, and external information was incorporated as constraints on the model parameters with additional restrictions on the Tucker3 core matrix.

FLUORESCENCE

Fluorescence data are often modeled using multi-way analysis due to its intrinsic relation to the PARAFAC model. Ross & Leurgans [1995] discuss in detail the theoretical rationale for using the PARAFAC and the Tucker2 model in fluorescence spectroscopy. Lee *et al.* [1992] used fluorescence excitation emission data of samples containing different amounts of proton acceptors and at different pH values to investigate the steady-state fluorescence of *N*-acetyl-L-tyrosinamide under different conditions in model systems. By using PARAFAC with nonnegativity constraints they were able to resolve the measured excitation–emission matrices and substantiate and develop earlier hypotheses regarding the basic configuration and characteristics of tyrosine. Ross *et al.* [1991] also used a similar setup to resolve a spectral data set (8 samples × 19 emissions × 28 excitations) of pigment complexes in pea thylakoids using PARAFAC with nonnegativity constraints. Millican and McGown [1990] used a rank annihilation-like algorithm to resolve single-sample two-component mixtures of different amounts of anthracene, 9,10-diphenylanthracene, and 1,3,5,8-tetraphenylpyrene. By introducing fluorescence lifetime besides excitation and emission resolution of minor components was significantly improved. Karukstis *et al.* [1995] characterized the steady-state room temperature fluorescence of 2-(*p*-toluidino)naphtalene-6-sulfonate in various solvents using the PARAFAC model on data sets of size 5–7 × 18 × 9. Russell *et al.* [1988] used fluorescence excitation–emission data sets of varying sizes to resolve systems containing platinum, palladium, and rhodium porphyrins even in the presence of quite severe spectral overlapping. In another paper direct trilinear decomposition was used for resolving fluorescence excitation and emission spectra of porphyrins in several animal dental calculus deposits [Ferreira *et al.* 1995]. The modeled data was of size 7 × 251 × 30. Three-way PLS and PARAFAC analysis were compared for the analysis of polycyclic aromatic hydrocarbons in water samples by Beltran *et al.* [1998b]. Data were obtained with a chromatographic system with a fast-scanning fluorescence spectra detector set to record short-time chromatograms containing several unresolved peaks. Resolution of closely eluting compounds was achieved, thus making a complete physical separation unnecessary. The procedure was applied to tap water samples (spiked at 0.10 and 0.20 mg/l levels) with good results, similar to those obtained with a HPLC system with a conventional fluorescence detector. Baunsgaard *et al.* [2001] investigated beet thick juice and cane final evaporator syrup to elucidate the nature of the color components in cane and beet syrup (47 samples × 222 emission × 21 excitations). Fluorescence excitation–emission landscapes resolved into individual fluorescent components with PARAFAC modeling were used as a screening method for colorants, and the method was validated with size exclusion chromatography using a diode array UV–vis detector. Bro [1999] showed how chemically meaningful models of complicated processes was possible by using fluorescence spectroscopy screening and PARAFAC modeling (data of size 265 × 571 × 7). Important fluorophores could be identified and related quantitatively to important quality parameters. da Silva & Novais [1998] measured mixtures of salicylic, gentisic and salicyluric acid by synchronous molecular fluorescence spectroscopy as a function of pH (between 2 and 11). A PARAFAC model was used to decompose the 12 × 41 × 21 data into the pure analyte spectra, concentration and pH profiles. Jiji *et al.* [2000] used PARAFAC on data from a field-portable, cuvette-based, single-measurement, excitation–emission matrix fluorometer. From the data (30 × 50 × 68/115) determination and resolution of three polynuclear aromatic hydrocarbons (PAHs) and of two carbamate pesticides was possible and allowed parts-per-trillion to parts-per-billion detection limits in aqueous solutions. Ling *et al.* [2001] resolved naphthalene, 1-naphthol and 2-naphthol whereas Martins *et al.* [1999] used PARAFAC and fluorescence spectroscopy to resolve the nonsteroidal anti-inflammatory drug Piroxicam in solution. The data used were of size 5 × 27 × 115. Tan *et al.* [2000] did a similar study (data size: 32 × 21 × 16) on a system of simultaneous degradations of chlorophyll a and b extracted with other interferents from fresh spinach. Xie *et al.* [2002] studied the competitive interaction with DNA of daunorubicin, being present in the clinical anti-tumor drug daunoblastina, and the fluorescence probe ethidium bromide.

10.3 Second-Order Calibration

Background of the field

Calibration is a method in analytical chemistry in which an instrumental response is related to a concentration of a chemical in a sample. Once this relationship is established the instrument can be used to measure the concentration of the analyte calibrated for in a future sample [Booksh & Kowalski 1994].

A very useful distinction between types of calibration is made by Sanchez and Kowalski [1988a, 1988b]. They distinguish between zeroth order, first order and second-order calibration, depending on whether the instrumental response of a single sample generates a scalar (zeroth order array), a vector (first order array) or a matrix (second-order array). The nomenclature zeroth, first, second-order comes from tensor algebra, a generalization of matrix algebra. Examples of zeroth order instruments are a pH sensor or an ion-selective electrode. A typical example of a first order instrument is a spectrometer (near-infrared, visual, Raman) operated over a certain wavelength range. Examples of second-order instruments are hyphenated instruments, such as liquid chromatography–diode array UV (ultraviolet) detection or GC–MS (gas chromatography with mass spectral detection) and excitation–emission fluorescence.

A further distinction in second-order instrumentation is based on the mathematical properties of the matrix resulting from a measurement. If the instrument is such that every analyte ideally contributes with a rank one to the instrumental (matrix) response, then this is called a rank-one second-order instrument. The term 'bilinear' is sometimes used in the literature to describe this property. This terminology is not strictly correct, because a 'bilinear' form of a matrix $\mathbf{X}$ is $\mathbf{X} = \mathbf{AB}'$ holds for any matrix and does not necessarily mean that the rank of $\mathbf{X}$ is one (see also Elaboration 10.1). Not all second-order instruments are of the rank-one type, e.g. tandem mass spectrometry is not a rank-one second-order instrument [Smilde *et al.* 1994].

ELABORATION **10.1**

Bi- and trilinearity revisited

In Chapter 2, bi- and trilinearity were explained in a mathematical sense, but some relevant concepts will be repeated and related to the examples. Any data matrix can be fitted to any extent by a bilinear, e.g. PCA, model by using sufficiently many components. This does not require the data or the underlying model (e.g. a physico-chemical model) to be of any particular form. In the same sense, any three-way array can be fitted by a PARAFAC model regardless of the underlying model or data properties. The terms bi- or trilinearity are taken to mean *low-rank* bi- or trilinear as bi- or trilinearity in itself is meaningless from a data analysis point of view (Figure 10.21). If the underlying model is (low-rank) bilinear, the measured data are approximately (low-rank) bilinear and a (low-rank) bilinear model can be meaningfully fitted in the ideal case. The same is valid for trilinearity.

With data coming from hyphenated instruments, low-rank bilinearity of a single-sample measurement can often be assumed. A hypothesized model of the data from one sample

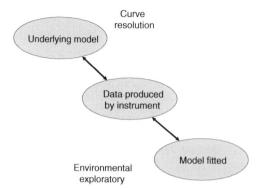

Figure 10.21. Bi- and trilinearity have at least three definitions: in the underlying model (mathematical structure plus possible distributional assumptions), in the data and in the fitted model. It is perfectly possible to fit a (low-rank) bi- or trilinear model without having the data or the underlying model assumed (low-rank) bi- or trilinear. Two possible scenarios are shown where focus is primarily on the fitted model and on the underlying model respectively.

can then be written as

$$\mathbf{X} = k_1 \mathbf{a}_1 \mathbf{b}_1' + k_2 \mathbf{a}_2 \mathbf{b}_2' + \cdots + \mathbf{E}$$

where $\mathbf{a}_1, \mathbf{a}_2, \mathbf{b}_1, \mathbf{b}_2$ etc. ($\mathbf{a}_r$ = unit chromatograms, $\mathbf{b}_r$ = unit spectra, k_r = concentrations) follow from the underlying model (Lambert–Beer).

When the underlying model is low-rank bilinear, the data matrices produced are often close to this bilinear model, except for noise and nonlinearities. In such a case, a low-rank bilinear model can be fitted, for example, by curve resolution methods. When certain added conditions such as selective channels, nonnegative parameters etc. are fulfilled, unique models can sometimes be obtained [Gemperline 1999, Manne 1995].

With a stack of low-rank bilinear data matrices such as the one above, a three-way array can be built. The underlying model of such an array is *not* necessarily low-rank trilinear if for example there are retention time shifts from sample to sample. However, the data can be still be fitted by a low-rank trilinear model. If the deviations from the ideal trilinearity (low-rank – one component per analyte) are small, directly interpretable components can be found. If not, components are still found in the same sense as in PCA, where the components together describe the data, although no single component can necessarily be related to a single analyte.

Trilinearity in the underlying model can be assumed, e.g., when the output from a chromatograph is subjected to excitation–emission fluorescence measurement in the low-concentration range. In this case, the underlying model is often low-rank trilinear, the data produced for one sample are trilinear and a PARAFAC model separates the data uniquely into unit chromatograms, unit excitation spectra and unit emission spectra as A-, B- and C-loadings.

It is also possible to take any three-way array $\underline{\mathbf{X}}$, and fit a PARAFAC or Tucker model for purely exploratory purposes. In this case, the PARAFAC loadings can be used to speculate about underlying trilinearity. With environmental data this way of working is often used. This amounts to exploring the data for unknown environmental profiles. Usually the data are only in part bi- or tri-linear with significant noise, and some nonlinearity and the guessed underlying model is hence speculative.

In the field of calibration the concept of an *interferent* is very important.[1] A useful terminology is as follows: an *analyte* (sometimes also called *analyte of interest*) is a compound in the sample for which quantification is needed, and an *interferent* is another compound in the sample that gives a contribution to the instrumental response but for which quantification is not needed. For a zeroth order instrument an interferent makes calibration impossible. The traditional way of dealing with interferents in analytical chemistry is to pretreat the sample such that only the analyte contributes to the signal of the zeroth order instrument. Extraction, separation, and or selective reagents are often used for this purpose.

The power of a first order instrument is that it can deal with interferents, as long as such interferents are in the calibration set. That is, the first order instrument can be calibrated for the analyte in the presence of interferents and the concentrations of such interferents do not even have to be known. If the analyte is present in a future sample and an interferent which was not 'seen' in the calibration set (i.e., it is a new interferent) the concentration estimate of the analyte is wrong, but the calibration model diagnoses this by indicating the sample as an outlier. These properties of first order instruments are very powerful and explain their use in, e.g., process analysis and multivariate calibration where sample pretreatment would take too much time.

For a second-order instrument there is an even stronger property with respect to interferents. Even in the case of a new interferent in a future sample, accurate analyte concentration estimates are possible. This is called the *second-order advantage* and is a very powerful property indeed. The first publications on second-order calibration were from Ho and co-workers [Ho *et al.* 1978, Ho *et al.* 1980, Ho *et al.* 1981], who used rank annihilation to quantify the analyte. Unfortunately, in order to obtain the second-order advantage, the second-order instrument has to operate in a very reproducible way, which is not always the case. Hence, up to now, it is not widespread in analytical practice.

There are also third order instruments available, e.g., GC–GC–MS, LC–MS/MS or time-resolved fluorescence excitation–emission. Such instruments also have the second-order advantage, but the problems of reproducibility become even more pronounced.

To set the stage for the examples, the most simple rank-one second-order problem is described in mathematical terms and its relation to three-way analysis is established. The standard sample (i.e. the sample that contains a known amount of analyte) is denoted $\mathbf{X}_1(J \times K)$ and the mixture (i.e. the unknown sample containing an unknown amount of analyte and possibly some interferents) is called $\mathbf{X}_2(J \times K)$. It is assumed that the first instrumental direction (e.g. the liquid chromatograph) is sampled at J points, and the second instrumental direction (e.g. the spectrometer) at K points. Then the rank-one second-order calibration problem for the case of one analyte and one interferent can be presented as

$$\mathbf{X}_1 = c_{1,1}\mathbf{a}_1\mathbf{b}_1^{\mathrm{T}} + \mathbf{E}_1$$
$$\mathbf{X}_2 = c_{2,1}\mathbf{a}_1\mathbf{b}_1^{\mathrm{T}} + c_{2,2}\mathbf{a}_2\mathbf{b}_2^{\mathrm{T}} + \mathbf{E}_2 \tag{10.4}$$

where $\mathbf{a}_1$ and $\mathbf{a}_2$ are the pure responses of the analyte and interferent, respectively, in the

---

[1] Matrix effects are also important, but these are not explicitly considered here.

first instrumental direction (e.g. pure chromatograms); $\mathbf{b}_1$ and $\mathbf{b}_2$ are the pure responses of the analyte and interferent, respectively, in the second instrumental direction (e.g. pure spectra); $c_{1,1}$ and $c_{2,1}$ are the concentrations of the analyte in the standard and the mixture, respectively, and $c_{2,2}$ is the unknown concentration of the interferent in the mixture. The matrices $\mathbf{E}_1$ and $\mathbf{E}_2$ represent residuals and contain model errors (e.g. deviations from the Lambert–Beer law) and instrumental noise.

Depending on the situation, there are several unknowns in Equation (10.4). In its most general case the unknowns are $\mathbf{a}_1, \mathbf{a}_2, \mathbf{b}_1, \mathbf{b}_2, c_{2,1}$ and $c_{2,2}$. The standard $\mathbf{X}_1$ has pseudo-rank one (see Chapter 2, Section 6), and, hence, can be decomposed in its contributions $\mathbf{a}_1$ and $\mathbf{b}_1$. Note that there is still intensity ambiguity because $\mathbf{a}_1\mathbf{b}_1^{\mathrm{T}}$ equals $\mathbf{a}_1\alpha\alpha^{-1}\mathbf{b}_1^{\mathrm{T}}$, for any nonzero α. The quantification of the analyte requires the estimation $c_{2,1}$ and sometimes it is also convenient to obtain estimates of $\mathbf{a}_1, \mathbf{a}_2, \mathbf{b}_1$ and $\mathbf{b}_2$ for diagnostics purposes. This is the 'curve resolution' aspect of second-order calibration.

Equation (10.4) can be written as a three-way problem by stacking $\mathbf{X}_1$ and $\mathbf{X}_2$ on top of each other generating the first mode ($i = 1,2$) of a three-way array:

$$\mathbf{X}_1 = \begin{bmatrix} \mathbf{a}_1 & \mathbf{a}_2 \end{bmatrix} \begin{bmatrix} c_{1,1} & 0 \\ 0 & 0 \end{bmatrix} \begin{bmatrix} \mathbf{b}_1^{\mathrm{T}} \\ \mathbf{b}_2^{\mathrm{T}} \end{bmatrix} + \mathbf{E}_1 = \mathbf{A}\mathbf{D}_1\mathbf{B}^{\mathrm{T}} + \mathbf{E}_1$$

$$\mathbf{X}_2 = \begin{bmatrix} \mathbf{a}_1 & \mathbf{a}_2 \end{bmatrix} \begin{bmatrix} c_{2,1} & 0 \\ 0 & c_{2,2} \end{bmatrix} \begin{bmatrix} \mathbf{b}_1^{\mathrm{T}} \\ \mathbf{b}_2^{\mathrm{T}} \end{bmatrix} + \mathbf{E}_2 = \mathbf{A}\mathbf{D}_2\mathbf{B}^{\mathrm{T}} + \mathbf{E}_2 \qquad (10.5)$$

$$\mathbf{C} = \begin{bmatrix} c_{1,1} & 0 \\ c_{2,1} & c_{2,2} \end{bmatrix}$$

where $\mathbf{A}, \mathbf{B}, \mathbf{D}_1$ and $\mathbf{D}_2$ are implicitly defined. Equations (10.5) show that this second-order calibration problem is a PARAFAC problem (see Chapter 4, Section 4.1). Hence, any PARAFAC algorithm can be used to solve for the unknown parameters. Traditionally, generalized rank annihilation was used to solve this problem [Lorber 1984]. An illustration of the generalized rank annihilation method for this situation is given in the Appendix 10.A.

Checking the Kruskal conditions (see Chapter 5, Section 2) for this problem gives

$$k_{\mathrm{A}} + k_{\mathrm{B}} + k_{\mathrm{C}} \geq 2R + 2$$
$$2 + 2 + 2 \geq 2 \times 2 + 2 \qquad (10.6)$$

which are exactly fulfilled, assuming that $\mathbf{a}_1$ and $\mathbf{a}_2$ are linearly independent; $\mathbf{b}_1$ and $\mathbf{b}_2$ are linearly independent; and $c_{2,2}$ is not zero. All these assumptions are reasonable (if $c_{2,2}$ is zero, then there is no interferent, and the problem becomes trivial) and generally valid in calibration situations. Hence, solving Equation (10.5) gives unique solutions for the unknown parameters $\mathbf{a}_1, \mathbf{a}_2, \mathbf{b}_1, \mathbf{b}_2$ (up to a constant, see above) and $c_{2,1}$. The concentration of the interferent $c_{2,2}$ cannot be established because only the shapes of $\mathbf{a}_2$ and $\mathbf{b}_2$ are unique, not their absolute magnitudes.

There are more complicated situations possible, e.g.: (i) a standard containing several analytes, (ii) mixtures containing more than one interferent, (iii) multiple standards, (iv) multiple mixtures or combinations of those four. The properties of these second-order calibration problems can be deduced by writing them in a PARAFAC structure (as in Equation (10.4)) and examining their properties. Two such examples are given in Appendix 10.B (cases (ii) and (iii) above). Appendix 10.B also contains a simple example showing the mechanics of second-order calibration. A detailed mathematical treatment of the different second-order calibration problems is given by Kiers and Smilde [1995].

Background of the example

The second-order calibration example shown next is from the field of environmental analytical chemistry. A sensor was constructed to measure heavy metal ions in tap and lake water [Lin *et al.* 1994]. The two heavy metal ions Pb^{2+} and Cd^{2+} are of special interest (the analytes) and there may be interferents from other metals, such as Co^{2+}, Mn^{2+}, Ni^{2+} and Zn^{2+}. The principle of the sensor is described in detail in the original publication but repeated here briefly for illustration. The metal ions diffuse through a membrane and enter the sensor chamber upon which they form a colored complex with the metal indicator (4-(2-pyridylazo) resorcinol PAR) present in that chamber. Hence, the two modes (instrumental directions) of the sensor are the temporal mode related to the diffusion through the membrane, and the spectroscopic mode (visible spectroscopy from 380 to 700 nm). Selectivity in the temporal mode is obtained by differences in diffusion behavior of the metal ions (see Figure 10.22) and in the spectroscopic mode by spectral differences of the complexes formed. In the spectroscopic mode, second-derivative spectra are taken to enhance the selectivity (see Figure 10.23). The spectra were measured every 30 s with a resolution of 1 nm from 420 to 630 nm for a period of 37 min. This results in a data matrix of size 74 (times) $\times$ 210 (wavelengths) for each sample.

The Lambert–Beer law states that (i) doubling the concentration of the analyte doubles the absorbance at a specific wavelength and (ii) the contributions of the absorbing species are independent of each other. This is also sometimes abbreviated to linearity (i) and additivity (ii). A specific example is shown, taking Pb as the analyte, and Co as the only interferent. Assuming that the Lambert–Beer law holds in this case, then upon writing $\mathbf{a}_{Pb}$, $\mathbf{a}_{Co}$, $\mathbf{b}_{Pb}$ and $\mathbf{b}_{Co}$ for the temporal profiles and spectra of the analyte Pb and the interferent Co, respectively, the sensor response for the standard, $\mathbf{X}_{Pb}$, and mixture, $\mathbf{X}_2$, can be written as

$$\mathbf{X}_{Pb} = c_{Pb,Pb}\mathbf{a}_{Pb}\mathbf{b}'_{Pb} + \mathbf{E}_{Pb}$$
$$\mathbf{X}_2 = c_{2,Pb}\mathbf{a}_{Pb}\mathbf{b}'_{Pb} + c_{2,Co}\mathbf{a}_{Co}\mathbf{b}'_{Co} + \mathbf{E}_2 \tag{10.7}$$

which is clearly a second-order calibration problem. The actual problem solved is slightly more complicated because multiple interferents are present in the mixture. Hence, this comes down to problem (10.24) (see Appendix 10.C) and the analyte concentrations can be obtained uniquely.

Building and using the models

A set of standards was prepared containing five different concentrations of Pb^{2+} (see first column of Table 10.3). Also a set of mixtures containing a fixed amount of interferents (all interferents Co^{2+}, Mn^{2+}, Ni^{2+} and Zn^{2+} present at 0.5 μM each), and varying amounts of Pb^{2+} (again at five concentrations) was available. Each standard was used to calibrate each mixture, resulting in 25 calibration models with the structure of Equation (10.5). Generalized rank annihilation was used to fit the PARAFAC model to obtain the concentrations of the analyte Pb^{2+} in the mixtures. The results are reported in Table 10.3.

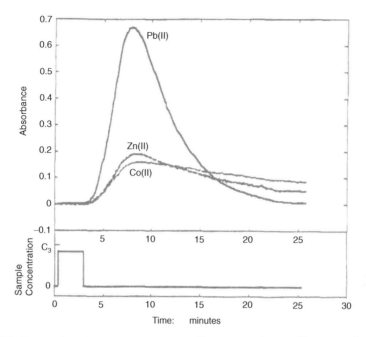

Figure 10.22. Temporal profiles of different ions: Pb^{2+} (1.5×10^{-6} M); Co^{2+} (3×10^{-6}); Zn^{2+} (3×10^{-6}). Reprinted with permission from Lin ZH, Booksh KS, Burgess LW, Kowalski BR, *Analytical Chemistry*, 1994, **66**, 2552–2560. Copyright (1994) American Chemical Society.

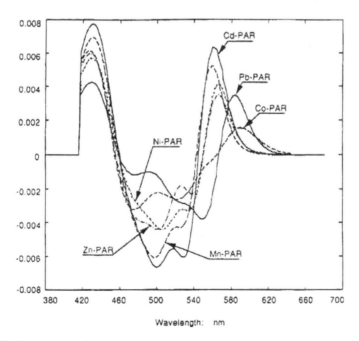

Figure 10.23. Second-derivative spectra of metal ion–PAR chelates. Reprinted with permission from Lin ZH, Booksh KS, Burgess LW, Kowalski BR, *Analytical Chemistry*, 1994, **66**, 2552–2560. Copyright (1994) American Chemical Society.

Table 10.3. Relative errors of Pb^{2+} prediction: the difference between the estimated and true value, expressed relative to the true value (in %)

Standards, Pb^{2+}, 10^{-6} M	Samples, Pb^{2+}, 10^{-6} M, Plus Interferents				
	0.5	1.5	2.5	3.5	4.5
0.5	−44.5	−7.9	−59.8	119.6	155.0
1.5	−52.2	−11.7	−5.9	20.8	22.7
2.5	−61.6	−25.0	−6.4	7.7	9.1
3.5	−66.2	−32.0	−13.7	2.8	4.7
4.5	−70.2	−37.7	−19.7	−0.5	3.6

The results are extensively discussed in the original paper. The high prediction errors for the 'low analyte concentration mixtures' may be due to a competition effect of the ions passing through the membrane. The relative large concentrations of interferents hamper the diffusion of the analyte. Hence, the additivity part of the Lambert–Beer law breaks down. Linearity is also a problem, as exemplified by the off-diagonal data in Table 10.3, where the mixtures contain different concentrations than the standards. Summarizing, there are deviations from the Lambert–Beer law and the quality of the calibration model 10.7 depends strongly on the validity of that law.

The generalized rank annihilation method also gives estimates of the temporal profile and the spectrum of the analyte Pb^{2+}. These can be used for qualitative analysis or to check the validity of the model. As can be seen in Figure 10.24, the estimated profiles resemble the pure ones very accurately, giving confidence in the estimation results. There are different ways to improve the results. Noting that the rising (=first) part of the temporal profiles is not particularly informative of the analyte and interferents, it is possible to truncate the data by discarding this rising part. The results of those calibrations are shown in Table 10.4. The remaining high prediction errors at low concentrations are due to competition effects in the membrane and cannot be corrected by manipulating the data.

In order to improve the calibration results, an analyte concentration in the mixture close to the analyte concentration in the standard would be favorable. This is not a practical solution, but an alternative is to use multiple standards. All five standards are used now. The calibration equations become slightly more complicated (i.e. a combination of cases (ii) and (iii) from above, see Appendix 10.C), but end up in a relatively straightforward three-way model. Generalized rank annihilation cannot be used anymore, since there are more than two slices in the three-way array. The authors used direct trilinear decomposition to solve this problem but an ordinary PARAFAC-ALS algorithm can also be used. The results are presented in Table 10.5 and show improvements at low concentrations.

Some concentration nonlinearities can be treated by a simple extension of the trilinear model. If there are concentration nonlinearities, then the model estimates *apparent* concentrations in **C**. Making a simple nonlinear calibration curve between the values in **C** and the *actual* concentrations in the standards gives the solution since the *apparent* concentration of the analyte in the mixture can then be transformed to the *actual* concentration domain [Booksh *et al.* 1994].

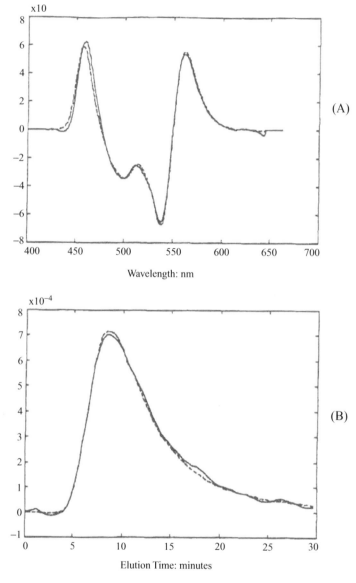

Figure 10.24. Comparison of the spectral and temporal profile of Pb^{2+} with the calculated ones with GRAM. (A) Spectra; (B) temporal profiles; ———, measured; _ _ _; calculated.

Summary and remarks

Second-order calibration is a set of methods developed to deal with second-order instrumental data: instruments that generate a matrix of measurements for a single sample. Second-order calibration problems can be written in the form of a PARAFAC model, which facilitates understanding the properties of the problem and of the solution.

Table 10.4. Relative errors of Pb^{2+} prediction (truncated data; see also Table 10.3)

Standards, Pb^{2+}, 10^{-6} M	Samples, Pb^{2+}, 10^{-6} M, Plus Interferents				
	0.5	1.5	2.5	3.5	4.5
0.5	−57.0	−23.3	−7.8	−2.1	−3.0
1.5	−50.4	−11.9	−4.1	5.0	3.8
2.5	−55.4	−19.7	−1.3	−0.9	−0.3
3.5	−56.0	−25.4	−2.2	−1.2	−1.3
4.5	−55.8	−19.3	0.0	1.5	1.7

Table 10.5. Relative errors of Pb^{2+} prediction (truncated data) using multiple standards

Standards, Pb^{2+}, 10^{-6} M	Samples, Pb^{2+}, 10^{-6} M, Plus Interferents				
	0.5	1.5	2.5	3.5	4.5
All	−14.6	−9.3	−11.1	−3.2	−4.6

Second-order calibration depends critically on the underlying assumptions of linearity. One of the problems, e.g., in using a hyphenated technique such as chromatography–spectroscopy, is the poor reproducibility of the retention time axis. This destroys the trilinearity of the calibration model. For certain situations, this can be counteracted by using different types of three-way models (e.g. PARAFAC2 [Bro *et al.* 1999]), but in most of the cases extensive (manual) preprocessing is needed to linearize the data. This renders second-order calibration a less practical technique for such data.

SOME OTHER APPLICATIONS IN THE FIELD

Second-order calibration has been used in flow injection analysis [Saurina and Hernandez-Cassou 2001] where Nørgaard and Ridder were the first to show the benefits and drawbacks of three-way methods applied to such data [Nørgaard & Ridder 1994a, Nørgaard & Ridder 1994b]. Samples containing three different though similar acids were injected into a FIA system with a pH profile induced over the sample plug. The size of the data was $12 \times 101 \times 89$. The differences in pK_a gave rise to conditions under which a second-order method should be able to quantify the analytes. However, due to problems arising from the fact that there is no dispersion of the sample constituents, the predictions using only one sample for calibration were not satisfactory. Their results were later elaborated and compared with other techniques [Reis *et al.* 2000, Smilde *et al.* 1999].

Wilson *et al.* [1989] used rank annihilation on MS/MS spectra of samples containing warfarin, different hydroxywarfarins, and phenylbutazone. The rank annihilation was also compared to and outperformed different ordinary curve resolution and regression techniques. Li *et al.* [1994] showed how GRAM can be used to estimate low concentrations of hydrocortisone in urine using LC–UV data. They noted that using the original rank annihilation algorithm of Lorber inaccurate results were obtained.

Xie *et al.* [1996] compared the merits of second-order calibration for quantifying binary mixtures of *p*-, *o*-, and *m*-aminobenzoic acids and orciprenaline reacting with diazotized sulfanilamide in a kinetic UV–vis study. The samples were measured at 31 times and at different numbers of wavelengths.

Ho *et al.* [1978, 1980] exemplified second-order calibration with simple one- and two-component mixtures of perylene and anthracene and showed that they were able to determine the concentration of one analyte in the presence of the other using only one pure standard and using fluorescence excitation–emission landscapes of size 30×30. They also used samples of a six-component polynuclear aromatic hydrocarbon to show the same principle in a more complex matrix. Gui *et al.* [1995] used direct trilinear decomposition on a $9 \times 40 \times 89$ array for quantifying initial concentrations of two components (glycine and glutamine) based on the kinetic development of fluorescence in a thin-layer chromatographic system. Poe and Rutan [1993] compared the use of generalized rank annihilation, peak height, peak area, and adaptive Kalman filter for quantifying analytes in a separation of polycyclic aromatic hydrocarbons using reversed phase liquid chromatography coupled with fluorescence detection. Each sample was measured at 256 wavelengths every second for 90 s. Generalized rank annihilation outperformed the other methods, but was found to be more sensitive to retention time shifts. Wilson *et al.* [1989] showed that the rank annihilation method could satisfactorily predict concentrations of different sugars in mixtures dissolved in D_2O measured by (256×256) 2D NMR. In a recent application of GC $\times$ GC, GRAM was successfully used to calibrate volatile and semi-volatile components [Sinha *et al.* 2003].

10.4 Multi-way Regression

Background of the field

Multi-way regression provides another route for building multi-way calibration and regression models. Whereas second-order calibration is intrinsically related to the special uniqueness properties of PARAFAC, multi-way regression models are similar to their two-way analogues. This implies that indirect relations can be handled if the measured data correlate indirectly to the responses. It also implies that many samples are usually required for building a robust model because the model is based on identifying a subspace unique to the sought response.

In the following, several approaches to constructing a regression model will be tested on some simple example data sets. In these examples, traditional second-order calibration is not possible, as there is no causal or direct relationship between specific measured variables and the responses. The purpose of this example is mainly to illustrate the relative merits of different calibration approaches on different types of data, whereas the model building (residual and influence analysis) itself is not treated in detail.

Example data sets

FLUORESCENCE DATA

The quality of refined sugar is given by several parameters. One of the important parameters is *color* and is measured in the laboratory. It is interesting to investigate if this color can be predicted directly from fluorescence measurements of dissolved sugar. In order to investigate this, 268 sugar samples were measured by fluorescence spectroscopy. The 268 samples were sampled every eighth hour for approximately 3 months. The quality parameter color was determined and a model sought for predicting color from fluorescence.

The emission spectra of 268 sugar samples dissolved in phosphate buffered water were measured at seven excitation wavelengths (excitation 230, 240, 255, 290, 305, 325 and 340 nm, emission 365–558 nm, 4.5 nm intervals). Hence the data can be arranged in a $268 \times 44 \times 7$ three-way array or a 268×308 matricized two-way matrix. The color was determined by a standard wet-chemical procedure as described by Nørgaard [1995].

Models (PARAFAC, Tucker3, N-PLS) of the fluorescence data are built. For PARAFAC and Tucker3, the scores of the model of the fluorescence data are used as independent variables which are related to the color by multiple linear regression [Bro 1997, Geladi *et al.* 1998]. In N-PLS, the regression model is determined simultaneous with finding the scores.

The quality of the models was determined by test set validation with the first 134 samples (corresponding to the first 1.5 months of the sampling period) in the calibration and the last 134 in the test set. This systematic approach was adopted in order to test the predictions on future samples. It is known that the chemistry of the process may change over time, so this test provides a more realistic measure of the quality of the models than if a random sampling had been chosen.

SENSORY DATA

In the data set BREAD, five breads made and were assessed by a sensory panel of eight people using a fixed vocabulary of 11 different attributes. The data are arranged in a $10 \times 11 \times 8$ array and a model is sought of these data in order to be able to predict the composition (salt in this case) in unknown samples.

The quality of the models was determined by leave-one-sample-out cross-validation due to the low number of samples. Due to the presence of replicates, the error measure is an indication of the reproducibility of the model. These sensory data have a poor signal-to-noise ration compared to the fluorescence data above [Gurden *et al.* 2001].

Building a regression model

The following models were investigated (all using centered data): PARAFAC regression, tri-PLS as well as two-way unfold-PLS on centered data, Tucker3 regression, and two-way unfold principal component regression (PCR). The Tucker3 regression was performed using the same number of components in each mode and using the scores (sample loadings) for the regression step. PCR was performed using the successively estimated score vectors from an unfold-PCA model for the regression model. The results from the calibration models

Table 10.6. The residual variance (in percentages of total variance) from test-set validation for the sugar data and for cross-validation for the bread data set. The results are given for *X* and *y*.

Number of Components	PARAFAC	Tucker	NPLS	PLS	PCR	
Sugar						
1	1.10	1.10	2.49	2.52	1.05	⎫
2	0.68	0.68	2.95	3.03	0.66	⎪ Residual *X*
3	0.28	0.31	3.08	3.20	0.32	⎬ variance
4	0.28	0.26	3.46	3.41	0.25	⎪
5	0.13	0.12	3.54	3.49	0.11	⎭
1	2.77	2.77	2.63	2.61	2.76	⎫
2	2.47	2.47	2.27	**2.32**	2.50	⎪ Residual *y*
3	**2.35**	2.41	**2.10**	3.80	2.40	⎬ variance
4	2.34	**2.16**	3.25	3.18	**2.20**	⎪
5	2.47	2.43	2.25	4.64	2.42	⎭
Bread						
1	25.70	25.70	8.34	8.68	26.19	⎫
2	22.18	22.18	13.50	12.21	19.94	⎪ Residual *X*
3	19.73	19.41	17.33	17.93	19.11	⎬ variance
4	18.53	20.09	18.35	18.34	18.89	⎪
5	17.59	17.48	19.95	18.96	17.84	⎭
1	7.87	7.87	5.23	4.97	7.29	⎫
2	5.35	5.36	2.29	3.13	5.24	⎪ Residual *y*
3	3.12	2.96	**1.23**	**2.11**	3.05	⎬ variance
4	**1.94**	**1.77**	1.21	2.08	2.94	⎪
5	2.29	1.50	1.30	2.05	**2.33**	⎭

are shown in Table 10.6. For the sugar data set, only few differences appear to be present between the different models. All are capable of describing nearly 100 % of the variation in both **X** and **y**. This is typical for data that have a high signal-to-noise ratio. *N*-PLS and PLS explain less of the variance in **X** but provide similar predictions to the remaining methods using fewer components. In general, the multi-way models contain only a fraction of the number of parameters that the matricized models do. For example, a one-component PARAFAC model contains 51 loading parameters and a one-component PCA model 308. Thus, if the matricized models do not provide better predictions, the trilinear models may be preferred. In this case, e.g. *N*-PLS provides a good model using relatively few parameters. Alternatively, the PARAFAC model may be preferred, if the chemical interpretation of its unique solution is important for the acceptance or use of the model. The fluorescence data illustrates that, when a trilinear model is adequate, there is no advantage of using more complex models.

For the bread data, the results are a little different. First of all the difference between cross-validated and fitted description (not shown) of **X** is more distinct because the number of samples is smaller, making the model more prone to overfit and because the data are also more noisy. The matricized models fit **X** much better than the multilinear models. For calibration with a three-component model, the residual variation in the X-block is 14 % for PARAFAC and 7 % for PCA (not shown). However, for cross-validation, the corresponding numbers read 19 % for the X-block in both cases. Hence, the additional fit of **X** provided by the matricized model is *only* overfit. This is also reflected in the predictions of salt where the matricized models are systematically worse than their multi-way counterparts.

Conclusion

Regression models between a multi-way array **X** and a response **y** can be made in many ways. The need for latent variables can be solved by decomposing **X** by PARAFAC, Tucker3, or PCA (after matricizing). This gives PCR-like models where **y** is not used in calculating the latent variables. Alternatively PLS models can be made where the decomposition of **X** is influenced by **y**. The field of regression between multi-way arrays has a huge potential for model development and for application.

When data can be assumed to be approximately multilinear there is little if any benefit in matricizing the data before analysis. Even though the two-way models describe more variation per definition, the increased modeling power does not necessarily provide more predictive models in terms of modeling either the independent or the dependent variables. Even when the data do not approximately follow a multilinear model (e.g. sensory data), the multilinear models can be preferred if the possible bias in having too simple an X-model is counteracted by the smaller amount of overfit.

Using PARAFAC for regression as shown here for the fluorescence data has the potential for simultaneously providing a model that predicts the dependent variable, and uniquely describes which latent phenomena are crucial for describing the variations in the dependent variable.

10.5   Process Chemometrics

Background of the field

PROCESS MONITORING VERSUS PROCESS CONTROL

Process monitoring is an area of engineering concerned with monitoring (chemical) processes. Process monitoring can be performed for several reasons, e.g. process optimization, process investigation or detection of deviating behavior of running processes. If processes are monitored for detecting deviating behavior and several variables are considered simultaneously, then the term multivariate statistical process control (MSPC) is often used. MSPC has its roots in statistical process control, which goes back to Shewhart [1931] who introduced the *control chart*. The term 'control' is confusing, as MSPC has little to do with automated process control using feedback control or model predictive control. A nice discussion on the relationships between MSPC and automated process control is given by Box and Kramer [1992]. To avoid confusion, the term 'process monitoring' will be used throughout this chapter indicating all activities on monitoring processes for detecting abnormal behavior.

The basic philosophy of process monitoring is that there are two types of variation in a process: common cause variation and variation due to special causes. The first type of variation is encountered when the process is running under normal operating conditions. If special events are happening, such as sensor failure, catalyst degradation, controller breakdown, then variation will show up that is different from common variation. The purpose of monitoring the process is three-fold: (i) detecting if there is special variation, (ii) locating where and when the special variation arises, and (iii) assigning a cause to the special variation. The second question is especially important if a whole plant is monitored.

There are two basic steps in the above mentioned process monitoring procedure. In phase 1, data from the process under normal operating conditions are obtained. From these data control charts are calculated. In phase 2, the daily operation of the process, measurements are taken and plotted in the control charts. It is then decided whether the process is still working under normal operating conditions or not. The first phase is called the training phase and the second phase the operational or test phase.

There are two fundamentally different ways of producing products: in a continuous fashion and batch-wise [Fogler 1992]. Cooking potatoes is typically a batch operation: the potatoes are put in a pan, water is added and the potatoes are cooked until they are done. A batch process is operated according to a recipe. In a continuous process, a product is made continuously, and such a process is operated at a certain set point (working point or steady-state). These two ways of producing products have their own process monitoring methods, although the general strategy is the same.

A short description of monitoring continuous processes for deviating behavior will be given, because some parts of batch process monitoring are taken from that area. Monitoring batch processes is discussed in detail as three-way analysis is heavily involved in that area.

MONITORING OF CONTINUOUS PROCESSES

For a continuous process, it is possible to use the concept of a 'set point' [Stephanopoulos 1984]. If the process is working under normal operating conditions, the average quality of the product should equal that of the product made under set point conditions, otherwise the process is off target. Hence, all the fluctuations of the product quality around this set point can be considered reflecting the common variation if the process is working under normal operating conditions. Based on this idea it is straightforward to build control charts, such as the Shewhart charts [Crosier 1988, MacGregor 1988].

In statistical process control a single quality characteristic is monitored. Multiple quality characteristics are mostly measured simultaneously and these characteristics are typically correlated. Multivariate process monitoring takes this correlation into account.

One of the pioneers of multivariate process monitoring was Jackson [1959], who used these methods at Eastman Kodak. The key concept of multivariate process monitoring of continuous processes is Hotelling's T^2 statistic [Hotelling 1931], which is a generalization of the student-t statistic for multivariate measurements [Anderson 1971]. The T^2 statistic measures the distance of a new measurement to the measurements taken under normal operating conditions corrected for the correlation structure in the latter. This corrected distance is closely related to the Mahanalobis distance [Anderson 1971, Mardia *et al.* 1979]. If the T^2 value of a new measured quality characteristic is too large (as compared to a 95 % or 99 % control limit) then the process is moving out of normal operating conditions.

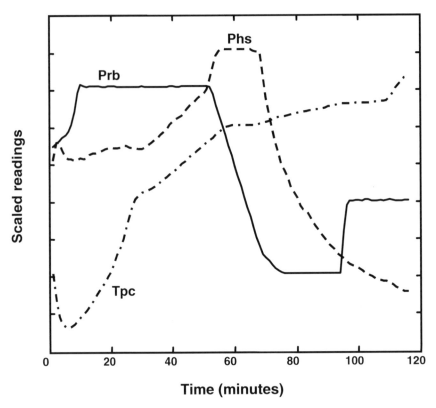

Figure 10.25. Typical trajectories of process variables in one run of a batch process (Prb and Phs are pressures and Tpc is a temperature).

FROM PRODUCT QUALITY MEASUREMENTS TO PROCESS VARIABLE MEASUREMENTS

Traditionally, product quality is based on direct product quality measurements. Such measurements are usually infrequent and tedious to perform. In modern process industry, many process variables are measured on line, such as pressure, temperature, flows, liquid levels, concentrations. It is also possible to monitor such variables in order to verify the quality of the process. The underlying assumption is that if all the process variables have a constant behavior then the product quality will still be on specification. The process variables are used detecting common variation or special variation. Hence, the full machinery of multivariate process monitoring for detecting deviating process behavior can also be applied in this situation.

BATCH PROCESS MONITORING

The process monitoring methods for detecting deviating behavior of continuous processes cannot be used directly for batch processes since the latter do not operate at a set point. Figure 10.25 shows some typical measured trajectories of process variables in a batch polymerization [Boqué & Smilde 1999, Dahl *et al.* 1999]. These trajectories show (highly) nonlinear behavior and reflect the nonsteady-state operation of batch processes. In Figure

10.25 the trajectories shown for the pressures and temperature are measured for a batch run which produced a good end product. It is not possible to use the concept of set point and investigate the variation around this set point, as was the procedure for continuous processes. For batch processes it is possible to use the concept of 'ideal trajectory'. The assumption is that if all the process variables follow their ideal time trajectories then the batch run will produce good quality products. Stated otherwise, there is not a set point for a process variable (independent of time, as in continuous processes) but a whole collection of set points: for each time point one set point. This collection of set points for a process variable is called an ideal trajectory [Nomikos & MacGregor 1995, Smilde 2001, Smilde *et al.* 2001].

It would be possible to estimate the ideal trajectories by running a large number of normal operating batches and calculating the mean of the trajectories of each process variable. This would also allow establishing what kind of variation there is around such ideal trajectories. Note, however, that the correlation between process variables is not taken into account in such a procedure. Hence, a different route is taken and explained in the following section.

THREE-WAY MODELS OF BATCH PROCESS VARIATION

In a series of papers, Nomikos and MacGregor established an approach for multivariate monitoring of batch processes [Nomikos & MacGregor 1994, Nomikos & MacGregor 1995]. Their basic idea is the following. Suppose that for a batch process under study J process variables are measured at K regularly spaced time points. For batch run i the process measurements are collected in $\mathbf{X}_i$ ($J \times K$). In the ideal case, all the process variables would follow their ideal trajectory for each batch run. Hence, if $\mathbf{X}_1$ is compared to $\mathbf{X}_2$ the difference would be zero. In practice, under normal operating conditions, there will be differences. If the differences are large, this indicates special variation. A systematic way of investigating differences between all batch runs $i = 1, \ldots, I$ is by stacking all $\mathbf{X}_i$ on top of each other, obtaining the three-way array $\underline{\mathbf{X}}$ ($I \times J \times K$). Three-way analysis methods can then be used to model $\underline{\mathbf{X}}$.

The standard method of analyzing $\underline{\mathbf{X}}$, proposed by Nomikos and MacGregor, is by matricizing $\underline{\mathbf{X}}$ to $\mathbf{X}$ ($I \times JK$) and performing an ordinary PCA, i.e. a Tucker1 model (see Chapters 2 and 4). Alternative models, such as Tucker3 and PARAFAC, are also applicable. The Tucker1 and Tucker3 models will be illustrated with an industrial example.

Background of the example

The example is taken from a polymerization batch process and has also been referred to previously by Dahl *et al.* [1999] and Kosanovich *et al.* [1996]. The dataset consists of 50 batches from which eight process variables are measured over approximately 120 time intervals. From this set of batches, two quality variables on the final product were also available. Both process and quality variables are listed in Table 10.7. The reactor in this chemical process (see Figure 10.26) converts the aqueous effluent from an upstream evaporator into a polymer product. The reactor consists of an autoclave and a cooling/heating system. It also has a vent to control the vapor pressure in the autoclave. The recipe specifies reactor and heat source pressure trajectories through five stages.

Table 10.7. Process variables measured during the batch run and quality variables corresponding to the properties of the resulting polymer

Process Variables	Quality Variables
(1) Polymer center temperature (T_{pc})	(1) Molecular weight
(2) Polymer side temperature (T_{ps})	(2) Titrated ends
(3) Vapor temperature (T_{va})	
(4) Reactor body pressure (P_{rb})	
(5) Heat source supply temperature (T_{hs})	
(6) Heat source jacket vent temperature (T_{hj})	
(7) Heat source coil vent temperature (T_{hc})	
(8) Heat source supply pressure (P_{hs})	

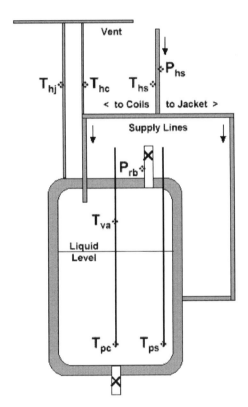

Figure 10.26. The reactor of the example process. For an explanation of the symbols, see Table 10.7.

The first problem to solve relates to the fact that each batch run will have a different duration. Formally, each batch run $\mathbf{X}_i$ has dimensions $J \times K_i$, where J is the number of monitored process variables and K_i is the number of time points monitored (the duration of the batch). In practice, all K_is are different ($i = 1, \ldots, I$). Hence, an array $\underline{\mathbf{X}}$ of size $I \times J \times K$ cannot be built. Below, an approach for handling this problem is discussed.

Alternative approaches are for example dynamic time warping [Kassidas *et al.* 1998], PARAFAC2 [Wise *et al.* 2001] or matricizing $\underline{\mathbf{X}}$ differently [Wold *et al.* 1998].

Building and using the model

INITIAL CALCULATIONS

The number of measurements at each stage of the batch process varies through the different batches. For every batch, the number of time points was counted at each stage, and then the mean values over all the batches were calculated. Next, every variable in a particular batch was interpolated by a linear function according to the number of data points at each stage for that batch. Finally, for every batch the same number of data points was extracted from the linear function at each stage. This number was taken to be the mean value for all the batches, but any other fixed number can be chosen. In this way, a total number of 116 time intervals resulted for every batch (9, 43, 22, 20 and 22 time intervals at each stage, respectively) [Boqué & Smilde 1999]. Figure 10.25 shows the operating profiles for some of the variables measured in the process. The reactor body pressure profile clearly shows the five different stages into which the process is divided.

Initial calculations showed that three batches were suspect and did not belong to normal operation conditions. This was detected by looking carefully at the process variable trajectories of the individual batches, hence from looking at the raw data.

These batches were removed from the data and kept apart for monitoring purposes (see later). The resulting data set $\underline{\mathbf{X}}$ ($47 \times 8 \times 116$) was preprocessed by centering across the batch mode. This way of centering batch data removes the nonlinear time trajectories of all process variables [Nomikos & MacGregor 1994]. After this centering procedure, the data were further processed by scaling each slab associated with a process variable to sum of squares one (scaling within the variable mode). Hence, each slab $\mathbf{X}_j$ ($I \times K$) has sum of squares of one after this scaling.

BUILDING THE MODELS

A pseudo-rank analysis on the three matricized matrices $\mathbf{X}_{(I \times JK)}$, $\mathbf{X}_{(J \times IK)}$ and $\mathbf{X}_{(K \times IJ)}$ was performed and the results are shown in Table 10.8. This table shows that the same number of components always explains the highest percentage of the variation in $\mathbf{X}_{(J \times IK)}$ (*J*-mode, variables), the second highest percentage in $\mathbf{X}_{(K \times IJ)}$ (*K*-mode, times) and the lowest percentage in $\mathbf{X}_{(I \times JK)}$ (*I*-mode, samples).

The topic of selecting the number of components in batch process monitoring models has not been addressed fully in the literature. Different alternatives have been suggested, e.g., cross-validation, broken stick rules etc. [Nomikos & MacGregor 1995]. The purpose of the batch process model is to monitor new batches and it seems therefore reasonable to base the number of components of the model on the ability of the model to detect deviating batches. This alternative has not been fully explored yet, but a related discussion can be found elsewhere [van Sprang *et al.* 2002]. There are also practical issues involved such as interpretability of the model and limiting the number of components to facilitate fault diagnosis. Hence, a practical approach is taken here with respect to the number of components.

Table 10.8. Cumulative percentages of explained variation (%SS) for the polymerization batch data

Number of Components	% I-mode	% J-mode	% K-mode
1	21.1	34.9	24.4
2	38.0	55.8	44.2
3	48.4	68.4	57.5
4	57.6	79.4	66.3
5	64.4	87.0	73.3
6	69.4	92.2	77.2

In the original approach, Tucker1 models are used for modeling the batch data [Nomikos & MacGregor 1994]. That is, the batch process data $\underline{\mathbf{X}}$ ($I \times J \times K$) are matricized to $\mathbf{X}_{(I \times JK)}$ and an ordinary PCA-model is built for $\mathbf{X}_{(I \times JK)}$. Hence, the first column of Table 10.8 is relevant for this model. Taking too many components in the model makes the diagnostics for fault detection more difficult because noise is added to the model. Taking too few components, the model does not capture the main systematic variation and the residuals carry systematic variation. Taking three components seems to be a good compromise as these components explain 48.4 % of the variation in $\underline{\mathbf{X}}$ (Nomikos and MacGregor also reported the use of three components for a similar data set capturing about 55 % of the variation [Nomikos & MacGregor 1995]). The Tucker1 model of $\underline{\mathbf{X}}$ is

$$\mathbf{X} = \mathbf{TP}' + \mathbf{E} \qquad (10.8)$$

where (as in ordinary PCA), the columns of $\mathbf{P}$ ($JK \times 3$) are orthonormal and the columns of $\mathbf{T}$ ($I \times 3$) are orthogonal.

The scores of the batches on the three principal components were calculated and Figure 10.27 shows the scores of the first principal component against the second one. This plot does not show deviating batches, nor large 'holes' or irregularities in the data. Other score plots did not show deviating behavior either. Batches 7, 8, 9 and 11 might seem a bit outlying, but a more detailed analysis (including checking the quality variables) showed that they were normal operation batches.

Another reasonable model for the batch process data is a Tucker3 model. The crucial step in building the Tucker3 model is to establish the size of the core-array $\underline{\mathbf{G}}$, that is, how many components to retain in each mode. The same comments with respect to the choice of the number of components are relevant here. Again, a practical approach is to strike a balance between complexity (a high number of components limits interpretability and fault diagnosis) and amount of explained variation. This results in a (3,2,3) Tucker3 model. This model explains 37.5 % of the variation in $\underline{\mathbf{X}}$. Table 10.9 summarizes the amounts of explained variations of alternative Tucker3 models.

Taking one component extra in the batch direction, i.e. a (4,2,3) model, does not contribute much to the explained variation, whereas deleting one component in the batch direction, i.e. a (2,2,3) model, reduces the explained variation considerably. Similar reasoning for the other modes shows that the (3,2,3) model is a reasonable choice. More sophisticated methods could be used to establish the optimal Tucker3 model complexity as explained in

Table 10.9. Percentages explained variation in the batch process data using different Tucker3 models

Tucker3 Model	Percentage Explained
3,2,3	37.5
2,2,3	31.0
4,2,3	40.6
4,3,3	43.8
3,3,3	38.3
4,3,4	45.6
3,2,4	38.0

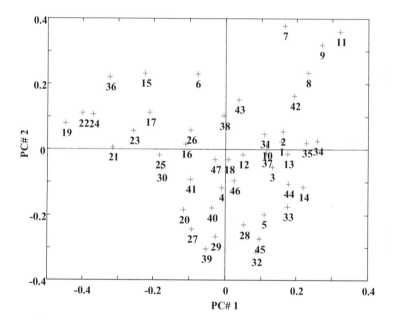

Figure 10.27. Score plot of a Tucker1 model for the polymerization data.

Chapter 7. The Tucker3 model of $\underline{X}$ becomes

$$\mathbf{X} = \mathbf{TG}(\mathbf{C} \otimes \mathbf{B})' + \mathbf{E} \tag{10.9}$$

where all matrices $\mathbf{T}$ ($I \times 3$), $\mathbf{B}$ ($J \times 2$) and $\mathbf{C}$ ($K \times 3$) are orthogonal, as is standard in Tucker3 models. Note that for convenience the same symbols $\mathbf{T}$ and $\mathbf{E}$ are used in Equations (10.8) and (10.9), but the content of the matrices is different.

The scores of the batches on the Tucker3 (3,2,3) model for the first and second components are shown in Figure 10.28. Minor clustering is observed, but other than that the data set seems homogeneous. Apart from rotations, there are some differences between Figures 10.27 and 10.28. This shows that Tucker1 and Tucker3 models are different models reflecting different properties of the data.

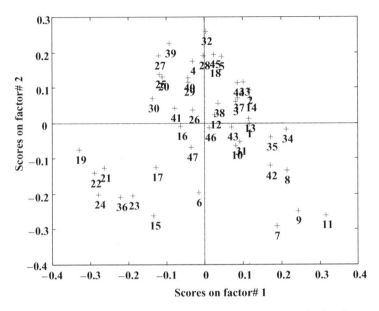

Figure 10.28. Score plot of a Tucker3 model for the polymerization data

Dahl *et al.* [1999] reported two-factor degeneracy PARAFAC models with three or more components for this data set. It was therefore decided not to use a PARAFAC model.

Using the models for on-line monitoring

There are two aspects of interest in the models produced so far. These are the approximation of the data (e.g. $\mathbf{TP}'$) and the residuals. For PCA, the matrix $\mathbf{P}$ represents the common variation in all batches and the matrix of scores $\mathbf{T}$ shows the behavior of the individual batches with respect to this common variation. The residuals show to what extent a batch fits the model. Hence, if the measurements of a new batch are available then two important sets of numbers can be calculated: its scores and residuals. Suppose that the data of the new batch can be represented as $\mathbf{X}_{new}$ ($J \times K$), and in its matricized version as $\mathbf{x}_{new}$ ($JK \times 1$). In order to find the scores and residuals of these data, it is assumed that they are preprocessed using the offsets and scales found during the model building. The scores and residuals of this new batch on the Tucker1 model are obtained as

$$\mathbf{t}_{new} = \mathbf{P}'\mathbf{x}_{new}$$
$$\mathbf{e}_{new} = \mathbf{x}_{new} - \mathbf{P}\mathbf{t}_{new}$$

$$(10.10)$$

where use is made of the orthogonality of $\mathbf{P}$. For the Tucker3 model, the equations look more complicated but the scores and residuals for this model are obtained in a similar way using properly preprocessed data

$$\mathbf{t}_{new} = (\mathbf{G}\mathbf{G}')^{-1}\mathbf{G}(\mathbf{C} \otimes \mathbf{B})'\mathbf{x}_{new}$$
$$\mathbf{e}_{new} = \mathbf{x}_{new} - (\mathbf{C} \otimes \mathbf{B})\mathbf{G}'\mathbf{t}_{new}$$

$$(10.11)$$

where use is made of the orthogonality of $\mathbf{B}$ and $\mathbf{C}$.

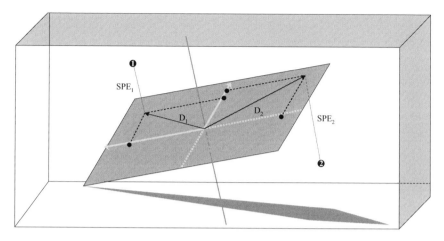

Figure 10.29. A pictorial representation of the D and SPE statistic.

The new batch can now be compared to the normal operating ones by comparing $\mathbf{t}_{new}$ with the $\mathbf{T}$s of the normal operating data and, and, likewise, comparing the average size of $\mathbf{e}_{new}$ with the normal operating residuals. A formal way of doing this is by using the test statistics D and SPE. A pictorial presentation of these statistics is given in Figure 10.29 where two new batches are projected onto the model. The D statistic measures the position of the new batch within the plane (the projection) spanned by the loadings (the model plane) and is calculated using the new scores $\mathbf{t}_{new}$. The SPE statistic measures the average size of the residuals and is calculated using $\mathbf{e}_{new}$. Approximate distributions can be derived for the D and SPE statistic and control charts can be based on this distribution [Box 1954, Nomikos & MacGregor 1995, Tracy *et al.* 1992]. Upon choosing a critical level (mostly α is set to 5 % corresponding to a 95 % control chart), the batch is out of control if either the D or SPE diagnostics (or both) exceed their respective limits.

On-line monitoring of a batch means that control signals are generated while the batch is running. That is, already during a batch run an indication can be obtained about the behavior of the batch. This means that if a fault is detected, the batch can be corrected by undertaking control actions. Automated process control is difficult for batch processes due to the nonlinear behavior of such processes. Hence, it is important to have a good on-line monitoring tool to enable the process operators to undertake action to save a batch.

There is a problem in on-line batch process monitoring which becomes clear when studying Equation (10.11): for calculating new scores and residuals for the ongoing batch, the vector $\mathbf{x}_{new}$ ($JK \times 1$) has to be available. This is by definition not the case for a batch that is not yet finished. Different strategies can be adopted to tackle this problem [Nomikos & MacGregor 1995], of which the 'current deviation' approach seems the most suitable [van Sprang *et al.* 2002]. In that approach, the future values of the process variables in the ongoing batch are filled-in with the values at the current time point in the batch. This approach is also adopted in this example.

Two of the batches earlier considered as suspect were subjected to an on-line monitoring analysis. The results for batch A are given in Figures 10.30 and 10.31. Both sets of charts show a similar behavior of batch A. In the beginning the batch is out of control but after a while it behaves according to normal operating conditions. Then at about time point 68 it

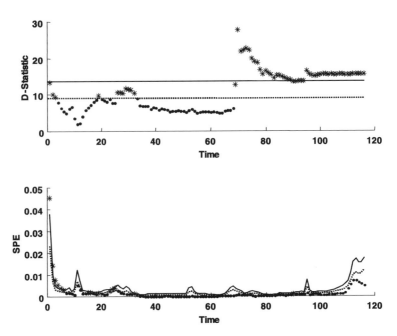

Figure 10.30. On-line *D* (upper) and *SPE* (lower) charts for batch *A* using the Tucker1 model. Legend: solid line, 99 % limit; dashed line, 95 % limit; points, time-resolved values within the 95 % limit; stars, time-resolved values outside the 95 % limits.

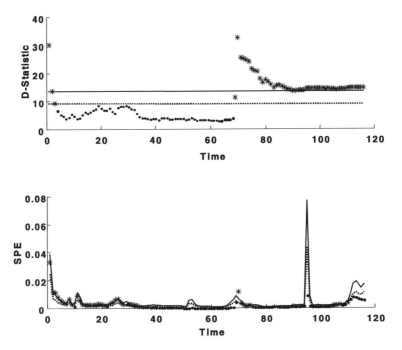

Figure 10.31. On-line *D* (upper) and *SPE* (lower) charts for batch A using the Tucker3 model. Legend: see Figure 10.30.

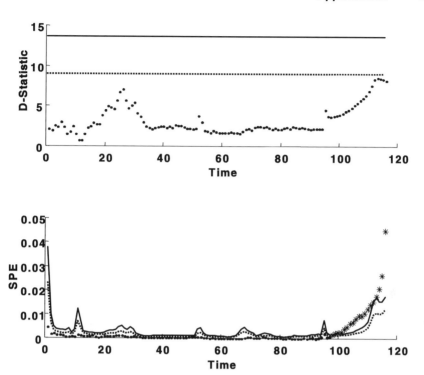

Figure 10.32. On-line *D* (upper) and *SPE* (lower) charts for batch *B* using the Tucker1 model. Legend: see Figure 10.30.

really becomes out of control, especially in the *D* chart and to some extent in the Tucker3 *SPE* chart. Contribution plots should indicate the problems of this batch [Miller *et al.* 1998, Westerhuis *et al.* 1999, Westerhuis *et al.* 2000a].

Note that the confidence limits for the *SPE* chart with the Tucker3 model (Figure 10.26) have a spike around time point 96. This might indicate a poor fitting batch in the normal operating set and should be checked in a further analysis. The alarms in the beginning of the batch should be interpreted with care, since the filling-in procedure used renders the results for the start of the batch rather unstable. Hence, these alarms might be false alarms.

A similar analysis was performed for batch B (see Figures 10.32 and 10.33). Batch B starts to get out-of-control around time point 97–98. This is clearly seen in the Tucker1 *SPE* chart (Figure 10.32 (lower)) and the Tucker3 *SPE* chart (Figure 10.33 (lower)). At a later point in time, also the Tucker3 *D* chart signals. Again a further analysis using contribution plots reveals the error of this batch.

Summary and remarks

The analysis above shows how three-way models can be used to model batch processes and serve as a basis for control charts. As of yet, there is no clear indication of what three-way models are best for modeling batch data. From the example above it is clear that both a

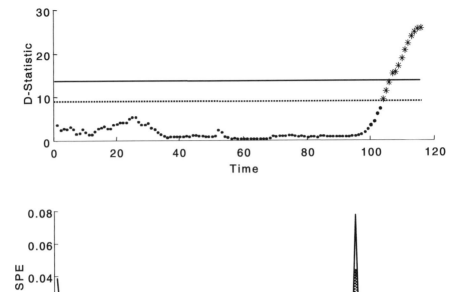

Figure 10.33. On-line D (lower) and SPE (upper) charts for batch B using the Tucker3 model. Legend: see Figure 10.30.

Tucker1 and a Tucker3 model have merits. There are indications that a PARAFAC model is not always suitable [Dahl *et al.* 1999], although they do have also monitoring capabilities in some cases [Louwerse & Smilde 2000, Wise *et al.* 1999]. The adequacy of different models also depends strongly on the type of measurements used to monitor the batch process. If spectroscopic measurements are available, then PARAFAC or restricted Tucker3 models might be appropriate due to the law of Lambert and Beer [Gurden *et al.* 2001, Westerhuis *et al.* 2000b].

A rigorous analysis of the relative merits of different models for a certain batch process also depends on the goal of the batch process model. If the goal is monitoring new batches then false alarms (type I errors), nondetects (type II errors) and diagnostic capability are very important. The three-way models should then be compared with respect to their performance on these criteria. There are unclear issues such as the degrees of freedom of the distribution of the D statistic which only depends on the number of batches (I) and the number of factors in the batch mode (R). This seems strange as the scores in the batch mode in Tucker1, Tucker3 and PARAFAC models are based on different degrees of freedom, yet, if the same number of components are used in the batch mode, these differences are not reflected in the limits of the D statistic.

Another way of dealing with batch process data is by matricizing $\underline{\mathbf{X}}$ ($I \times J \times K$) such that the J-mode (process variable) is left intact [Wold *et al.* 1998]. This results in an $\mathbf{X}$

($IK \times J$) matrix which can be modeled with a PCA model. Hence, this gives also a Tucker1 model, but with a different mode kept intact as compared to earlier. It should be realized that there are large differences between the Tucker1 models used by Wold and MacGregor.

The Tucker1 models used by Wold do not have to use filling-in procedures for new batches. This is an advantage when compared to the Tucker1 models used by MacGregor, although experience has shown that the current deviations filling-in procedure performs rather well [van Sprang *et al.* 2002]. Note, however, that large differences in batch lengths give unstable control chart limits in Wold's Tucker1 models.

The centering of the Tucker1 model used by MacGregor ensures that the main nonlinear dynamics of the data are removed. Hence, it can be considered as a linearization step, which is a common tool in engineering, and this is a very sensible preprocessing step when linear models are used subsequently. The centering of the Tucker1 model used by Wold is performed over two modes simultaneously, which might introduce extra multilinear components (see Chapter 9). Hence, multilinear models may have problems with Wold's arrangement of the data.

The model parameters $\mathbf{P}_M$ ($JK \times R$) in the Tucker1 model used by MacGregor capture the correlation between the process variables but also the correlation in time (even their 'interactions': changes of correlations between process variables in time). The model parameters $\mathbf{P}_W$ ($J \times R$) in the Tucker1 model used by Wold only capture the correlation between the process variables averaged over all time points. Hence, the dynamics of the batch process are captured in either $\mathbf{P}_M$ for MacGregor's arrangement or in the $\mathbf{T}_W$ for Wolds arrangement. This results in a completely different behavior of the control charts. For the Tucker1 model used by MacGregor, the *SPE* chart usually signals an upset, whereas in the Tucker1 model used by Wold this is usually done by the *D* chart. Again, the relative merits of both arrangements should be established for a particular application keeping in mind the type I error, the type II error and the diagnostic capabilities. For a more detailed discussion of these issues, see Smilde [2001] and van Sprang *et al.* [2002].

Some other applications in batch process monitoring

Batch process monitoring has been used by Albert and Kinley to study a fermentation process [Albert & Kinley 2001]. The merits of matricizing $\underline{\mathbf{X}}$ ($I \times J \times K$) to $\mathbf{X}_{(I \times JK)}$ or $\mathbf{X}_{(IK \times J)}$ are discussed. For the analysis, 144 batches were collected, with 17 on-line variables recorded hourly (one fermentation takes about 140 h to complete). The final yield was also available to judge the performance of the fermentation. There were 65 batches available of high yield and these were used to build the batch process model using five components. A detailed discussion on selecting the number of components and the amount of explained variation of these components is not provided. The 44 batches of low yield were monitored with this model according to the strategy of Nomikos and MacGregor [1995]. In 38 % of the cases, the control charts based on the model could signal a low yield batch. A univariate analysis (studying the individual process variables trajectories) signals only 20 % of the low yield batches.

In the area of polymerization processes, case studies of batch process monitoring based on the Nomikos and MacGregor scheme were reported. In an emulsion batch polymerization, 46 batches were available each having 106 measurements taken during the batch run of 20 process variables [Neogi & Schlags 1998]. The measured quality parameter for a finished batch was product viscosity. Variables were selected on the basis of their significance with respect to variation in product viscosity, and a six-component Tucker1 model was built using 13 variables. This model explained 67 % of the process variation. No information is provided regarding the rationale of using six components. The model was capable of detecting a faulty batch, and pointing to the cause of the deviating behavior. In an application of a suspension polymerization process making PVC, 15 on-line process variables were measured at 60 points in time [Tates *et al.* 1999]. After removing outlying batches, 69 batches were available for batch process modeling. A Tucker1 model was built, and cross-validation indicated that two components were sufficient. This two-component model explained 38 % of the variation in the data. Subsequent monitoring of two suspect batches using this model detected these batches as deviating. Contribution plots were used to diagnose the faults.

10.6 Exploratory Analysis in Chromatography

Background of the field

Chromatography is by far the most widely used analytical separation method [Harris 1997]. There are different kinds of chromatography, of which liquid (LC) and gas chromatography (GC) are the most abundant. Liquid chromatography can again be divided into normal-phase and reversed-phase chromatography. The principle of chromatography is that a mixture of solutes is injected into a column that contains a stationary phase. The mobile phase transports this mixture through the column and the difference in the affinities of a solute for the stationary and mobile phases determines the time it takes to elute. Elution is usually expressed as a retention factor (see below). The names LC and GC arise because in LC the mobile phase is a liquid and in GC the mobile phase is a gas. *Normal-phase* LC has a polar stationary phase and an apolar mobile phase, while in *reversed-phase* LC this is reversed.

There are different ways in which LC or GC analysis can lead to three-way data. These were already briefly described in the introduction. If chromatographic measurements are performed on different stationary phases, at different mobile phase compositions for different solutes (chemicals), then a three-way array of retention factors (or of peak-widths, asymmetry factors etc.) is generated. Such arrays can be studied to explore the differences and similarities between stationary phases and/or solutes.

Chromatographic instruments can also be combined with spectroscopic instruments, e.g., LC can be coupled with a diode-array UV spectrometer. This generates, per sample, a two-way matrix of measured absorbances. For multiple samples, this generates a three-way array. This array can be analyzed with several goals: (i) finding the concentrations of unknowns in mixtures; this is called second-order calibration as explained in Section 10.3; (ii) finding underlying profiles of the chemical compounds, e.g. finding pure component UV spectra; this is called curve-resolution (Section 10.1). Sometimes (i) and (ii) are combined.

Background of the example

An example will be given on the use of three-way models to explore chromatographic data. The data have the structure of *stationary phases* × *mobile phases* × *solutes* and the purpose of the analysis is to understand the differences between different stationary phases and

Table 10.10. The solutes used in the LC example

Name	Abbreviation
Acetophenone	ACP
n-Butyl-4-aminobenzoate	BAB
Ethyl-4-hydroxybenzoate	EHB
2-Phenylethanol	PE
Toluene	TOL
n-Propyl-4-hydroxybenzoate	PHB
Ethyl-4-aminobenzoate	EAB
Methyl-4-hydroxybenzoate	MHB

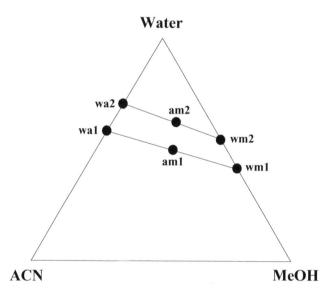

Figure 10.34. The mixture design used in the LC example. Legend: all fractions are reported as water/acetonitrile/methanol fractions; wm1 = 0.47/0.00/0.53, wm2 = 0.55/ 0.00/0.45, wa1 = 0.62/0.38/0.00, wa2 = 0.70/ 0.30/0.00, am1 = 0.54/0.19/0.27, am2 = 0.63/0.15/0.22.

analytes. Hence, it is a purely exploratory endeavor, trying to find interpretable patterns that give insight into the chromatography.

This example is from reversed-phase chromatography [Smilde 1990, Smilde *et al.* 1990, Smilde & Doornbos 1991, Smilde & Doornbos 1992]. A data set was constructed from the retention factors of eight aromatic compounds measured on six different types of stationary phases at six mobile-phase compositions. The mobile phase consisted of water, acetonitrile and methanol in different ratios, chosen according to a mixture design [Cornell 1981]. This design is shown in Figure 10.34. The six stationary phases had different functionalities, i.e., the bonded phase contained methyl (C1), hexyl (C6), octyl (C8), octadecyl (C18), cyanide (CN) or phenyl (PHE) end-groups. The solutes used are reported in Table 10.10 together with their abbreviations.

This resulted in a three-way array $\underline{\mathbf{X}}$ ($6 \times 6 \times 8$) of retention factors with mode one holding the stationary phase, mode two the mobile phase and mode three the solutes. The purpose of the analysis was two-fold: (i) to investigate the differences between the types of stationary-phase material and (ii) to develop a calibration model for transfer of retention factors from one stationary phase to another one. Only the first part of the investigations will be discussed.

There is a priori knowledge about this data set. The main source of variation between the stationary phases is due to differences in hydrophobicity. The C18 phase is the most hydrophobic one whereas the CN and phenyl phases are the least hydrophobic. The C6 and C8 phases are slightly less hydrophobic than the C18 phase and the C1 phase takes an intermediate position between C18 and CN. There may also be differences in selectivity[2] between the types of stationary phases. Chemically, C6 and C8 are similar, but C6 and CN are quite different and may show a different selectivity. The column packings may also differ in efficiency, i.e. yield different peak widths, but this is not reflected in a retention factor. Hence, two sources of variation are expected between the stationary phases.

For reasons to be explained below, all retention factors k were transformed logarithmically to $\ln k$. Due to the design used, it is possible to check for nonlinear mixing behavior of the mobile phase, i.e. nonlinear variation of $\ln k$ with composition. Such behavior is indeed present: the mean $\ln k$ of toluene on C1 at mobile phase wm1 and wa1 is 1.38, corresponding to a k value of 3.98. The $\ln k$ value of toluene on C1 at mobile phase am1 (the 0.5:0.5 mixture of wm1 and wa1) is 1.10, corresponding to a k value of 3.01. This is a clear difference, which indicates that nonlinearities are present. Hence, there are at least three sources of variation in the mobile-phase mode: the relative amounts of the two organic modifiers and nonlinear mixing behavior.

The solutes are all substituted benzenes, but have chemical differences. There are members of two different homologous series in the data set (MHB, EHB, PHB and EAB, BAB) that differ mainly in hydrophobicity, but selective effects can be expected for the other solutes.

The nonlinear mixing behavior may also depend on the type of stationary phase and on the solute: for the C18 phase less pronounced differences are expected. Hence, interactions can be expected between the modes.

Building and using models

INITIAL CALCULATIONS

From the raw retention values, retention factors can be calculated, which are dimensionless numbers correcting for variations in instrumental instabilities such as variation in flowrate. The retention factor is defined as the ratio of the net retention time ($t_r - t_u$) and the retention time of an unretained peak (t_u), where t_r is the retention time of the solute. This is standard practice in liquid chromatography [Christian & O'Reilly 1986]. Moreover, taking the logarithms of these retention factors ideally linearizes the data, e.g., with respect to fractions of organic modifier in the mobile phase composition [Schoenmakers 1986].

---

[2] Selectivity means that the variations in retention are different for different solutes.

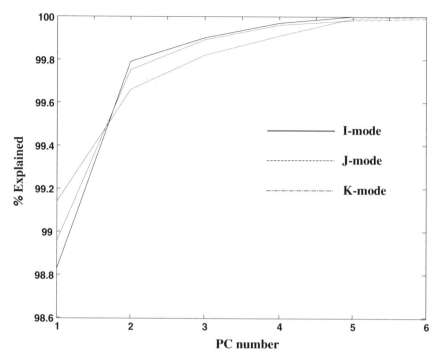

Figure 10.35. The percentage of explained variation of the three PCA models of the different matricized arrays. Legend: I-mode is a PCA on $\mathbf{X}_{(I \times JK)}$ etc.

Retention factors are measured with an approximate constant relative error and this translates to a constant absolute error after the logarithmic transformation [Atkinson 1985]. Hence, after these transformations, it is reasonable to apply multilinear models assuming a constant error.

After the above-mentioned transformations, the data were mean centered across the stationary phase mode, because investigating the differences between stationary phases was the primary goal of the analysis. No scaling was applied, because all the values were measured in the same units and were of the same order of magnitude.

The first analyses performed were principal component analyses on the three matricized arrays $\mathbf{X}_{(I \times JK)}$, $\mathbf{X}_{(J \times IK)}$ and $\mathbf{X}_{(K \times IJ)}$ to get an impression of the complexity in the different modes. The explained variation of these three PCA models is shown in Figure 10.35. For models with two principal components, all explained variations are already high. The highest one is for the stationary-phase (I) mode, and the lowest one for the mobile-phase (J) mode. This is in agreement with the a priori knowledge (see earlier), where two sources of variation were expected between stationary phases and three between the mobile phases. All in all, two or three components seem sufficient to describe the systematic variation in all the models.

Because the interest is primarily in the differences between the stationary phases, the PCA results of $\mathbf{X}_{(I \times JK)}$ are discussed in some more detail. The root-mean-squared error (RMSE) of the one-, two- and three-component PCA models go down from 0.08, via 0.04 to 0.02. These values were calculated as the square root of the residual sum of squares divided by

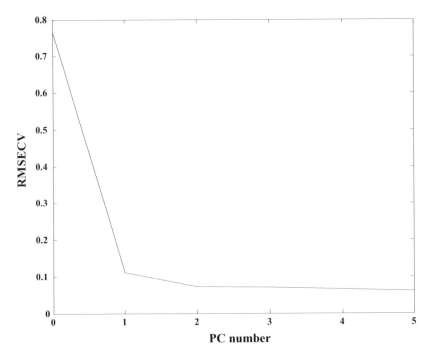

Figure 10.36. Cross-validation results of $\mathbf{X}_{(I \times JK)}$. Legend: RMSECV stands for root-mean-squared error of cross-validation and represents the prediction error in the same units as the original measurements.

the total number of elements in the array. A cross-validation was performed on the matrix $\mathbf{X}_{(I \times JK)}$ and the results are shown in Figure 10.36. The RMSECV value is calculated as the square root of the prediction error sum of squares divided by the total number of elements in the array. This figure indicates that probably two principal components are sufficient to describe the data.

A score plot of this PCA model is shown in Figure 10.37(a). The scores on the first principal component show the ordering of the stationary phases according to their known hydrophobicity, i.e., $CN < PHE \sim C1 < C6 \approx C8 < C18$. The second principal component is more difficult to interpret, but could show differences in selectivity. This second component, though, only explains 0.96 % of the variation.

The corresponding loading plot of this PCA model is shown in Figure 10.37(b). The loadings of the first principal component are all positive reflecting the interpretation of the first component as a hydrophobicity scale: for all solute/mobile phase combinations retention values increase in the range CN, PHE, C1, C6, C8 and C18. The loadings of the second component are more difficult to interpret.

The RMSE value of 0.0351 of the two component model can be translated (roughly) to a relative error of 3.5 % in the k values, which is a reasonable error for chromatography. The RMSE values can also be calculated per stationary phase and per variable (combination of one solute and one mobile phase composition). The results are shown in Figure 10.38. The variation on the stationary phases CN and C18 is explained the best. This was expected

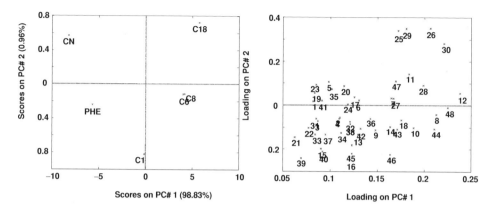

Figure 10.37. (a) Scores and (b) loadings of the first two components of a PCA model on $\mathbf{X}_{(I \times JK)}$.

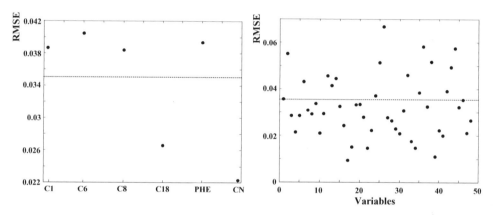

Figure 10.38. Residual plots of (a) the stationary phases and (b) all mobile-phase/solute combinations. Legend: RMSE stands for root-mean-squared error (see text) and the dotted line indicates the mean RMSE for the whole data set.

as these stationary phases 'span' the hydrophobicity range (also see the scores plot). From Figure 10.38. there are no indications of outlying stationary phases and/or solute/mobile-phase combinations.

PARAFAC ANALYSIS

The next step in the analysis was to perform a PARAFAC analysis of the data set $\underline{\mathbf{X}}$ where only centering across the stationary-phase mode was used. Going from a one-component PARAFAC model to a three-component model, the RMSE value dropped from 0.093, via 0.056 to 0.042. Already for a two-component model problems were encountered indicated by a high Tucker congruence coefficient (see Chapter 5) of -0.99, which was observed to decrease with increasing number of iterations. For a three-component model, this number was -0.98 for two of the components. These numbers indicate that the solutions might be degenerate and not meaningful. The reason for this degeneracy will become clear when discussing the Tucker results. The loading plots for the two-component PARAFAC model

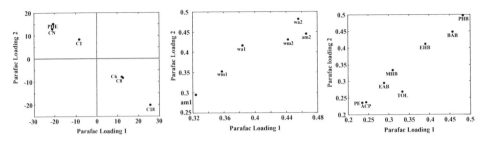

Figure 10.39. Loading plots of the two-component PARAFAC model of **X**. (a) the stationary-phase mode, (b) the mobile-phase mode and (c) the solute mode.

illustrate the degeneracy: severe collinearity between the two components (Figure 10.39). The degeneracy also seriously hampered the cross-validation of the PARAFAC model, which was hence not pursued further.

One way to deal with degeneracy problems is too force orthogonality between the factors of the first PARAFAC mode. This resulted in models going from RMSE values of 0.093 (one component), via 0.060 (two components) to 0.056 (three components). Due to the orthogonality restrictions, the RMSE values are higher than the corresponding values of the unrestricted model (except for the one-component model, of course). The loading plot of the stationary-phase mode for the two-component orthogonal PARAFAC model is shown in Figure 10.40. Clearly, the high correlation has disappeared and the first component again describes the hydrophobic differences, as earlier in the PCA model (see above). As there were indications that a Tucker3 model was more appropriate for this data, the orthogonal PARAFAC model was not investigated further.

TUCKER3 ANALYSIS

A Tucker3 analysis was performed on **X**, where again the data were only mean centered across the stationary-phase mode. Different models were tried up to and including three components in all directions. For these models the RMSE values were calculated; these are shown in Figure 10.41. This figure also shows the RMSECV values as a result of a cross-validation of all the above-mentioned models. Cross-validation was performed by dividing the data set in 15 segments thereby leaving out 19 or 20 elements per segment ($6 \times 6 \times 8/15 = 19.2$). Figure 10.41 contains a lot of information and deserves a detailed discussion. All RMSECV values are larger than the corresponding RMSE values, as should be, since fit is mostly better than prediction. Comparing a model, say, a (1,3,3) model, to a more complex model (say, a (2,3,3) model), the RMSE always goes down, which is also expected. Comparing the (1,1,1) component model with the ones having one more components in two of the modes[3] (e.g. (1,2,2)) consistently gives a lower RMSECV. Hence, the (1,1,1) model is too simple.

Models having two components in the stationary-phase mode can be compared to 'neighboring' models having one extra stationary-phase component (e.g. comparing (2,2,2) with (3,2,2)). There are four such comparisons and they all indicate that going from two to three components in the stationary-phase mode does not improve the model. Similar comparisons

---

[3] Note that a (1,1,2) Tucker3 model is not relevant since it always has exactly the same fit as the (1,1,1) model (see Chapter 5).

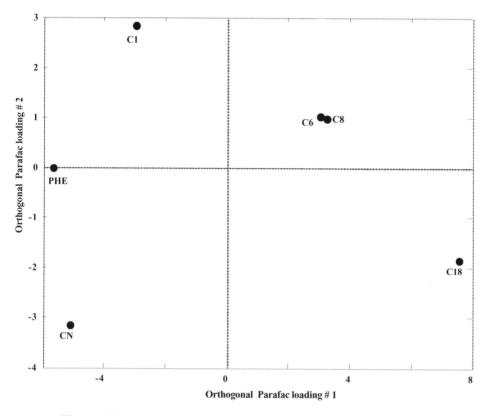

Figure 10.40. Loadings for the two-component orthogonal PARAFAC model.

show that for the mobile phase going from two to three components improves the model (three out of four comparisons suggest this) and, likewise, in the solute mode there is also evidence that going from two to three components improves the model (three out of four comparisons suggest this).

As a (3,3,3) component model already had a low RMSE (0.0396) and the size of the data set is small, the analysis was not pursued for more than three components. The conclusion is that a (2,3,3) Tucker3 model is very reasonable for these data with an RMSE of 0.043, which corresponds to 99.69 % of explained variation. This choice of the Tucker3 model also corresponds with the a priori knowledge of this data (see earlier). The loading plots for this model are shown in Figure 10.42. In the stationary-phase loadings of the (2,3,3) Tucker3 model, the hydrophobicity scale is clearly reflected. These loadings are very similar to the PCA loadings (Figure 10.37) and were discussed there. The mobile-phase loadings clearly show the effect of nonlinear mixing behavior. Although the mobile-phase compositions are arranged according to an experimental design, this design is not reflected in the loadings (e.g., am1 is not in between wa1 and wm1). The solute loadings show a pattern for the members of the homologous series: MHB, EHB and PHB are on a line parallel to the first loading and likewise EAB, BAB. This again supports the assumption that the first loading represents hydrophobicity because in going from MHB to PHB the solute becomes more hydrophobic and will be affected more by the hydrophobicity of the stationary phase. Hence, the loadings on component one will increase for MHB to PHB. Preliminary calculations

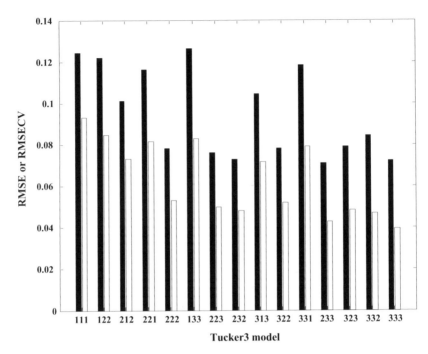

Figure 10.41. RMSE (open bars) and RMSECV (filled bars) values of some relevant Tucker3 models of **X**.

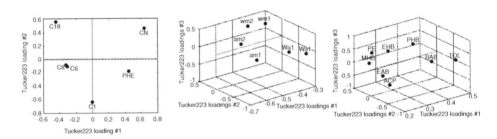

Figure 10.42. Loading plots of the (2,3,3) Tucker3 model of **X**. (a) Stationary-phase loadings, (b) mobile-phase loadings and (c) solute loadings.

(not shown) indicated that the solute toluene has the largest variation in the data set: it is the most affected by the hydrophobic differences between the stationary phases. This is also reflected in a high loading of toluene on the first component.

The elements of the core-array of the (2,3,3) Tucker3 model are given in Table 10.11. The Tucker3 model is calculated in the standard way (orthonormal loadings). Hence, the squared elements of the core-array represent the amount of variation explained by the corresponding loadings. These values are expressed in percentages of the total variation to be explained and add up to the 99.69 % of the total model.

The values of the core-array clearly show a dominant first factor, which represents the hydrophobicity of the stationary phases. The second largest value is the (2,1,2) element,

Table 10.11. The core-array of the (2,3,3) Tucker3 model of $\underline{\mathbf{X}}$. The values between brackets are the percentages of explained variation of that component

	M.Ph. 1	M.Ph. 2	M.Ph. 3
Sol. 1 St.Ph. 1	12.8896	−0.0035	0.0124
	(98.52 %)[a]	(0 %)	(0 %)
Sol. 1 St.Ph. 2	0.0048	0.7022	−0.2449
	(0 %)	(0.29 %)	(0.04 %)
Sol. 2 St.Ph. 1	−0.0064	−0.5118	−0.2225
	(0 %)	(0.16 %)	(0.03 %)
Sol. 2 St.Ph. 2	0.9525	0.0610	−0.1488
	(0.54 %)	(0 %)	(0.01 %)
Sol. 3 St.Ph. 1	−0.0040	−0.2440	−0.2244
	(0 %)	(0.04 %)	(0.03 %)
Sol. 3 St.Ph. 2	−0.1740	0.1150	0.1085
	(0.02 %)	(0.01 %)	(0.01 %)

[a] Values were rounded to the second decimal place. Values below 0.005 % are indicated by 0 %.

which corresponds to the second component in the stationary-phase mode, the first component in the mobile-phase mode and the second component in the solute mode. Also the (2,2,1) element has a contribution. The Tucker analysis also sheds light on the degeneracy problem encountered in the PARAFAC analysis. It is known that a Tucker3 structure underlying a three-way array can cause a PARAFAC degeneracy [Kruskal *et al.* 1989]. This is probably the case here.

Summary and remarks

In this example it is shown that a Tucker3 model provides a good description of the variation in the data. The PARAFAC model is problematic for these data and the Tucker1 model mixes up two of the modes. Core rotations for the Tucker3 model could be used to enhance the interpretation of the Tucker3 model. Another possibility is to rotate the loading matrix of the mobile-phase mode to reflect the underlying design more appropriately [Carroll *et al.* 1980].

Some other applications in this field

The earliest application of the Tucker3 model in chemistry is from the work of Spanjer [1984] and de Ligny *et al.* [1984] working with chromatographic applications. In this application two three-way data sets are analyzed. The first data set contains measured normal-phase liquid chromatographic retention factors of 19 solutes, measured on six different stationary phases at two mobile-phase compositions. This results in a $19 \times 6 \times 2$ array. The solutes were monosubstituted benzenes and polycyclic aromatic compounds. The second data set is a $39 \times 3 \times 2$ array. It contains retention values of 39 more complicated solutes, viz.,

Table 10.12. Analysis of variance results for the second data set

	P^a	Q^a	R^a	$NPAR^b$	Df^c	s^d
ANOVA				42	171	0.58
Tucker3	1	1	1	42	171	0.38
	2	2	2	84	129	0.21
	3	2	2	122	91	0.08
	3	3	2	126	87	0.07
	4	3	2	164	49	0.05

[a] P, Q and R refer to the number of Tucker3 components in the solute, mobile-phase and stationary-phase mode, respectively.
[b] NPAR stands for 'number of parameters'.
[c] Df stands for degrees of freedom and are calculated according to de Ligny *et al.* [1984]. Note, however, the difficulties in defining degrees of freedom for three-way models.
[d] The value s is the standard deviation of the residuals and is calculated by dividing the sum of squared residuals by Df and taking the square root.

monosubstituted phenols, anilines and pyridines measured on three stationary phases at the same two mobile-phase compositions as the first data set. This second data set will be discussed in some detail.

The retention factors were expressed as logarithms of net retention volumes (values similar to retention factors, see above). This transformation again linearizes the data. The different models for each data set were compared with respect to their standard deviations of residuals. This is shown in Table 10.12. An analysis of variance (ANOVA) model was fitted to the data. No interactions were assumed, because this would enlarge the number of parameters to be estimated. Even the simplest Tucker3 model outperforms this ANOVA model in terms of the s value. The best Tucker3 model is the (4,3,2) model, giving a very low s value. For this model there is no reduction in the mobile-phase and stationary-phase modes, as the number of components in these modes equals the number of levels. The results obtained can be compared with the results obtained with the Snyder equation in chromatography [de Ligny *et al.* 1984].

A special feature of the analysis of de Ligny *et al.* is that they have used an expectation maximization algorithm for handling missing data (see Chapter 6). Using the calculated model parameters, it is possible to provide predictions of the missing values. Moreover, error variances of these predictions are provided based on local linearizations of the model around the parameters.

10.7 Exploratory Analysis in Environmental Sciences

Background of environmental three-way analysis

Environmental measurements are known for creating large data sets, because the data are taken at many locations and over long periods of time. Because of the expense in collecting the data there is also a need for measuring as many variables as possible once the samples are taken. This typically gives data sets having many chemical variables for many locations over many time units (see also Figure 10.4). The future trend is to increase the data flow with the use of automatic samplers/analyzers and with shortening of sampling periods to get better time resolution. Also better spatial resolution is a future trend as sampling equipment becomes more automated and easier to handle.

The processes underlying the variations in data taken in the environment are very complex and it is almost impossible to find a true linear underlying hard model that has created the content of each environmental sample. Environmental data are also noisy, more because of limitations of sampling than because of instrument noise. Therefore, the assumption of low-rank trilinear or bilinear variation is usually weak. Additionally, the analysis is complicated by the fact that the measured chemical variables may not all be relevant to the process studied. Often the relevance of the measured variables is a guess or due to an instrumentation compromise. Because of this, latent variables are useful for exploring the data. Many of the variables also suffer from detection limit problems. Three-way analysis solves many of the above-mentioned problems: it provides estimates of latent variables, it reduces the influence of noise, and it allows an insight into which measured variables are most important to the latent variables and which variables are correlated or unique. This may be sufficient to characterize an environmental problem, or if not to give feedback for organizing new measurements.

Environmental studies can be carried out in the atmosphere, in water, in sediments, on waste, in soil etc. It is possible to measure in the air, sediment, waste, soil and water directly or indirectly by collecting plants and animals from them. It is also possible to measure food prepared from animals living in a certain environment, e.g. enhanced levels of trace elements in milk, cheese or yoghurt. The environment may be defined on a number of different size scales. Sometimes large water and air masses are studied and sometimes limited indoor environments are under investigation. Sometimes local environmental problems are under study and sometimes global sources and sinks of chemicals are the subject. All this makes environmental studies very broad: any chemicals in any medium over any kind of time resolution can be of importance.

An example – the VENICE data

Venice is a historical city built on top of a lagoon. The beauty of the city is also its worst enemy: the water erodes the foundations of the buildings. The closeness to the water creates environmental concerns. Water quality is essential for the livelihood of the city. There are two main phenomena to be studied here: increase in turbidity by naval traffic, industrial activity and growth of algae. The example is based on monthly sampling over almost 4 years of water in 13 locations along the Venice lagoon. Eleven water quality variables (chlorophyll-a, solid material, transparency, fluorescence, turbidity, suspended matter, NH_4^+, NO_3^-, P, COD, BOD) were measured in 44 monthly samples giving a $(13 \times 11 \times 44)$ array. The VENICE data were initially analyzed by Leardi *et al.* [2000] and the pretreatment of the data was done as recommended by them: taking logarithms of variables 1–9 and 11 and mean centering and unit variance scaling of the 572×11 matrix over the 11 variables. After that the $(13 \times 11 \times 44)$ three-way array was built. Note that the centering part is not

Table 10.13. Selecting the PARAFAC rank for the VENICE data model. Numbers are rounded to the nearest integer. Component sizes are given as percentage sum of squares relative to the sum of squares of the data

Pseudorank	%SS Explained	Component Sizes (%)
1	20	20
2	35	20,15
3	45	39,38,16

in accordance with the general guidelines provided in Chapter 9. According to Chapter 9, the same centering and scaling could also be performed iteratively, providing a better fit to the data. This is not pursued here.

The VENICE model

The analysis of this array is done by PARAFAC modeling. Determination of the model rank in this exploratory study is done by increasing the rank until the loadings start to show degeneration. The model of rank three shows degeneration. This can be seen in the component sizes in Table 10.13, but even better in plots of the components.

Figure 10.43 shows the correlation of C-loadings one and two for a rank-three model. Figure 10.44 shows the A-loadings as line plots and indicates a strong similarity of the two largest components. Such similarities of loadings often indicate degenerate solutions and make the results difficult to interpret. Additional diagnostics also pointed to a two-component model being adequate and therefore, rank two is chosen for the final model. This model is repeated in Table 10.16.

Interpretation of the VENICE model

The analysis continues with an interpretation of the loadings of the rank-two model. The C-loadings represent time evolution and are preferably shown as line plots. Figure 10.45 shows these line plots. The first loading shows seasonality and a strong downward trend. The second one shows a weak downward trend. The first C-loading is the most interesting one to study because it shows seasonality.

The first loading in Figure 10.45 is almost exactly the same as the one in the original paper [Leardi *et al.* 2000] showing that the Tucker3 model with an almost superdiagonal core used in the paper and the PARAFAC model fitted here are equivalent for all practical purposes. The role of the variables is interpreted by looking at the B-loadings, this time in a scatter plot in Figure 10.46. The first loading direction (horizontal) shows negative loadings for the variables related to the presence of algae (fluorescence, chlorophyll-a, solids) and positive loadings for the ions that are considered nutrients for the algae. This axis may be called a *biological* axis. The second loading contrasts transparency with almost everything else in the water. It is thus an *anthropogenic* loading.

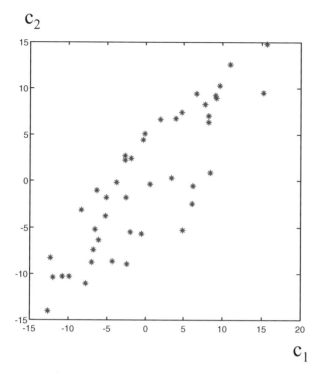

c_2

c_1

Figure 10.43. The two largest C-loadings of the rank three PARAFAC model show some correlation, which could be an indication of degeneracy and hence difficult interpretation.

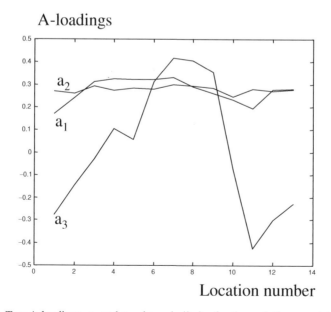

A-loadings

Location number

Figure 10.44. Two A-loadings, a_1 and a_2, show similarity for the rank three model. The model is degenerate.

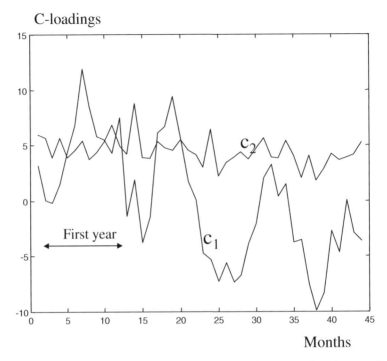

Figure 10.45. The C-loadings of the PARAFAC model of rank two. Trends and seasonality can be seen, especially in the first loading.

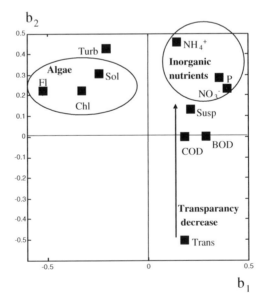

Figure 10.46. The B-loadings of the rank-two model show the importance of the variables. Chl = Chlorophyll-a, Sol = solid material, Trans = transparancy, Fl = fluorescence, Turb = turbidity, Susp = suspended matter. The scatter plot is made square, with equal scales and zero is indicated.

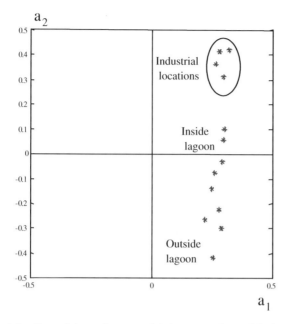

Figure 10.47. The A-loadings of the rank two model show groupings of the locations in the second component. The values of the second loading depend very closely on the distance to the open sea. A cluster of industrial locations is highlighted. The scatter plot is made square, with equal scales and zero is indicated.

The A-loadings in Figure 10.47 show the different locations with a clear trend and clustering in the second component. A cluster of industrial locations is indicated. Also locations inside and outside the lagoon are marked. The variability in the first loading is rather small. Now it is possible to interpret the ensemble of loadings. The first loading is biological. It is a contrast between algae and nutrient parameters (Figure 10.46). More algae means that more inorganic nutrients are used up, so these nutrients decrease. This loading is very seasonal (Figure 10.45), showing differences between winter and summer, but also a general increase in eutrophication over the years. The locations behave almost identically in this component. The second component is mainly reflecting transparency of the water (Figure 10.46) and general pollution and is assumed to be an anthropogenic loading. Transparency is high outside the lagoon (far from the city and coastline) and especially low in the industrial area (Figure 10.47). This effect has no seasonality and only a slight decrease over the years (Figure 10.45). Such a small time trend may not be significant.

The interpretation of the model is dependent on the variables used. Other ions or trace metals could have led to a different model and different interpretations. Therefore a well-motivated choice of variables is crucial. In most projects, this choice has to be modified after the first conclusions become available. Leardi *et al.* have left out temperature and salinity in advance in their work because these were not related to the goals of the study and would have given confusing large components with trivial interpretations. As with most environmental data, there is a rather large residual variation. This residual is not only noise, but it contains also systematic information that a PCA model would be able to explain in part, but the data set is probably too small to warrant such an effort.

Table 10.14. Examples of environmental three-way data arrays

Dataset	Subject	Array	Mode 1	Mode 2	Mode 3	Preprocessing	Goal of the Study
CRAB	Crabs/water	$48 \times 25 \times 3$	Individual/ location	Elements	Organ	Center/scale	Study the influence of mining effluents in water on the health of crabs
NIGER	Water/Niger	$10 \times 13 \times 22$ $(10 \times 15 \times 22)$	Location	Quality parameter	Time	Center/scale	Study the influence of industry on river water quality, seasonality
VENICE	Water/Venice	$13 \times 11 \times 44$	Location	Quality Parameter	Time	Logarithm/ center/scale	Study of the influence of industry on lagoon water quality, seasonality
SCENES	Air	$6 \times 18 \times 45$	Location	Elements	Time	Center/scale	Receptor modeling, checking source profiles
ALERT	Air	$11 \times 52 \times 24$	Year	Week	Ion/element	Scale	Receptor modeling, establishing source profiles, study of seasonal variations

With data of this type it would be possible to include sampling depth as a fourth mode to give a location × depth × chemical × time array.

Table 10.15. Chemical variables used in selected environmental examples

Dataset	Variables Studied
CRAB: 25 variables in crabs	Ag, Al, As, Ca, Cd, Co, Cr, Cu, Fe, K, Li, Mg, Mn Mo, Na, Ni, P, Pb, Se, Si, Sn, Ti, U, V, Zn
NIGER: 13 variables in water	Turbidity, pH, conductivity, Cl^-, hardness, alkalinity, O_2, BOD^a, COD^b, oils and fats, NH_4^+, NO_3^-, NO_2^-
VENICE: 11 variables in water	Chlorophyll-a, solid material, transparancy, fluorescence, turbidity, suspended matter, NH_4^+, NO_3^-, P, COD, BOD
SCENES: 18 variables in air	Al, Si, P, S, Cl, K, Ca, Ti, V, Mn, Fe, Cu, Zn, Br, Cd, Pb, organic C, elemental C
ALERT: 24 variables in air	Cl^-, Br^-, NO_3^-, SO_4^{2-}, H^+, Na^+ $NH4^+$, K^+, MSA^c, Mn, V, Al, Zn, Pb, Ca, Ti, I^-, In, Si, As, La, Sb, Sm, Sc

^{a} BOD: biological oxygen demand.
^{b} COD: chemical oxygen demand.
^{c} MSA: methanesulphonic acid.

Background of selected examples from the literature

In the literature examples described next, it is shown how authors have used three-way models and their parameters. Detailed environmental conclusions are equally important, but these can be found in the original papers and are not repeated here.

Three-way applications in environmental studies became important from the beginning of the 1990s. A few examples will highlight the possible use of three-way analysis in environmental studies. The VENICE example shown earlier is included with the models and conclusions of the authors for the sake of comparison. An overview is given in Table 10.14 and the variables used in different examples are given in Table 10.15.

An early study on crabs was carried out by Gemperline and coworkers [Gemperline *et al.* 1992, West *et al.* 1992]. They studied how mining influences water quality and how the water quality in its turn influences crab disease (CRAB data). The variables were element concentrations measured by inductively coupled plasma emission. They used a 48 × 28 × 3 array: 48 individuals × 28 elements × 3 organs. One group of crab individuals were healthy controls living in a clean environment. Three of the 28 elements were not used because they showed no variation at all in the data set. The crab example was also presented in Chapter 8.

Henrion studied water quality by chemical and physical parameters. The study included ten sampling locations along the Niger delta and 22 sampling occasions spread almost monthly over 2 years (NIGER data). They measured 13 water quality variables giving a 10 × 13 × 22 array [Henrion & Henrion 1995]. Five sampling stations were upstream from an industrial zone and five were downstream. The raw data are given in their book. A different study used 15 variables [Henrion *et al.* 1992].

Another study of water quality was published by Leardi *et al.* [2000] based on monthly sampling over almost 4 years of water in 13 locations along the Venice lagoon (VENICE data). Eleven water quality variables were measured in 44 monthly samples giving a $13 \times 11 \times 44$ array. The three water examples have heterogeneous variables measured in quite different scales. This can be appreciated by looking at the variables in Table 10.15. Therefore, centering and some appropriate scaling are needed. The authors also use logarithmic transformation due to skew distributions of most variables.

Because of the physical properties of water, most studies are local, concentrating on a part of a river, a lake, an estuary or a lagoon. Dispersion of polluting agents in water is often slow and only in one or two dimensions. In air, dispersion is in two or three dimensions and at higher speeds, so that larger areas have to be studied. Numerous air pollution studies are carried out by national agencies, where physical and chemical properties of air or particulate matter in air are measured in many locations at regular intervals. The air pollution is assumed to come from a number of sources (e.g. coal combustion, automotive exhausts, sea spray, metal smelters) and by a complex dispersion process it is collected at different locations. The dispersion process is different for each sampling occasion because of meteorological factors, mainly wind speeds and directions [Xie *et al.* 1999a, Xie *et al.* 1999b, Xie *et al.* 1999c, Zeng & Hopke 1992a, Zeng & Hopke 1992b]. A systematic study of the dispersion process is called receptor modeling.

ELABORATION **10.2**

Receptor modeling

Receptor modeling works on a number of simple assumptions:

- There are a limited number of emitting sources located at fixed points and each source has a well-defined emission profile (a fingerprint defined by chemical composition in elements, cations, anions, organic chemicals).
- Dispersion mixes the contributions from the sources.
- At the sampling station a linear combination of source contributions is measured.

With many sampling stations and sampling periods it is possible to do a curve resolution giving the number of sources, the 'pure' source profiles and the contribution of each profile to each sample [Hopke 1991]. This curve resolution problem was earlier solved by factor analysis, but lately three-way solutions have been applied. The establishment of source profiles relies a lot on factor rotation. Bilinearity is very much dependent on the dispersion process.

A simulation study using artificial sources was performed by Zeng & Hopke [1990]. They showed that a Tucker3 model could give the correct core size, but that a rotation was needed to get interpretable loadings. This was done by varimax rotations of the core and related rotations of the loadings of the profiles.

Zeng & Hopke [1992a] studied air particulate matter collected on filters by mainly X-ray fluorescence analysis. The eleven sampling locations were in four states in the southwestern USA. The three-way array consisted of measurements of concentrations at 6 locations × 18 species × 45 sampling times (SCENES data) . Some locations and sampling times had to be left out because of missing data. The data were centered and scaled as in the VENICE example, because the variables had huge concentration differences. A simplification of the

Table 10.16. Three-way models for some selected examples. Rounding to integers is applied to the last column

Dataset	Model	3-Way Rank	Rotation	%SS Explained
CRAB	Tucker3	$4 \times 5 \times 2$	No	70
NIGER	Tucker3	$2 \times 2 \times 2$	Core	30
	PARARAC	2	No	29[a]
VENICE	Tucker3	$2 \times 2 \times 2$	Core	35
	PARARAC	2	No	35[a]
SCENES	Tucker 3	$3 \times 4 \times 5$	Varimax	63
ALERT	PARAFAC	2	Yes, profiles	≈ 90
	(+PCA)	(+5)		

[a] A new calculation on data received from the authors.

receptor model was made by defining a small number of location zones and dispersion regimes. This makes it possible to get a Tucker3 core of size: emission sources × emission zones × meteorological regimes.

The articles of Hopke *et al.* [1999] and Xie *et al.* [1999c] have measurements of arctic air in one location (ALERT data) over 52 weeks a year during 11 years. The analysis by mainly instrumental neutron activation analysis, ion chromatography and inductively coupled plasma gave concentrations of 24 species, both cations and anions. Missing weeks were replaced by geometric means of present weeks. The week mode is expected to give information on seasonality and the year mode would give annual differences. An important goal is to identify source profiles. The model is built using the multilinear engine [Paatero 1999].

ENVIRONMENTAL MODELS AND THEIR INTERPRETATION FOR
THE LITERATURE EXAMPLES

Table 10.16 shows some of the models used for environmental analysis in the literature. Tucker models with a small core are popular. It is obvious that environmental data are noisy and that the models do not explain a very large percentage of the total sum of squares.

The interpretation of the CRAB data [Gemperline *et al.* 1992] is mainly performed by three-dimensional joint plots of the Tucker3 components and by visualization of the core. This core is used without rotation. One of the joint plots is shown in Chapter 8, Figure 8.37, now as a two-dimensional plot. This plot represents a plane of the core with a high correlation to the gill. It shows groupings in healthy site 1, healthy site 2 and diseased site 2. The healthy references of site 1 form a dense cluster, meaning low variability. The healthy and diseased crabs from site 2 show up in separate spread out clusters, with a gradient towards the healthy references of site 1. The diseased crabs are spread out towards a cluster of elements, some of which are typical for the local clay that is dispersed in the water because of the influx of industrial effluents. Results for other data and detailed interpretations are in the paper.

The VENICE data were explained earlier. The authors used a Tucker3 model that gave very simlar results as the PARAFAC model. For the NIGER data, Henrion & Henrion [1995] calculated a $2 \times 2 \times 2$ core and rotated it to a more interpretable (attempted superdiagonal) solution. The choice of the small core was simply made for plotting purposes. Scatter plots of Tucker3 loadings and superimposed plots are used. A clustering in the locations can be related to selected variables and the seasonality is explained. The first rotated component explains seasonality and the second one explains site differences upstream and downstream from the industrial zone. The seasonality is due to the rain season and influences ionic content. The site differences are mainly related to oxygenation that is reduced downstream from the industrial site.

For the SCENES data Zeng and Hopke [1992a] find a Tucker 3 core of four emission sources, three immission zones and five meteorological regimes. The loadings are varimax rotated and a rotated core is given. Identified sources are soil, urban dust, vegetative combustion and Cu. The authors admit that relating the empirically found regimes to meteorology is difficult. The paper also has an interesting appendix with nomenclature definitions.

Xie *et al.* [1999b] use the multilinear engine to find a model with five two-way factors and two three-way factors for the ALERT data. Because of the use of five PCA factors, the model explains a high percentage of the total sum of squares. The seven factors need rotation to become interpretable. For the two-way factors, the variation over the 11 years and the seasonality over the weeks are interpreted. The authors also take up rescaling of the factors. The interpretation of the element factors needs a backscaling to ng m^{-3} to make the interpretation more meaningful.

Summary and remarks

Many environmental measurements in water and air give location × chemical (property) × sampling time three-way arrays. Sometimes locations and sampling times have to be removed because of missing data or results below the detection limit. This removal can severely damage the three-way array by leaving out columns, rows or tubes. The chemicals/elements/ions measured have widely different concentration ranges necessitating mean centering and unit variance scaling over the variables. Sometimes skew distributions suggest a logarithmic transformation. The environmental studies give many variables. In some cases (XRF, ICP) the variables are determined by the method of analysis chosen and not by careful selection. As a result, many measured variables are not useful in the models.

Model building using PARAFAC or Tucker3 models gives parsimonious models that are easily interpreted. The three-way rank or Tucker 3 core gives useful information about the data under study. Because of the noise in the data, large residuals are to be expected. Sometimes two PARAFAC or Tucker3 loadings suffice to understand the main information in the environmental problem under study. Different numbers of components are investigated until a reasonable amount of the sum of squares is explained or until more loadings become difficult to interpret. This may for example happen when the PARAFAC model degenerates by the addition of one more PARAFAC component.

Interpretation of the Tucker3 core is very important and sometimes core rotations are needed. Some authors rotate in order to simplify the core, while others see the need for a varimax rotation of the loadings. These rotations are very subjective and require extensive background knowledge. The loadings are used in line or scatter plots and allow the detection of outliers, groupings and trends. Joint plots are used frequently.

Environmental data come in a wide variety and what is expected from them also varies a lot. This makes data pretreatment, model building and model interpretation very dependent on background knowledge, more than, e.g., in fluorescence or MSPC data where general goals are formulated more easily.

10.8 Exploratory Analysis of Designed Data

Background

Sometimes two of the modes of a three-way array are formed by a two-way ANOVA layout in qualitative variables and the three-way structure comes from measuring a continuous variable (e.g. a spectrum) in each cell of the ANOVA layout. In such cases each cell of the ANOVA has a multitude of responses that not even MANOVA can handle. When quantitative factors are used, one or two modes of the three-way array may result from an experimental design where the responses are noisy spectra that behave nonlinearly. Such data are treated in this section.

In the cases described above, the three-way data usually have no strong underlying assumptions of low-rank bilinearity but this does not make three-way analysis meaningless. If it is possible to put the data in a three-way array ($I \times J \times K$) it is often possible to model that array with PARAFAC or Tucker3 models in an exploratory way. The determination of the model dimension supplies important information about the data and the use of the A-, B- and C-loadings in plots gives an insight into outliers, clustering, trends and other meaningful properties of the data. Some examples are given in Table 10.17. They have in common that the dimensions of the modes, except the spectral one, are rather small. In Table 10.18, a summary is given of the preprocessing used for these data. Additional details are given in the following text.

The BATCH background

The batch synthesis data [Geladi & Forsström 2002] is concerned with the production of isoamyl acetate by an esterification from isoamyl alcohol and acetic acid in a boiling reaction mixture. An acid catalyst is needed and paratoluenesulphonic acid is used. Water is removed by azeotropic distillation with benzene for moving the equilibrium towards the ester.

The simplified reaction is:

$$C_5H_{11}OH + CH_3CO-OH \overset{H^+}{\rightleftharpoons} C_5H_{11}O-COCH_3 + H_2O(C_6H_6)\uparrow$$

Eleven batch reactions were carried out as a 3^2 design in molar ratio of the reagents and amount of catalyst, with two extra center points. Vis–NIR transflectance spectra of the batches were taken with a fiberoptic probe every 3 min over almost 2 h giving 40 points in time. The Vis–NIR spectra contain 1050 wavelengths, every 2 nm from 400 to 2498 nm. The data form an $11 \times 40 \times 1050$ three-way array. This is a typical small × small × large array with a spectral mode as the large mode. Figure 10.48 gives the raw spectra of one batch. The

Table 10.17. Some examples of designed three-way arrays

Name	Array	Mode 1	Mode 2	Mode 3	Goal
XPS	6 × 15 × 90	Pressure	Exposure time	Electron energy	Detect chemical bonds, follow surface reactions as a function of pressure
MALT	11 × 6 × 351	Batch	Day	Wavelength	Follow malting kinetics, detect intermediates, cluster batches
PEAT	5 × 7 × 21	Peat type	Treatment	Particle area	Compare types, treatments, check treatment efficiency
SEEDS	6 × 10 × 1050	Mother	Father	Wavelength	Compare mothers and fathers of seeds
SYNT	14 × 13 × 701	Batch	Time	Wavelength	Follow reaction kinetics, identify reagents, intermediates, product
BATCH	6 × 40 × 776	Batch	Time	Wavelength	Follow reaction kinetics, identify reagents, intermediates, product

Table 10.18. Some construction principles and preprocessing techniques used for the three-way arrays in Table 10.17

Name	Construction Principle	Preprocessing
XPS	2-way ANOVA in modes 1 and 2	Background correction
MALT	Incomplete 2-way ANOVA for batches	None
PEAT	2-way ANOVA in modes 1 and 2	None
SEEDS	2-way ANOVA in modes 1 and 2	Smoothing/derivative
SYNT	Factorial design in mode 1	Smoothing/derivative
BATCH	Factorial design in mode 1	Smoothing/derivative

region 400–1000 nm shows no peaks, just background. The region above 2200 nm shows noise from using a fiber optic probe. Between 1000 and 2200 nm there are some peaks that change during the reaction. Very often, wavelength ranges are removed before the final model is made.

Near infrared spectra are known to be difficult to interpret and to exhibit nonlinear behaviour. For example, the spectra of pure analytes can change in mixtures. This invalidates Beer's law. For the interpretation of the results, spectra of pure chemicals and known mixtures are also collected. The goal of the example is to check whether a PARAFAC decomposition of the three-way array allows the interpretation of the loadings as a function of what happens in the reaction.

The example is rather complicated. It consists of a preliminary PARAFAC model on the whole data. This shows a block effect that is identified as leading to a reduction of the design. After that the data are pretreated by smoothing the first derivative, cropped in

Absorbance

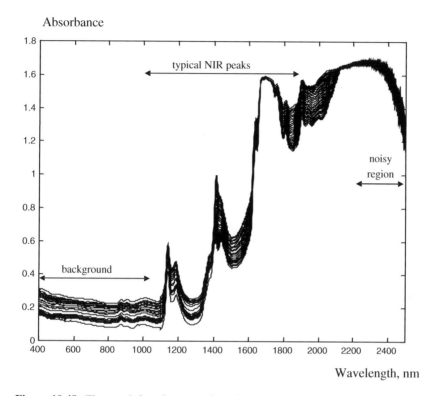

Wavelength, nm

Figure 10.48. The near infrared spectra of one batch (40 times) of the BATCH example.

wavelength range and then a definite PARAFAC model is made. For the interpretation of the spectral loadings, PCA models of the loadings together with the pure spectra need to be made.

The BATCH model

A PARAFAC model of the noncentered $11 \times 40 \times 1050$ array can be made. A model of rank one explains 99 % of the sum of squares. The A-loading of this model is shown in Figure 10.49. There is clearly a block effect in the batches. After five batches, something has changed. The C-loading is given in Figure 10.50 and this loading can explain what the cause of the block effect is. The interpretation is not easy, but comparison with spectra of pure chemicals and known mixtures makes it possible to get an indication of what is wrong.

It is shown in [Geladi & Forsström 2002] that the block effect can be attributed to a contamination of the fiber optic probe with benzene and ester. This contamination makes the original experimental design useless and only the last six batches are used for further analysis. The reduced experimental design is shown in Figure 10.51. The loading in Figure 10.50 shows that the spectra have high absorbance values above 1200 nm. In such a case, it is very common, especially when centering is not used, that the first component explains this 'offset' spectrum common to all samples. As this spectrum constitutes the main fraction

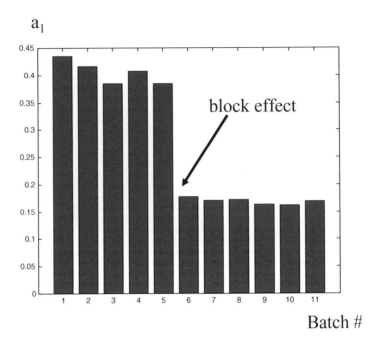

Figure 10.49. The A-loading of a rank one PARAFAC model.

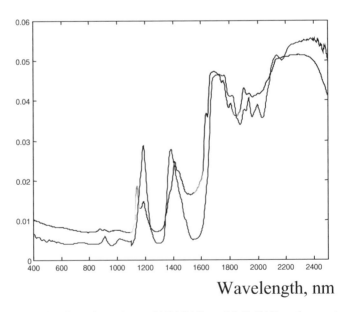

Figure 10.50. The C-loading of a rank one PARAFAC model. Solid line: the spectrum of ester in benzene. Dashed line: the C-loading.

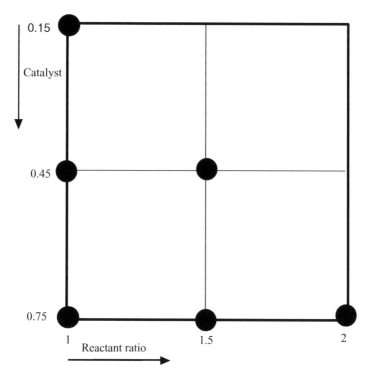

Figure 10.51. The reduced design in molar ratio and catalyst.

of the variance in the data, the percentage explained by this component can be exceedingly high (typically above 99 %). In this case, the percentage explained is more than 99 %. Superficially this is counterintuitive because the component is not contributing much to explaining the variation *between* samples but rather explains the variation in common. However, it is a simple consequence of the structure of the data and does not imply an incorrect model.

This common shape, however, may often be due to baselines and other non-chemical variation which is not necessarily interesting to model. In order to avoid modeling this variation, derivatives of the spectra can be used. A Savitzky–Golay smoothing derivative is used with a window of 27, fourth order fitting and first derivative calculation. Some first derivative spectra are shown in Figure 10.52. From the area below 1200 nm (mainly background) and above 2300 nm (noise from glass fiber), the figure shows that some wavelength regions can be left out. Cropping to the region 1000–2198 nm gives the final array the size 6 × 40 × 600. The spectra are in derivative form and this takes away the intrinsic nonnegativity of the spectra because contrary to raw spectra, the first derivatives have negative parts. This is no problem for the PARAFAC analysis, but it becomes problematic for the interpretation of the loadings. This is so because of the sign uncertainty problems as presented in Chapter 8. The C-loadings (corresponding to the spectral mode) can be changed in sign together with a sign change for A-loadings or B-loadings without a change in model fit resulting from it (see Chapter 8). This can give interpretation problems for the signs of less important (smaller) loadings.

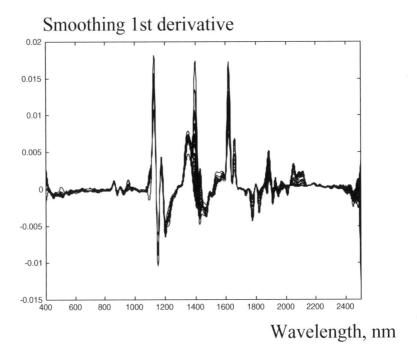

Figure 10.52. The smoothing first derivative spectra for one batch.

For the selection of model dimensionality, PARAFAC models of increasing rank are fitted and compared. The component sizes of models of rank one to five are given in Figure 10.53. The interpretation of this figure is simple: components with sizes larger than the total sum of squares point to degenerate or otherwise unstable models. The components do not come in size order from the particular algorithm used. It is good practice to reorder the loadings (A, B and C) after size.

Rank one or rank four is the best choice based on this type of evaluation. For each model rank tested, the appearance of the loadings as well as check for degeneracy and compliance with background knowledge was assessed. The rank-four PARAFAC model explains 99 % of the sum of squares and was chosen to be the most appropriate model. The four components explain 71 %, 31 %, 16 % and 7 % of the sum of squares of the data (the sum can exceed 100 % due to the nonorthogonality). The C-loadings have to be used to identify the chemical composition of each component. This is done by comparing with spectra of pure chemicals and known mixtures. These have to undergo the same smoothing-derivative and cropping procedures as the three-way array.

The BATCH model interpretation

Visual comparison of loadings and pure chemical/mixture spectra is not easy. It is possible to do a singular value analysis of the matrix containing the ten pure chemical/mixture spectra and the four C-loadings. This analysis requires all spectra and loadings to be normalized to the same norm because shapes are compared, not sizes. The singular value decomposition

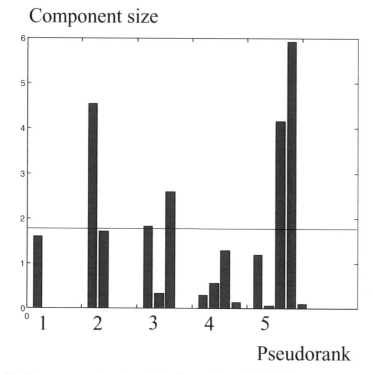

Figure 10.53. The component sizes in models of increasing rank. The horizontal line is the total sum of squares of the data set.

is done without any further pretreatment. The singular values of this analysis are shown in Figure 10.54.

The broken stick or elbow method indicates that four components are significant. These explain 90 % of the total sum of squares in the data. The score plots for these components can be studied in more detail. Figure 10.55 shows the scatter plot of the first two components. In the plot, chemicals, mixtures and C-loadings are identified. The third C-loading is an outlier. The loadings c_1, c_2 and c_4 are on a line, with c_4 between c_1 and c_2. The chemicals and mixtures dissolved in benzene are all close to the c_1–c_2 line. The chemicals and mixtures not dissolved in benzene are further away. This is to be expected because the reaction was carried out in benzene. In order to make a better interpretation, Figure 10.55 is shown again as Figure 10.56, with some names removed for clarity. Figure 10.56 shows that the loading c_1 resembles the alcohol in benzene and the reaction mixture in benzene. The conclusion is that the first loading represents the reaction mixture. The second PARAFAC, c_2, component resembles benzene, ester in benzene and acetic acid in benzene. The conclusion is that the second PARAFAC component is the final product. The closeness to acetic acid has to do with the excess of this reagent used in some experiments. Benzene is never used up, so the final product would have spectral features of it. The fourth PARAFAC component is between c_1 and c_2. The interpretation is that this small component is a compensation for the fact that different batches have different reaction rates.

The PARAFAC loading c_3 can not be explained by any of the pure chemical/mixture spectra. The c_3 loading is shown in Figure 10.57. There is a strong peak at 1400 nm. This

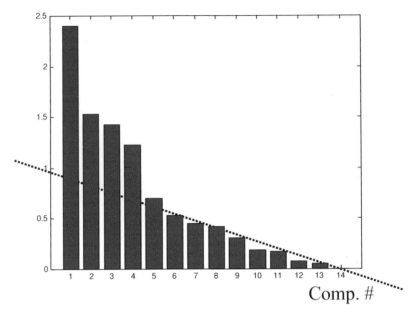

Figure 10.54. The singular values for the 14 × 600 matrix. There are four significant components. The dashed line is used in the elbow method to find significant components.

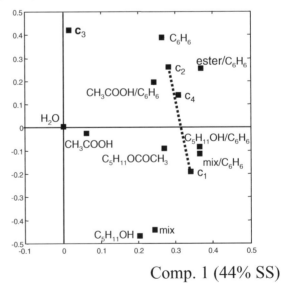

Figure 10.55. The score plot of components one and two for the 14 × 600 matrix. Pure chemicals, mixtures and PARAFAC C-loadings are indicated. c_4 is on a line between c_1 and c_2.

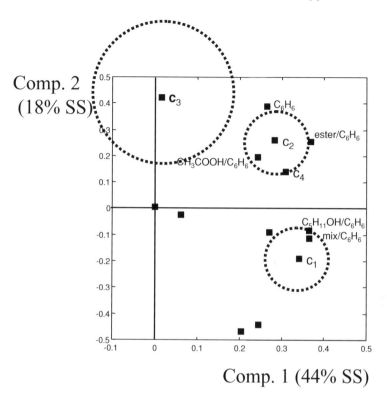

Figure 10.56. The score plot of components one and two for the 14×600 matrix. Pure chemicals and mixtures in benzene and PARAFAC C-loadings are indicated. $\mathbf{c}_3$ has no close neighbours that explain its composition. $\mathbf{c}_1$ represents the starting reaction mixture. $\mathbf{c}_2$ represents the final product, the excess of acetic acid and the remaining benzene.

peak also exists in the acetic acid spectra, as shown in Figure 10.58. The raw spectra show that there is a peak broadening around 1400 nm, probably caused by a quick protonation of the acetic acid when the catalyst is added. The order of the components is different from that in the original work [Geladi & Forsström 2002] because a different wavelength range is used, but the interpretation of the components is the same. The third and fourth component from the singular value decomposition of the 14×600 matrix corroborate these conclusions and do not add much extra, so they are not shown here.

Figure 10.59 shows the first and second B-loadings as a scatter plot, giving the time profiles in the reaction. There is first a quick reaction with the catalyst. Then the reaction slows down and changes direction to slowly move towards the endpoint that is never reached. The reaction was not finished after 117 min. The B-loadings show that the reaction goes in two stages, with a first quick protonation of acetic acid by the catalyst and then the reaction product is slowly used up and the ester is formed.

The interpretation of the smaller B-loadings is problematic because of sign reversal problems. Figure 10.60 gives the sign-reversed third B-loading and Figure 10.61 gives the sign-reversed third C-loading. This is the way these loadings were presented in the original publications. Figure 10.60 shows a decrease in the product of Figure 10.61, the protonated

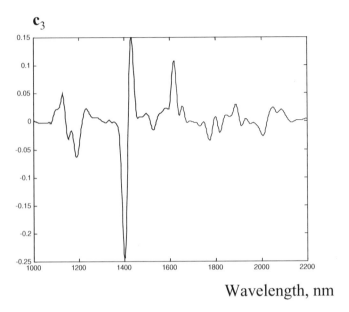

Figure 10.57. The c_3 loading. A strong peak at 1400 nm is visible.

Derivative spectra

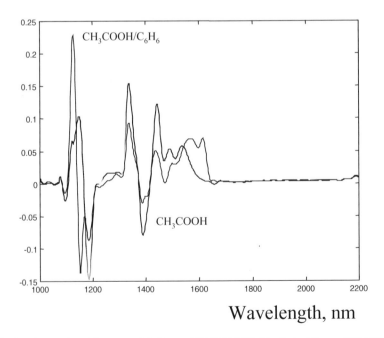

Figure 10.58. The acetic acid spectra: full line $= CH_3COOH/C_6H_6$, dashed line $= CH_3COOH$.

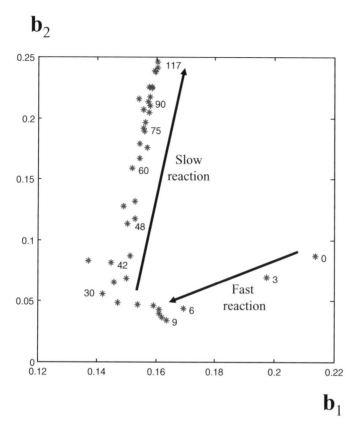

Figure 10.59. The $\mathbf{b}_1-\mathbf{b}_2$ scatter plot. Times indicated are in minutes. There is a fast first reaction, a change in direction and then a slow evolution towards the endpoint.

acetic aced. It is impossible to conclude only from mathematical or statistical arguments which sign reversal is right. Also chemical arguments are hard to use, because the product of component three is an intermediate that has no known chemical spectrum.

The third and second A-loadings are shown in Figure 10.62 as a scatter plot. The experimental conditions are indicated. There is a strong direction of increasing reagent ratio and a less pronounced direction of increasing amount of catalyst. The third component is protonation of acetic acid and this is dependent on the amount of catalyst used and the ratio of acetic acid and alcohol. The second component is formation of the ester, and this is also dependent on the ratio of acetic acid and alcohol. There is also an interaction between factors that makes the interpretation less straightforward.

This example shows that it is possible to explain the batch reactions through PARAFAC components and that the PARAFAC loadings can provide a chemical interpretation:

- Component 1: benzene + reagents decrease over time to a fixed level because the benzene is never used up and there is excess acetic acid.
- Component 2: ester that increases over time.
- Component 3: fast protonation of acetic acid by the catalyst giving a peak shape change.
- Component 4: a small compensation component explaining nonlinearities resulting from the different reaction rates of the batches.

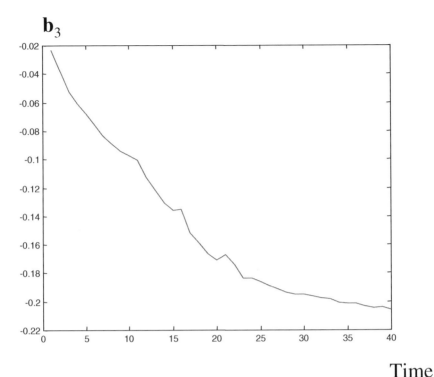

Figure 10.60. The sign-reversed third B-loading.

Background of selected literature examples

An overview of the examples in this section is given in Table 10.17, with an explanation of the design and pretreatment in Table 10.18. The examples are not exhaustive.

The XPS example of Do and McIntyre [1999] describes the oxidation of a pure aluminum surface in a controlled atmosphere of low pressure water vapor. The surface was measured by X-ray photoelectron spectroscopy. The experiment was a two-way ANOVA in the quantitative variables pressure and exposure time using six pressures and 15 nonlinearly spaced reaction times. The spectral variable consisted of 90 photoelectron energies, expressed in eV. The total array is 6 × 15 × 90. A nonlinear reaction of the photoelectron spectra to the pressure and exposure time was expected. Background correction based on physical background knowledge of the XPS spectra was used for preprocessing the data.

The MALT example of Allosio *et al.* [1997] is about monitoring a malting batch reaction by near infrared spectroscopy. The eleven batches form an incomplete two-way ANOVA in grain varieties and malting conditions. Malting takes 5 days and every day, also on day zero, the near infrared spectrum in 351 wavelengths was measured, giving an 11 × 6 × 351 array.

The PEAT example of Geladi *et al.* [2000] was introduced in Chapter 8 (Figures 8.11 and 8.13). It is made up of an ANOVA layout in five peat types (treatment A) and seven particle

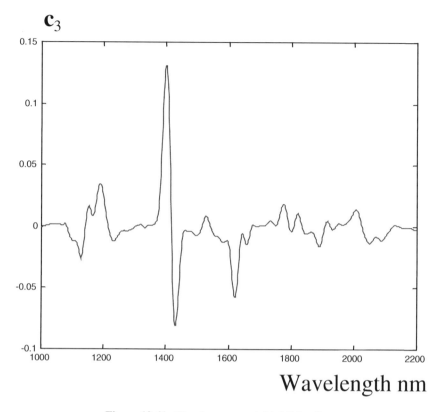

Figure 10.61. The sign-reversed third C-loading.

coagulation treatments (treatment B). Each combination of treatments gave a particle size distribution histogram in 21 bins. The array is $5 \times 7 \times 21$. The goal was to coagulate small particles in a slurry that would otherwise slow down a filtration process. The data were used without any pretreatment. Because of the possible errors in measurement of particle sizes large residuals were expected.

Another ANOVA example is given by Lestander *et al.* [2002]. Here the material was pine seeds with known clones for mothers and fathers giving a two-way ANOVA layout in the qualitative treatments mother (A) and father (B). For each cell in the ANOVA layout, visible and near infrared spectra (400–2498 nm) of the resulting seeds were measured in 1050 wavelengths (SEEDS data). The spectra were measured to determine if genetic information (genotype or phenotype) about the seeds showed up in them. This could then lead to the construction of equipment for quick automatic seed sorting. The array is $6 \times 10 \times 1050$. The spectra are averages of three replicates. The near infrared spectra contain too many responses to allow for an ANOVA or MANOVA calculation. Furthermore, the information is spread out in such spectra. The data were also subjected to principal component analysis of a 180×1050 array and a number of principal components were used in ANOVA calculations (each component at a time) and MANOVA calculations (all components together). The spectra were pretreated by taking a smoothing first derivative in the spectral mode.

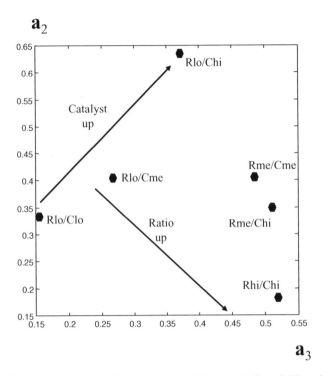

Figure 10.62. The a_3–a_2 scatter plot. Experimental conditions are indicated. There is a strong direction of increasing reagent ratio and a somewhat weaker direction of increasing amount of catalyst. An interaction of the factors is assumed.

The first batch data set from Geladi and Åberg [2001] was used in Chapter 8. It will be called SYNT. The production of isoamylacetate was carried out by esterification from isoamyl alcohol and acetic acid under reflux conditions. The reaction was catalyzed by a proton donor and a molecular sieve was used to remove the water, moving the reaction equilibrium towards the ester. Chemical synthesis was carried out in a number of batches according to an experimental design in reagent ratio, amount of molecular sieve and amount of catalyst. This was a 2^3 design with some center points and some runs replicated. Each batch was monitored by fiberoptic visible-NIR measurments (400–2500 nm) every 10 min during 2 h. This gives a $16 \times 13 \times 1050$ array. This array was later reduced to a $14 \times 13 \times 701$ size by removal of unreliable batches and deletion of empty or noisy wavelength ranges. It was impossible to use the raw spectra for building a three-way model and therefore smoothing first derivatives were used.

Model building and interpretation of examples

An overview of the models is given in Table 10.19.

The XPS [Do & McIntyre 1999] example uses the positive matrix factorization variant of the PARAFAC algorithm [Paatero 1997] and obtains three components with all-positive

Table 10.19. Three-way models used

Name	Model	3-way rank	%SS explained
XPS	PARAFAC[a]	3	Not in paper
MALT	PARAFAC[a]	4	Not in paper
PEAT	PARAFAC	4	80
	Tucker3	$4 \times 4 \times 4$ core	81
		$2 \times 4 \times 4$ core	80
SEEDS	PARAFAC	3	99
SYNT	PARAFAC	3	97
BATCH	PARAFAC	4	99

[a] Constrained.

values in all the loadings. The selection of model-rank and the percentage of the total sum of squares explained are not discussed in the paper. The loadings are interpreted as line plots. The spectral loadings allow the detection of chemical bonds. The time mode loadings indicate that pure Al is reacted away slowly at the surface, that oxide is formed quickly and that an intermediate state of hydride formation exists. The hydride formation is very nonlinear as a function of the pressure. A more detailed interpretation of the ANOVA structure is not given in the paper.

The MALT data [Allosio *et al.* 1997] are analyzed by PARAFAC with orthogonality constraints on the batch and wavelength modes. Four components can be extracted and their spectral loadings may be interpreted as intermediates and reaction products of the malting by looking at the line plots. The ANOVA structure is incomplete and cannot be interpreted easily. The paper is unclear about why four components are chosen or how much of the total sum of squares they explain.

For the PEAT example [Geladi *et al.* 2000] it is possible to extract four PARAFAC components from the $5 \times 7 \times 21$ data array and to interpret these loadings meaningfully. Also Tucker3 models were developed. Determination of the model-rank is done by increasing the number of PARAFAC components until the model becomes unstable or by cross-validation (see also Figures 7.5 and 7.6). The A-mode is not that structured and the Tucker3 core can easily be reduced from four to two A-mode components. It is difficult to explain more than 80 % of the sum of squares. As described in Chapter 8, line and scatter plots can be made useful and lead to the following conclusions: peat humification is more important than peat type, the treatments are different and some are better than others. This is also the subjective interpretation of the ANOVA, without formal hypothesis testing or interaction terms. ANOVA only provides significance of the within/between treatment differences for all treatments. The scatter plots in Chapter 8 allow for a more detailed interpretation of outliers, clusterings and gradients (Figures 8.14–8.18, 8.23–8.24, 8.26–8.27). Influence and residual plots for the example are given in Chapters 8 (Figures 8.38, 8.40–8.41).

The SEEDS example [Lestander *et al.* 2002] uses a principal component analysis of a mean-centered 180×1050 array and the principal components are used in the ANOVA analysis. Replicates of the spectra allow the calculation of interaction terms. Seven principal components explain 99.6 % of the total sum of squares. ANOVA of each principal component becomes quite complicated and therefore MANOVA can be made useful. This model

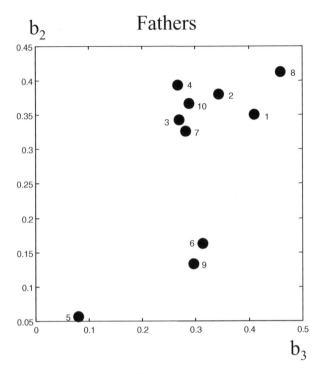

Figure 10.63. The scatter plot of the SEEDS loadings showing the clustering of the fathers.

gives high significance for the mother effect and less but still significant father and interaction effects. The ANOVA/MANOVA results only show whether treatments in a group are significantly different and do not show quantitative results. The PARAFAC analysis of the $6 \times 10 \times 1050$ array with first derivative spectra gives three components explaining 99 % of the total sum of squares.

The largest of these simply describes the average spectrum of the seeds. The second and third components can be used in scatter plots to show detailed information of outliers and clustering in the mother and father spaces. Figure 10.63 shows the scatter plot of b_3 and b_2 for the fathers. For the mothers, Figure 10.64 shows the scatter plot of a_3 and a_2. This shows that genetic information is present in the spectra. According to the MANOVA results, the mothers information should be more reliable than that of the fathers.

The SYNT example $14 \times 13 \times 701$ of Geladi & Åberg [2001] gives three PARAFAC components. Attempts to extract more components give unstable models and meaningless results. The loadings can be used in line or scatter plots. One of these is given in Figure 8.12. The wavelength mode loadings are not easily interpreted, as is usual for near infrared spectra, but comparison with the spectra of pure chemicals allows the interpretation of two loadings as those of starting material and final product. The corresponding time mode loadings show a decrease for the reagents and an increase for the final product. The batch mode loadings cannot be used to interpret the experimental design, but a clustering in fast and slow reactions is possible. A scatter plot of the time mode loadings shows that the

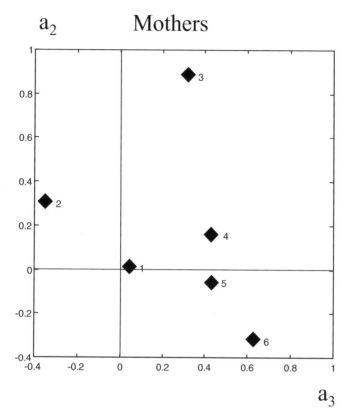

Figure 10.64. The scatter plot of the SEEDS loadings showing the relative positions of the mothers.

reaction moves in two stages: a fast first stage and a slow evolution in the second stage. The unidentified wavelength mode loading can be interpreted as an intermediate. The use of derivatives and of pure chemicals and known mixtures is crucial for the interpretation of the loadings. The residual plot for this example is given as bias and sum of squares in Figure 8.38.

Summary and remarks

Sometimes the modes of a three-way array are made up of ANOVA layouts or designed experiments. One of the modes is usually a 'spectral mode'. This gives arrays of the type small × small × large, where large stands for the spectral mode and small × small for the two-way ANOVA layout. In the case of a two-way ANOVA layout with spectral responses, the number of responses is too large to allow the classical ANOVA interpretation. Even MANOVA does not help in such cases. A reduction in responses can be achieved by PCA, but even better is using PARAFAC. While an ANOVA model would allow the interpretation of between/within distances, the line and scatter plots of PARAFAC (and Tucker3) loadings give a more detailed overview of outliers, clusters and gradients among the treatments. Stated otherwise, ANOVA/MANOVA gives a qualitative interpretation, the PARAFAC loadings of the array allow for a more quantitative interpretation. Selection of the model dimension is sometimes done by increasing the model rank until the model destabilizes or the loadings obtained become meaningless. This requires some subjective background knowledge.

Loadings are interpreted as line or scatter plots. The spectral mode loadings allow identification of chemical species. The loadings can also be used for interpretation of ANOVA layouts or experimental designs. In this way PARAFAC is an efficient data reduction tool for the space of the responses.

10.9   Analysis of Variance of Data with Complex Interactions

Background of the field

Many tools are available for analyzing experimentally designed data [Hoaglin & Welsch 1978, Latorre 1984, Rao 1973, Searle *et al.* 1992, Weisberg 1985]. Common to many of these approaches is that the estimated effects are treated as *additive*. This means that the effect of each factor is independent of the variation in other factors. In some situations, an additive model of main effects is not realistic because the factors do not affect the response independently. A well-working remedy to this is to allow interactions between the factors. Conceptually, traditional analysis of variance models start from main effects and seek to keep the number of interactions as low as possible and of the lowest possible order.

It was already noted by Fisher and MacKenzie [1923] that, even though a standard additive ANOVA model can describe a given data set, the model does not necessarily shed sufficient light on the underlying phenomena causing the variations in the data. Instead, they proposed a *multiplicative* model which was possible to fit in the two-factor case. Even though the model did not provide significantly better fit, the nature of the model was more in line with the nature and understanding of the data, and therefore more interpretable and informative. The usefulness of the multiplicative model as compared to a model where the interaction is modeled in the ordinary way is that the interaction between the two factors is *separated* into a multiplicative contribution from each factor.

In some situations, the multiplicative nature of the effects can be handled indirectly by a logarithmic transform of the response and at other times it may be beneficial to remain in the original response domain (e.g. because of the error structure or because of the structure of the model). Several authors have described different analysis of variance methods that include multiplicative terms for two-factor experiments [Gollob 1968, Hegemann & Johnson 1976, Heiser & Kroonenberg 1997, Mandel 1969, Mandel 1971]. Extending these ideas to more than two factors leads to a decomposition of a three- or higher-way array of responses into a tri- or multilinear PARAFAC-like model which is the starting point for three-factor GEMANOVA (GEneralized Multiplicative ANnalysis Of VAriance) [Bro & Jakobsen 2002, Bro 1998, Bro & Heimdal 1996, Heimdal *et al.* 1997]. If the data are higher order, e.g. a four-way array arising from a four-factor factorial design, a four-way quadrilinear model is used. In a study of enzymatic activity of polyphenol oxidase [Bro & Heimdal 1996], a five-factor experimental design was evaluated. For these particular data, it was found that the five-way array of responses of enzymatic activity, y_{ijklm}, could be well modeled by a five-way interaction term of the form $a_i b_j c_k d_l e_m$ and a simple and interpretable model was obtained.

The following example provides an illustration of some of the interpretational advantages of the GEMANOVA model but also illustrate that the inferential power of such models is still largely lacking.

Background of the example

Modified atmosphere packaging (MAP) is widely used to extend the shelf life of fresh meat. Gas flushing is used to replace the air surrounding the meat with an atmosphere containing elevated levels of oxygen and carbon dioxide. Normally an atmosphere containing 20–30 % CO_2 and 70–80 % O_2 is used for retail packaging of fresh beef. An elevated level of carbon dioxide is used to retard microbial growth and an elevated level of oxygen is used to prolong the color stability [Taylor 1972]. Color is the most important factor for consumer preferences when purchasing fresh meat. The attractive red meat color stems from the oxygenated derivative (oxymyoglobin) of the meat pigment. A high oxygen level stabilizes the color, but inevitably the pigment will oxidize to metmyoglobin resulting in an undesirable brown meat color. A high oxygen level is also expected to enhance other deteriorative processes in meat, such as lipid oxidation [Jakobsen & Bertelsen 2000, Zhao *et al.* 1994].

This study focuses on describing to what extent different storage and packaging conditions affect the color stability of fresh beef. Different levels of storage time, temperature, oxygen content in package headspace and extent of light exposure were investigated. Development of mathematical models describing color changes during storage can help identify the important factors and find critical levels for these factors. Further, such models can form the basis for designing an optimal modified atmosphere composition, also concerning other important quality deteriorative processes in meat.

Building a three-way model

DATA

Longissimus dorsi muscles of beef were used. Details of the sample and measurement preparation can be found in the original literature. Meat from three different animals was used and left and right parts from each animal were treated independently giving a total of six muscles. Storage time, temperature, time of light exposure and oxygen content in the package headspace (balanced with carbon dioxide) were varied (Tables 10.20 and 10.21). On days 0, 3, 7, 8 and 10 color was measured on the meat surface immediately after opening of the package using a Minolta Colorimeter CR-300 (Minolta, Osaka, Japan) measuring the so-called L, a, b coordinates. Red color was expressed as the a value and only this response value was used in the analysis. A high a value represents a red color of the meat sample. The measurement was repeated on five randomly selected locations for each sample and

Table 10.20. Level of factors in designed experiments for which a color response is measured

Variable	Levels
Storage time (days)	0, 3, 7, 8, 10
Temperature (°C)	2, 5, 8
O_2 content in headspace (%)	40, 60, 80
Exposure time to light (%)	0, 50, 100
Muscle no.	1, 2, 3, 4, 5, 6

Table 10.21. Other characteristics of the design

Variable	
No. of samples in full design	810
No. of samples in reduced design	324
Missing elements in array	60 %

the average used. On day zero there is no effect of storage and packaging conditions yet, and therefore all variable combinations are assigned the same color a value. This value is obtained as an average of measurements from all six muscles.

The full factorial design constitutes $5 \times 3 \times 3 \times 3 \times 6 = 810$ (storage × temperature × oxygen × light × muscle) combinations all to be measured in five replicates. Due to the limited amount of samples that can be taken from an individual muscle [Jakobsen & Bertelsen 2000], a reduced design was chosen using a modified quadratic D-optimal design with 324 settings (40 % of the 810 initial combinations).

The experimental error in the instrumental analysis can be used to evaluate the results from the model. The measurement error (incorporating both instrumental and sampling error) was determined by measuring at different locations on the same muscle five times. This yielded a pooled standard deviation of 2.2 (arbitrary units) and as the data used are averaged over the replicates, the standard deviation of this average is approximately $2.2/\sqrt{5} = 1.0$. This is the error approximately expected to be the lowest attainable estimated error in the ANOVA model.

RESULTS FROM TRADITIONAL ANALYSIS OF VARIANCE

A traditional analysis of variance was performed on the data. The factors are storage time (x_i), temperature (x_j), oxygen (x_k), light (x_l), and muscle (x_m) which is qualitative. The response color is held in y_{ijklm}. The data were scaled and centered and all cross-products and interactions added. After removing insignificant variables, the following ANOVA model was obtained through cross-validated multiple linear regression. The actual choice of model was based on the validation principles already outlined in Chapter 7 and more specific details

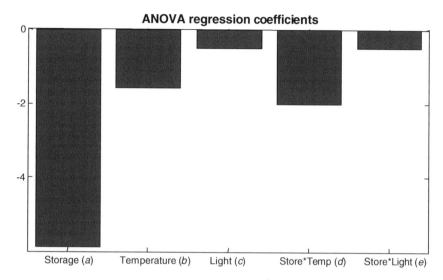

Figure 10.65. Significant effects in analysis of variance model of color. Effects are shown in terms of scaled and centered factors and response.

can be found in Bro and Jakobsen [2002].

$$y_{ijklm} = \mu + ax_i + bx_j + cx_k + dx_ix_j + ex_ix_l + \varepsilon_{ijklm}, \quad i = 1, \dots, I;$$
$$j = 1, \dots, J; k = 1, \dots, K; l = 1, \dots, L; m = 1, \dots, M \tag{10.12}$$

where

- color at storage level i, temperature j, light level k, oxygen level l, muscle m: y_{ijklm};
- overall offset: μ;
- storage main effect: a;
- temperature main effect: b;
- light main effect: c;
- interaction between storage and temperature: d;
- interaction between storage and light: e;
- residuals: ε_{ijklm}.

This model fits well ($r = 0.97$) and provides reasonable cross-validated errors (1.76). It is evident from the effects shown in Figure 10.65 that an increase in the level of any of the significant factors leads to a decrease in color. This is in agreement with expectations as color is known to deteriorate with increasing time, temperature and light exposure [Andersen *et al.* 1990, Jakobsen & Bertelsen 2000]. It is also evident that storage time has the largest effect on color in the given experimental domain and that light has a relatively low effect. Oxygen is excluded from the model because it does not have any significant effect. This is somewhat surprising because general practice is to maintain a high level of oxygen ($>70\,\%$) because lower levels are assumed to lead to excessive degradation of quality/color. Muscle does not have any significant effect either.

Determining the appropriate GEMANOVA model

As an alternative to the traditional ANOVA model, a multilinear model was investigated. Even though the original set of responses makes an eight-way array (storage × temperature × oxygen × light × animal × left–right muscle × replicate × color *L/a/b*), the averaged data used for analysis is a five-way array where the animal and the left–right muscle mode have been concatenated into one mode and only *a* color is used. The *a* color is thus a function of storage time, temperature, oxygen, light, and muscle.

In establishing the GEMANOVA model, it is necessary to determine which effects to include. This is done by first evaluating different alternatives to a one-component five-way PARAFAC model. From cross-validation results, it is determined that an adequate and simple GEMANOVA model for these data, is given by

$$y_{ijklm} = a_i b_j c_k d_l + e_m + \varepsilon_{ijklm}, \quad i = 1, \ldots, I; j = 1, \ldots, J;$$
$$k = 1, \ldots, K; l = 1, \ldots, L; m = 1, \ldots, M \tag{10.13}$$

where the notation is as before [Bro & Jakobsen 2002]. In Figure 10.66, the predictions obtained from cross-validation are shown. Note, that this model has a completely different structure from that of the ANOVA model. Hence, these two models although of similar predictive quality have widely different structures. The choice of which to use is thus

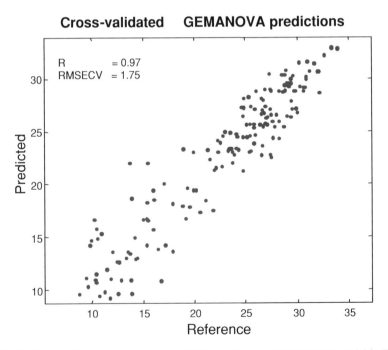

Figure 10.66. Results of leave-one-out cross-validation using the GEMANOVA model in Equation (10.13).

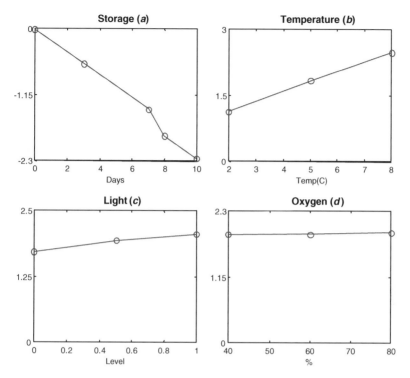

Figure 10.67. Parameters from multiplicative effect in GEMANOVA model. The estimated response at specified levels of the four factors equal the product of the corresponding effects plus the muscle term, which varies between 31 and 33.

mainly based on which one that offers the most understandable description of the variation in the data.

In order to interpret the model, the parameters are depicted in Figure 10.67. Here, the parameter a_i, $i = 1, \ldots, 5$ holds the storage effect, b_j, $j = 1,2,3$ is the temperature effect, c_k, $k = 1,2,3$ is the light effect and d_l, $l = 1,2,3$ is the oxygen effect. The separate muscle effects, e_m, are found to be identical to the zero-time starting color. The multiplicative term in Equation (10.13), $a_i b_j c_k d_l$, is zero, thus absent, at time zero due to a_1 being zero. Thus, the initial level is given specifically by the separate muscle effect. Every muscle has a starting color level of approximately 32. The estimated muscle effect is almost identical to the initial color measured. For all other settings of the factors, the estimated response is simply the starting level plus the product of the four effects read from ordinates in Figure 10.67. Some important points deduced from the plot are:

- Any change in color is negative; hence a decrease from the starting point, because the product of the four parameters $a_i b_j c_k d_l$ consists of one negative number (storage – upper left) and three positive numbers.
- The changes are relative. For example, in going from temperature 2 °C to 8 °C it is seen that the loading b_j increases from 1.2 to 2.4 Therefore, the temperature effect will be

that the overall decrease is twice as high at $8\,^{\circ}\mathrm{C}$ as it is at $2\,^{\circ}\mathrm{C}$, regardless of all other factors.

- The relative effect of temperature is linear.
- The effect of light is small, but there seems to be a small increase (hence decrease in color) with amount of light. The significance of this effect is treated below.
- The effect of oxygen is insignificant, thus supporting the surprising conclusion from the traditional analysis of variance, that even values as low as 40 % oxygen do not significantly alter the color development during storage.

The model shows that, apart from starting level, which is sample specific, the functional shape of the decrease as a function of the storage time, temperature and light is a simple one-component PARAFAC model $a_i b_j c_k$. This multiplicative part holds for all samples, whereas the starting level is specific.

Summary

Multiplicative multilinear models can be used for modeling ANOVA data. Such multilinear models can be interesting, for example, in situations where traditional ANOVA interactions are not possible to estimate. In these situations GEMANOVA can be a feasible alternative especially if a comparably simple model can be developed. As opposed to traditional ANOVA models, GEMANOVA models suffer from less developed hypothesis testing and often the modeling is based on a more exploratory approach than in ANOVA.

Appendix 10.A: An Illustration of the Generalized Rank Annihilation Method

An illustration of the generalized rank annihilation solution of a second-order problem is given to show some of its properties. Suppose that the problem is the following:

$$
\begin{aligned}
\mathbf{X}_1 &= \mathbf{a}_1 \mathbf{b}_1^{\mathrm{T}} \\
\mathbf{X}_2 &= c\mathbf{a}_1 \mathbf{b}_1^{\mathrm{T}} + \mathbf{a}_2 \mathbf{b}_2^{\mathrm{T}}
\end{aligned}
\tag{10.14}
$$

where for simplicity the error terms are dropped and the concentrations of the analyte in the standard and the interferent in the mixture are absorbed in the a profiles. Also assume that both $\mathbf{X}_1$ and $\mathbf{X}_2$ are (2×2) matrices (can always be done, see truncation GRAM in Chapter 6). The following generalized eigenvalue problem results

$$
\mu \mathbf{X}_2 \mathbf{z} = \lambda \mathbf{X}_1 \mathbf{z}
\tag{10.15}
$$

which has to be solved to obtain an estimate of c. The matrix $\mathbf{X}_2$ is 2×2 and has rank 2, hence, it is of full rank and its null-space contains only the zero vector. The matrix $\mathbf{X}_1$ is also 2×2, but has rank one, hence, the dimension of its null-space is one. This gives the following solutions for nonzero $\mathbf{z}$ of Equation (10.15):

$$
\begin{aligned}
&\text{(i)}\quad \mu = \lambda = 0\\
&\text{(ii)}\quad \mu = 0 \text{ and } \mathbf{X}_1\mathbf{z} = 0\\
&\text{(iii)}\quad \mu \neq 0 \text{ and } \lambda \neq 0
\end{aligned}
\tag{10.16}
$$

of which (iii) is the only nontrivial solution (note that the solution $\lambda = 0$ and $\mathbf{X}_2\mathbf{z} = 0$ is not possible for nonzero $\mathbf{z}$ and full rank $\mathbf{X}_2$). For this solution, Equation (10.15) can be written as

$$
\begin{aligned}
\mu \mathbf{X}_2\mathbf{z} &= \lambda \mathbf{X}_1\mathbf{z}\\
\Rightarrow \mu\left(c\mathbf{a}_1\mathbf{b}_1^{\mathrm{T}} + \mathbf{a}_2\mathbf{b}_2^{\mathrm{T}}\right)\mathbf{z} &= \lambda\left(\mathbf{a}_1\mathbf{b}_1^{\mathrm{T}}\right)\mathbf{z}\\
\Rightarrow \mu\left(c\mathbf{a}_1\mathbf{b}_1^{\mathrm{T}} - (\lambda/\mu)\mathbf{a}_1\mathbf{b}_1^{\mathrm{T}}\right)\mathbf{z} &= -\mu\left(\mathbf{a}_2\mathbf{b}_2^{\mathrm{T}}\right)\mathbf{z}\\
\Rightarrow \left(c\mathbf{a}_1\mathbf{b}_1^{\mathrm{T}} - (\lambda/\mu)\mathbf{a}_1\mathbf{b}_1^{\mathrm{T}}\right)\mathbf{z} &= -\left(\mathbf{a}_2\mathbf{b}_2^{\mathrm{T}}\right)\mathbf{z}\\
\Rightarrow (c - (\lambda/\mu))\left(\mathbf{b}_1^{\mathrm{T}}\mathbf{z}\right)\mathbf{a}_1 &= -\left(\mathbf{b}_2^{\mathrm{T}}\mathbf{z}\right)\mathbf{a}_2
\end{aligned}
\tag{10.17}
$$

and under the following conditions

$$
\begin{aligned}
&\text{(i)}\quad \mathbf{a}_1 \neq \alpha\mathbf{a}_2\\
&\text{(ii)}\quad \mathbf{b}_2^{\mathrm{T}}\mathbf{z} = 0\\
&\text{(iii)}\quad \mathbf{b}_1^{\mathrm{T}}\mathbf{z} \neq 0
\end{aligned}
\tag{10.18}
$$

the solution of problem (10.17) is $c = \lambda/\mu$, which gives the desired concentration. The conditions in Equation (10.18) are very interpretable from a calibration point of view. Condition (i) means that the temporal profiles of the analyte and the interferent should not have the same shape, which points to the selectivity in the temporal domain. Condition (ii) means that the solution vector $\mathbf{z}$ has no component in the interferent direction (i.e., is orthogonal to the interferent) and condition (iii) shows that $\mathbf{z}$ should have a component in the direction of the analyte. Both conditions (ii) and (iii) are related to the concept of net analyte signal in first order calibration [Lorber et al. 1997].

Appendix 10.B: Other Types of Second-Order Calibration Problems

This appendix contains examples of other types of second-order calibration, as mentioned in the main text.

Case (ii) second-order calibrations

The case in which the mixture contains two interferents and one analyte can be described as follows

$$
\begin{aligned}
\mathbf{X}_1 &= c_{11}\mathbf{a}_1\mathbf{b}_1^{\mathrm{T}} + \mathbf{E}_1\\
\mathbf{X}_2 &= c_{21}\mathbf{a}_1\mathbf{b}_1^{\mathrm{T}} + c_{22}\mathbf{a}_2\mathbf{b}_2^{\mathrm{T}} + c_{23}\mathbf{a}_3\mathbf{b}_3^{\mathrm{T}} + \mathbf{E}_2
\end{aligned}
\tag{10.19}
$$

where all symbols are the same as in Equation (10.4) and a_3, b_3, c_{23} represent the first order instrumental profile, second-order instrumental profile and the concentration of the second interferent, respectively. Rewriting Equation (10.19) as a PARAFAC model gives

$$X_1 = [\, a_1 \quad a_2 \quad a_3 \,] \begin{bmatrix} c_{11} & 0 & 0 \\ 0 & 0 & 0 \\ 0 & 0 & 0 \end{bmatrix} \begin{bmatrix} b_1^T \\ b_2^T \\ b_3^T \end{bmatrix} + E_1$$

$$X_2 = [\, a_1 \quad a_2 \quad a_3 \,] \begin{bmatrix} c_{21} & 0 & 0 \\ 0 & c_{22} & 0 \\ 0 & 0 & c_{23} \end{bmatrix} \begin{bmatrix} b_1^T \\ b_2^T \\ b_3^T \end{bmatrix} + E_2 \qquad (10.20)$$

$$C = \begin{bmatrix} c_{11} & 0 & 0 \\ c_{21} & c_{22} & c_{23} \end{bmatrix}$$

where again the matrices A, B, and C are implicitly defined. The Kruskal condition for this PARAFAC model reads

$$k_A + k_B + k_C \geq 2R + 2$$
$$2 + 2 + 1 < 2 \times 2 + 2 \qquad (10.21)$$

and these are not fulfilled because the k-rank of C has dropped to one due to the presence of two linearly dependent columns in C (column 2 and 3). Hence, unique results are not guaranteed. Fortunately, partial uniqueness is present for such PARAFAC models, which means that a_1, b_1 and c_{21} are uniquely solvable [Harshman 1970]. This shows the real power of second-order calibration.

Case (iii) second-order calibrations

The case in which there are multiple standards and the mixture contains one interferent and one analyte can be described as follows:

$$X_1 = c_{11} a_1 b_1^T + E_1$$
$$X_2 = c_{21} a_1 b_1^T + E_2 \qquad (10.22)$$
$$X_3 = c_{31} a_1 b_1^T + c_{32} a_2 b_2^T + E_3$$

which can be written as the following PARAFAC model:

$$X_1 = \begin{bmatrix} a_1 & a_2 \end{bmatrix} \begin{bmatrix} c_{11} & 0 \\ 0 & 0 \end{bmatrix} \begin{bmatrix} b_1^T \\ b_2^T \end{bmatrix} + E_1 = AD_1B^T + E_1$$

$$X_2 = \begin{bmatrix} a_1 & a_2 \end{bmatrix} \begin{bmatrix} c_{21} & 0 \\ 0 & 0 \end{bmatrix} \begin{bmatrix} b_1^T \\ b_2^T \end{bmatrix} + E_2 = AD_2B^T + E_2$$

$$X_3 = \begin{bmatrix} a_1 & a_2 \end{bmatrix} \begin{bmatrix} c_{31} & 0 \\ 0 & c_{31} \end{bmatrix} \begin{bmatrix} b_1^T \\ b_2^T \end{bmatrix} + E_3 = AD_3B^T + E_3$$

(10.23)

$$C = \begin{bmatrix} c_{11} & 0 \\ c_{21} & 0 \\ c_{31} & c_{32} \end{bmatrix}$$

where again the matrices A, B, D_1, D_2 and D_3 are implicitly defined. The Kruskal condition of this model are fullfilled, i.e., $k_A + k_B + k_C = 2 + 2 + 2 = 6 \geq 2 \times 2 + 2$.

Appendix 10.C: The Multiple Standards Calibration Model of the Second-Order Calibration Example

In the example, five standards $X_1, \ldots, X_5$ were available and one mixture X_6. Hence, the resulting PARAFAC model is:

$$X = A(C \odot B)^T + E$$
$$X^T = \begin{bmatrix} X_1^T & X_2^T & X_3^T & X_4^T & X_5^T & X_6^T \end{bmatrix}$$ (10.24)
$$E^T = \begin{bmatrix} E_1^T & E_2^T & E_3^T & E_4^T & E_5^T & E_6^T \end{bmatrix}$$

in which A contains the temporal profiles of the analyte and the four interferents; B contains the spectra of the analyte and the four interferents and C contains the concentrations:

$$C = \begin{bmatrix} c_{1,Pb} & 0 & 0 & 0 & 0 \\ c_{2,Pb} & 0 & 0 & 0 & 0 \\ c_{3,Pb} & 0 & 0 & 0 & 0 \\ c_{4,Pb} & 0 & 0 & 0 & 0 \\ c_{5,Pb} & 0 & 0 & 0 & 0 \\ c_{6,Pb} & c_{6,Co} & c_{6,Mn} & c_{6,Ni} & c_{6,Zn} \end{bmatrix}$$ (10.25)

and the elements of C are self-explanatory (the zeros indicate that the interferents are not present in the standards). Obviously, the Kruskal condition is not fulfilled (see case (ii)), but also in this case partial uniqueness holds and a_{Pb}, b_{Pb} and $c_{6,Pb}$ can be recovered.

REFERENCES

Albert S, Kinley RD, Multivariate statistical monitoring of batch processes: an industrial case study of fermentation supervision, *Trends in Biotechnology*, 2001, **19**, 53–62.

Allosio N, Boivin P, Bertrand D, Courcoux P, Characterisation of barley transformation into malt by three-way factor analysis of near infrared spectra, *Journal of Near Infrared Spectroscopy*, 1997, **5**, 157–166.

Amrhein M, Reaction and flow variants/invariants for the analysis of chemical reaction data. Ph.D. thesis, École Polytechnique Fédérale de Lausanne, 1998.

Amrhein M, Srinivasan B, Bonvin D, Schumacher MM, On the rank deficiency and rank augmentation of the spectral measurement matrix, *Chemometrics and Intelligent Laboratory Systems*, 1996, **33**, 17–33.

Andersen CM, Bro R, Practical aspects of PARAFAC modelling of fluorescence excitation-emission data, *Journal of Chemometrics*, 2002,

Andersen HJ, Bertelsen G, Skibsted LH, Colour and colour stability of hot processed frozen minced beef. Results from chemical model experiments tested under storage conditions, *Meat Science*, 1990, **28**, 87–97.

Anderson TW, *An Introduction to Multivariate Statistical Analysis*, John Wiley & Sons, Inc., New York, 1971.

Andersson CA, Arndal A, Combining HPSEC, fluorescence and multiway chemometrics to resolve and identify potential colour precursors in sugar process streams, in *Conference on Dairy and Food Science*, Copenhagen, LMC, 1999.

Andersson CA, Bro R, Improving the speed of multi-way algorithms: Part I. Tucker3, *Chemometrics and Intelligent Laboratory Systems*, 1998, **42**, 93–103.

Andersson CA, Direct orthogonalization, *Chemometrics and Intelligent Laboratory Systems*, 1999, **47**, 51–63.

Andersson CA, Henrion R, A general algorithm for obtaining simple-structure of core arrays in N-way PCA with application to fluorometric data, *Computational Statistics and Data Analysis*, 1999, **31**, 255–278.

Andrews DT, Wentzell PD, Applications of maximum likelihood principal component analysis: incomplete data sets and calibration transfer, *Analytica Chimica Acta*, 1997, **350**, 341–352.

Appellof CJ, Davidson ER, Strategies for analyzing data from video fluorometric monitoring of liqiud chromatographic effluents, *Analytical Chemistry*, 1981, **53**, 2053–2056.

Atkinson AC, *Plots, Transformations, and Regression: an Introduction to Graphical Methods of Diagnostic Regression Analysis,* Clarendon Press, Oxford, 1985.

Aubourg SP, Sotelo CG, Perez MR, Assessment of quality changes in frozen sardine (Sardina pilchardus) by fluorescence detection, *Journal of the American Oil Chemists' Society*, 1998, **75**, 575–580.

Barbieri P, Andersson CA, Massart DL, Predonzani S, Adami G, Reisenhofer E, Modeling bio-geochemical interactions in the surface waters of the Gulf of Trieste by 3-way PCA, *Analytica Chimica Acta*, 1999, **398**, 227–235.

Barnes RJ, Dhanoa MS, Lister SJ, Standard normal variate transformation and de-trending of near-infrared diffuse reflectance spectra, *Applied Spectroscopy*, 1989, **43**, 772–777.

Baunsgaard D, Factors affecting 3-way modelling (PARAFAC) of fluorescence landscapes, *Internal report, Dept. Dairy and Food Science, The Royal Veterinary and Agricultural University, Denmark*, 1999.

Baunsgaard D, Nørgaard L, Godshall MA, Specific screening for color precursors and colorants in beet and cane sugar liquors in relation to model colorants using spectrofluorometry evaluated by HPLC and multiway data analysis, *Journal of Agriculture and Food Chemistry*, 2001, **49**, 1687–1694.

Bax A, *Two Dimensional Nuclear Magnetic Resonance in Liquids*, Reidel, Boston, MA, 1982.

Beebe KR, Kowalski BR, An introduction to multivariate calibration and analysis, *Analytical Chemistry* 1987, **59**, 1007A–1017A.

Beebe KR, Pell R, Seasholtz MB, *Chemometrics: A Practical Guide,* John Wiley & Sons, Inc., New York, 1998.

Belsley DA, Kuh E, Welsch RE, *Regression Diagnositcs: Identifying Influential Data and Sources of Collinearity,* John Wiley and Sons, Inc., New York, 1980.

Belsley DA, Kuh E, Welsch RE, *Regression Diagnostics*, John Wiley & Sons, Inc., New York, 1980.

Beltrami E, Sulle funzioni bilineari, *Giornale di Matematiche ad Uso degli Studenti Delle Universita*, 1873, **11**, 98–106.

Beltran JL, Ferrer R, Guiteras J, Multivariate calibration of polycyclic aromatic hydrocarbon mixtures from excitation-emission fluorescence spectra, *Analytica Chimica Acta*, 1998a, **373**, 311–319.

Beltran JL, Guiteras J, Ferrer R, Three-way multivariate calibration procedures applied to high-performance liquid chromatography coupled with fast-scanning fluorescence spectrometry detection. determination of polycyclic aromatic hydrocarbons in water samples, *Analytical Chemistry*, 1998b, **70**, 1949–1955.

Booksh KS, Kowalski BR, Error analysis of the generalized rank annihilation method, *Journal of Chemometrics*, 1994, **8**, 45–63.

Booksh KS, Kowalski BR, Theory of analytical chemistry, *Analytical Chemistry*, 1994, **66**, 782A–791A.

Booksh KS, Lin Z, Wang Z, Kowalski BR, Extension of trilinear decomposition method with an application to the flow probe sensor, *Analytical Chemistry*, 1994, **66**, 2561–2569.

Boqué R, Smilde AK, Monitoring and diagnosing batch processes with multiway covariates regression models, *AIChE Journal*, 1999, **45**, 1504–1520.

Borisenko AI, Tarapov IE, *Vector and Tensor Analysis*, Dover Publications, New York, 1968.

Box G, Jenkins G, *Time Series Analysis, Forecasting and Control,* Holden-Day, San Francisco, CA 1976.

Box GEP, Draper NR, *Empirical Model Building and Response Surfaces*, John Wiley & Sons, Inc., New York, 1987.

Box GEP, Hunter WG, Hunter JS, *Statistics for Experimenters*, John Wiley & Sons, Inc., New York, 1978.

Box GEP, Kramer T, Statistical process monitoring and feedback adjustment - a discussion, *Technometrics*, 1992, **34**, 251–267.

Box GEP, Some theorems on quadratic forms applied in the study of analysis of variance problems. I. Effect of inequality of variance in the one-way classification, *The Annals of Statistics*, 1954, **25**, 290–302.

Brereton R (Ed.), *Multivariate Pattern Recognition in Chemometrics, Illustrated by Case Studies*, Elsevier, Amsterdam, 1992.

Bright FV, Modern molecular fluorescence spectroscopy, *Applied Spectroscopy* 1995, **49**, 14A–19A.

Bro R, Andersson CA, Kiers HAL, PARAFAC2 - Part II. Modeling chromatographic data with retention time shifts, *Journal of Chemometrics*, 1999, **13**, 295–309.

Bro R, de Jong S, A fast non-negativity-constrained least squares algorithm, *Journal of Chemometrics*, 1997, **11**, 393–401.

Bro R, Exploratory study of sugar production using fluorescence spectroscopy and multi-way analysis, *Chemometrics and Intelligent Laboratory Systems*, 1999, **46**, 133–147.

Bro R, Heimdal H, Enzymatic browning of vegetables. Calibration and analysis of variance by multiway methods, *Chemometrics and Intelligent Laboratory Systems*, 1996, **34**, 85–102.

Bro R, Jakobsen M, Exploring complex interactions in designed data using GEMANOVA. Color changes in fresh beef during storage, *Journal of Chemometrics* 2002, **16**, 294–304.

Bro R, Multi-way analysis in the food industry. Models, algorithms, and applications, Ph.D. thesis, University of Amsterdam (NL), http://www.mli.kvl.dk/staff/foodtech/brothesis.pdf, 1998.

Bro R, Multiway calibration. Multi-linear PLS, *Journal of Chemometrics*, 1996, **10**, 47–61.

Bro R, PARAFAC. Tutorial and applications, *Chemometrics and Intelligent Laboratory Systems*, 1997, **38**, 149–171.

Bro R, Sidiropoulos ND, Smilde AK, Maximum likelihood fitting using simple least squares algorithms, *Journal of Chemometrics*, 2002, **16**, 387–400.

Bro R, Smilde AK, Centering and scaling in component analysis, *Journal of Chemometrics*, 2003, **17**, 16–33.

Bro R, Smilde AK, de Jong S, On the difference between low-rank and subspace approximation: improved model for multi-linear PLS regression, *Chemometrics and Intelligent Laboratory Systems*, 2001, **58**, 3–13.

Brouwer P, Kroonenberg PM, Some notes on the diagonalization of the extended three-mode core matrix, *Journal of Classification*, 1991, **8**, 93–98.

Brown P, *Measurement, Regression and Calibration*, Clarendon Press, Oxford, 1993.

Budiansky B, Tensors, *Handbook of Applied Mathematics*, (Ed. Pearson, CE), Van Nostrand Reinhold, New York, 1974, pp. 179–225.

Burdick DS, An introduction to tensor products with applications to multiway data analysis, *Chemometrics and Intelligent Laboratory Systems*, 1995, **28**, 229–237.

Burdick DS, Tu XM, McGown LB, Millican DW, Resolution of multicomponent fluorescent mixtures by analysis of the excitation-emission-frequency array, *Journal of Chemometrics*, 1990, **4**, 15–28.

Burnham AJ, Viveros R, MacGregor JF, Frameworks for latent variable multivariate regression, *Journal of Chemometrics*, 1996, **10**, 31–45.

Cadima J, Jolliffe IT, Some comments on ten Berge&Kiers (1996). Optimality criteria for principal component analysis and generalizations, *British Journal of Mathematical and Statistical Psychology*, 1997, **50**, 365–366.

Campbell SL, Meyer CD, *Generalized Inverses of Linear Transformations*, Pitman, London, 1979.

Carroll JB, An analytical solution for approximating simple structure in factor analysis, *Psychometrika*, 1953, **18**, 23–38.

Carroll JD, Chang J, Analysis of individual differences in multidimensional scaling via an N-way generalization of 'Eckart–Young' decomposition, *Psychometrika*, 1970, **35**, 283–319.

Carroll JD, Pruzansky S, Kruskal JB, CANDELINC: A general approach to multidimensional analysis of many-way arrays with linear constraints on parameters, *Psychometrika*, 1980, **45**, 3–24.

Cattell R, The three basic factor-analytic research designs- their interrelations and derivatives, *Psychological Bulletin*, 1952, **49**, 499–452.

Cattell RB, The scree test for the number of factors, *Multivariate Behavioral Research*, 1966, **1**, 245–276.

Cattell RB, The three basic factor-analytic research designs – their interrelations and derivatives, *Psychological Bulletin*, 1952, **49**, 499–521.

Cattell RB, 'Parallel proportional profiles' and other principles for determining the choice of factors by rotation, *Psychometrika*, 1944, **9**, 267–283.

Christian GD, O'Reilly JE, *Instrumental Analysis*, Vol. 2, Prentice-Hall, Englewood Cliffs, NJ, 1986.

Christian GD, O'Reilly JE, *Instrumental Analysis*, 2nd edn, Prentice Hall, New Jersey, 1986.

Cleveland W, McGill M (Eds), *Dynamic Graphics for Statistics*, Wadsworth & Brooks/Cole, Belmont, CA, 1988.

Cleveland W, *The Elements of Graphing Data*, Wadsworth & Advanced Books and Software, Monterey, CA, 1985.

Cook RD, Added-variable plots and curvature in linear regression, *Technometrics*, 1996, **38**, 275–278.

Cook RD, Weisberg S, Characterizations of an empirical influence function for detecting influential cases in regression, *Technometrics*, 1980, **22**, 495–508.

Cook RD, Weisberg S, *Residuals and Influence in Regression*, Chapman and Hall, New York, 1982.

Cornelius PL, Seyedsadr M, Crossa J, Using the shifted multiplicative model to search for 'separability' in crop cultivar trials, *Theoretical and Applied Genetics*, 1992, **84**, 161–172.

Cornell JA, *Experiments with Mixtures. Designs, Models, and the Analysis of Mixture Data*, John Wiley and Sons, Inc., New York, 1981.

Critchley F, Influence in principal component analysis, *Biometrika*, 1985, **72**, 627–636.

Crosier RB, Multivariate generalizations of cumulative sum quality-control schemes, *Technometrics*, 1988, **30**, 291.

Currie LA, Nomenclature in evaluation of analytical methods including detection and quantification capabilities, *Pure and Applied Chemistry*, 1995, **67**, 1699–1723.

da Silva JCGE, Novais SA, Trilinear PARAFAC decomposition of synchronous fluorescence spectra of mixtures of the major metabolites of acetylsalicylic acid, *Analyst*, 1998, **123**, 2067–2070.

Dahl KS, Piovoso MJ, Kosanovich KA, Translating third-order data analysis methods to chemical batch processes, *Chemometrics and Intelligent Laboratory Systems*, 1999, **46**, 161–180.

Daubechies I, Recent results in wavelet applications, *Journal of Electronic Imaging*, 1998, **7**, 719–724.

Davies P, Coxon A, *Key Texts in Muldimensional Scaling*, Heinemann, London, 1982.

De Jong S, Kiers HAL, Principal covariates regression. Part 1. Theory, *Chemometrics and Intelligent Laboratory Systems*, 1992, **14**, 155–164.

de Jong S, Phatak A, Partial least squares regression, in *Recent Advances in Total Least Squares Techniques and Errors-in Variables Modeling*, van Huffel (Ed.), SIAM, Philadelphia, 1997, pp. 25–36.

De Jong S, Regression coefficients in multilinear PLS, *Journal of Chemometrics*, 1998, **12**, 77–81.

De Jong S, Ter Braak CJF, Comments on the PLS Kernel algorithm, *Journal of Chemometrics*, 1994, **8**, 169–174.

De Lathauwer L, Signal processing based on multilinear algebra, Ph.D. thesis, K.U. Leuven, 1997.

de Leeuw J, Young FW, Takane Y, Additive structure in qualitative data: An alternating least squares method with optimal scaling features, *Psychometrika*, 1976, **41**, 471–503.

de Ligny CL, Spanjer MC, van Houwelingen JC, Weesie HM, Three-mode factor analysis of data on retention in normal-phase high-performance liquid chromatography, *Journal of Chromatography*, 1984, **301**, 311–324.

Dempster AP, Laird NM, Rubin DB, Maximum likelihood from incomplete data via the EM algorithm, *Journal of the Royal Statistical Society Series B Statistical Methodology*, 1977, **39**, 1–38.

Digby PGN, Matrix algebra for data analysis, in *Food Research and Data Analysis*, Martens H and Russwurm H (Eds), Applied Science, London, 1983, pp. 435–472.

Do T, McIntyre NS, Application of parallel factor analysis and X-ray photoelectron spectroscopy to the initial stages in oxidation of aluminium, *Surface Science*, 1999, **435**, 136–141.

Dong D, McAvoy TJ, Nonlinear principal component analysis-based on principal curves and neural networks, *Computers in Chemical Engineering*, 1996, **20**, 65–78.

Draper NR, Smith H, *Applied Regression Analysis*, 3rd edn, John Wiley & Sons, Inc., New York, 1998.

Duchesne C, MacGregor JF, Jackknife and bootstrap methods in the identification of dynamic models, *Journal of Process Control,* 2001, **11**, 553–564.

Durell SR, Lee C, Ross RT, Gross EL, Factor analysis of the near-ultraviolet absorption spectrum of plastocyanin using bilinear, trilinear, and quadrilinear models, *Archives of Biochemistry and Biophysics*, 1990, **278**, 148–160.

Eastment HT, Krzanowski WJ, Cross-validatory choice of the number of components from a principal component analysis, *Technometrics*, 1982, **24**, 73–77.

Eckart C, Young G, The approximation of one matrix by another of lower rank, *Psychometrika*, 1936, **1**, 211–218.

Efron B, Gong G, A leisurely look at the Bootstrap, the Jackknife, and the cross-validation, *The American Statistician*, 1983, **37**, 36–48.

Efron B, Tibshirani RJ, *An Introduction to the Bootstrap,* Chapman & Hall, New York, 1993.

Efron B, *The Jackknife, the Bootstrap and Other Resampling Plans,* Society for industrial and applied mathematics, Philadelphia, PA, 1982.

Ewing GW, *Instrumental Methods of Chemical Analysis,* McGraw-Hill, New York, 1985.

Faber N, Meinders M, Geladi P, Sjöström M, Buydens L, Kateman G, Random error bias in principal component analysis. Part I. Derivation of theoretical predictions, *Analytica Chimica Acta*, 1995, **304**, 257–271.

Faber NM, Buydens LMC, Kateman G, Aspects of pseudorank estimation methods based on the eigenvalues of principal component analysis of random matrices, *Chemometrics and Intelligent Laboratory Systems*, 1994, **25**, 203–225.

Faber NM, Buydens LMC, Kateman G, Generalized rank annihilation. III: practical implementation, *Journal of Chemometrics*, 1994, **8**, 273–285.

Faber NM, Lorber A, Kowalski BR, Generalized rank annihilation method: Standard errors in the estimated eigenvalues if the instrumental errors are heteroscedastic and correlated, *Journal of Chemometrics*, 1997, **11**, 95–109.

Faber NM, On solving generalized eigenvalue problems using MATLAB, *Journal of Chemometrics*, 1997, **11**, 87–91.

Fearn T, On orthogonal signal correction, *Chemometrics and Intelligent Laboratory Systems*, 2000, **50**, 47–52.

Ferguson GA, The concept of parsimony in factor analysis, *Psychometrika*, 1954, **19**, 281–290.

Ferreira MMC, Brandes ML, Ferreira IMC, Booksh KS, Dolowy WC, Gouterman M, Kowalski BR, Chemometric study of the fluorescence of dental calculus by trilinear decomposition, *Applied Spectroscopy*, 1995, **49**, 1317–1325.

Fisher RA, MacKenzie WA, Studies in crop variation. II The manurial response of different potato variaties, *Journal of Agricultural Science*, 1923, **13**, 311–320.

Fogler HS, *Elements of Chemical Reaction Engineering*, Prentice-Hall, London, 1992.

Frank IE, Friedman JH, A statistical view of some chemometrics regression tools, *Technometrics*, 1993, **35**, 109–148.

Frank IE, Kowalski BR, A multivariate method for relating groups of measurements connected by a causal pathway, *Analytica Chimica Acta*, 1985, **167**, 51–63.

Friedman JH, Multivariate adaptive regression splines, *The Annals of Statistics*, 1991, **19**, 1–141.

Friedman JH, Stuetzle W, Projection pursuit regression, *Journal of the American Statistical Association*, 1981, **76**, 817–823.

Fujikoshi Y, Satoh K, Modified AIC and C-p in multivariate linear regression, *Biometrika*, 1997, **84**, 707–716.

Gabriel K, The biplot-graphic display of matrices with applications to principal component analysis, *Biometrika*, 1971, **58**, 453–467.

Gabriel KR, Least squares approximation of matrices by additive and multiplicative models, *Journal of the Royal Statistical Society Series B Statistical Methodology*, 1978, **40**, 186–196.

Galton F, Regression towards mediocrity in heredity stature, *Journal of Anthropological Institute*, 1885, **15**, 246–263.

Gardner WP, Shaffer RE, Girard JE, Callahan JH, Application of quantitative chemometric analysis techniques to direct sampling mass spectrometry, *Analytical Chemistry*, 2001, **73**, 596–605.

Geisser S, A predictive approach to the random effect model, *Biometrika*, 1974, **61**, 101–107.

Geladi P, Åberg P, Three-way modelling of a batch organic synthesis process monitored by near infrared spectroscopy, *Journal of Near Infrared Spectroscopy*, 2001, **9**, 1–9.

Geladi P, Bergner H, Ringqvist L, From experimental design to images to particle size histograms to multiway analysis. An example of peat dewatering, *Journal of Chemometrics*, 2000, **14**, 197–211.

Geladi P, Forsström J, Monitoring, of a batch organic synthesis by near-infrared spectroscopy: modeling and interpretation of three-way data, *Journal of Chemometrics*, 2002, **16**, 329–338.

Geladi P, Grahn H, Multivariate Image Analysis, in *Encyclopedia of Analytical Chemistry*, Meyers R (ed.), John Wiley & Sons, Ltd, Chichester, 2000, pp. 13540–13562.

Geladi P, Grahn H, *Multivariate image analysis*, John Wiley & Sons, Ltd, Chichester, 1996.

Geladi P, Hadjiiski L, Hopke P, Multiple regression for environmental data: nonlinearities and prediction bias, *Chemometrics and Intelligent Laboratory Systems*, 1999, **47**, 165–173.

Geladi P, MacDougall D, Martens H, Linearization and scatter-correction for near-infrared reflectance spectra of meat, *Applied Spectroscopy*, 1985, **39**, 491–500.

Geladi P, Manley M, Lestander T, Scatter plotting in multivariate data analysis. *Journal of Chemometrics*, 2003, **17**, 503–511.

Geladi P, Some recent trends in the calibration literature, *Chemometrics and Intelligent Laboratory Systems*, 2002, **60**, 211–224.

Geladi P, Xie YL, Polissar A, Hopke PK, Regression on parameters from three-way decomposition, *Journal of Chemometrics*, 1998, **12**, 337–354.

Gemperline P, Miller K, West T, Weinstein J, Hamilton J, Bray J, Principal component analysis, trace elements, and blue crab shell disease, *Analytical Chemistry*, 1992, **64**, 523A–532A.

Gemperline PJ, Computation of the range of feasible solutions in self-modeling curve resolution algorithms, *Analytical Chemistry*, 1999, **71**, 5398–5404.

Gemperline PJ, Miller KH, West TL, Weinstein JE, Hamilton JC, Bray JT, Principal component analysis, trace elements, and blue crab shell disease, *Analytical Chemistry*, 1992, **64**, 523A–532A.

Gerritsen MJP, Tanis H, Vandeginste BGM, Kateman G, Generalized rank annihilation factor analysis, iterative target transformation factor analysis, and residual bilinearization for the quantitative analysis of data from liquid chromatography with photodiode array detection, *Analytical Chemistry*, 1992, **64**, 2042–2056.

Gifi A, *Nonlinear multivariate analysis*, John Wiley & Sons, Ltd, Chichester, 1990.

Gill PE, Murray W, Wright MH, *Practical Optimization*, Academic Press, London, 1981.

Gollob HF, A statistical model which combines features of factor analytic and analysis of variance techniques, *Psychometrika*, 1968, **33**, 73–115.

Golub GH, van Loan CF, *Matrix Computations*, The John Hopkins University Press, Baltimore, MD, 1989.

Gower J, Multidimensional scaling displays, in *Research Methods for Multimode Data Analysis*, Law H, Snyder C, Hattie J, McDonald R (Eds), Praeger, New York, 1984, 592–601.

Gower JC, Generalized Procrustes Analysis, *Psychometrika*, 1975, **40**, 33–51.

Grung B, Manne R, Missing values in principal component analysis, *Chemometrics and Intelligent Laboratory Systems*, 1998, **42**, 125–139.

Gui M, Rutan SC, Agbodjan A, Kinetic detection of overlapped amino acids in thin-layer chromatography with a direct trilinear decomposition method, *Analytical Chemistry*, 1995, **67**, 3293–3299.

Guiteras J, Beltran JL, Ferrer R, Quantitative multicomponent analysis of polycyclic aromatic hydrocarbons in water samples, *Analytica Chimica Acta*, 1998, **361**, 233–240.

Guo Q, Wu W, Massart DL, The robust normal variate transform for pattern recognition with near-infrared data, *Analytica Chimica Acta*, 1999, **382**, 87–103.

Gurden SP, Westerhuis JA, Bijlsma S, Smilde AK, Modelling of spectroscopic batch process data using grey models to incorporate external information, *Journal of Chemometrics*, 2001, **15**, 101–121.

Gurden SP, Westerhuis JA, Bro R, Smilde AK, A comparison of multiway regression and scaling methods, *Chemometrics and Intelligent Laboratory Systems*, 2001, **59**, 121–136.

Gurden SP, Westerhuis JA, Smilde AK, Monitoring of batch processes using spectroscopy, *AIChE Journal*, 2002, **48**, 2283–2297.

Hanson RJ, Lawson CL, *Solving Least Squares Problems*, Prentice-Hall, Englewood Cliffs, NJ, 1974.

Harman HH, *Modern Factor Analysis*, University of Chicago Press, Chicago, 1967.

Harris D, *Quantitative Chemical Analysis*, 5th edn, W. H. Freeman, New York, 1997.

Harris RID, Judge G, Small sample testing for cointegration using the bootstrap approach, *Economics Letters*, 1998, **58**, 31–37.

Harshman R, Foundations of the Parafac procedure: models and conditions for an 'explanation' multi-modal factor analysis, *UCLA Working Papers in Phonetics*, Los Angeles CA, 1970, Vol. 16.

Harshman RA, De Sarbo WS, An application of PARAFAC to a small sample problem, demonstrating preprocessing, orthogonality constraints, and split-half diagnostic techniques, in *Research Methods for Multimode Data Analysis*, Law HG, Snyder CW, Jr, Hattie JA and McDonald RP (Eds), Praeger Special Studies, New York, 1984, 602–642.

Harshman RA, Determination and proof of minimum uniqueness conditions for PARAFAC1, *UCLA Working Papers in Phonetics*, 1972, **22**, 111–117.

Harshman RA, Foundations of the PARAFAC procedure: Models and conditions for an 'explanatory' multi-modal factor analysis, *UCLA Working Papers in Phonetics*, 1970, **16**, 1–84.

Harshman RA, How can I know it's real? A catalog of diagnostics for use with three-mode factor analysis and multidimensional scaling, in *Research Methods for Multimode Data Analysis*, Law HG, Snyder CW, Jr, Hattie J and McDonald RP (Eds), Praeger, New York, 1984, pp. 566–591.

Harshman RA, Lundy ME, Data preprocessing and the extended PARAFAC model, in *Research Methods for Multimode Data Analysis*, Law HG, Snyder CW, Jr., Hattie J, McDonald RP (Eds), Praeger, New York, 1984, pp. 216–284.

Harshman RA, Lundy ME, The PARAFAC model for three-way factor analysis and multidimensional scaling, *Research Methods for Multimode Data Analysis* (Eds Law, HG, Snyder, CW, Hattie, JA, and McDonald, RP), Praeger, New York, 1984, pp. 122–215.

Hegemann V, Johnson DE, On analyzing two-way anova data with interaction, *Technometrics*, 1976, **18**, 273–281.

Heimdal H, Bro R, Larsen LM, Poll L, Prediction of polyphenol oxidase activity in model solutions containing various combinations of chlorogenic acid, (−)-epicatechin, O2, CO2, temperature and pH by multiway analysis, *Journal of Agricultural and Food Chemistry*, 1997, **45**, 2399–2406.

Heiser WJ, Kroonenberg PM, Dimensionwise fitting in PARAFAC-CANDECOMP with missing data and constrained parameters, Internal Report PRM 97-01, University of Leiden, The Netherlands, 1997.

Helland IS, Næs T, Isaksson T, Related versions of the multiplicative scatter correction method for preprocessing spectroscopic data, *Chemometrics and Intelligent Laboratory Systems*, 1995, **29**, 233–241.

Helland IS, On the structure of partial least squares regression, *Communications in Statistics–Simula*, 1988, **17**, 581–607.

Henrion R, Andersson CA, A new criterion for simple-structure transformations of core arrays in *N*-way principal components analysis, *Chemometrics and Intelligent Laboratory Systems*, 1999, **47**, 189–204.

Henrion R, Body diagonalization of core matrices in three-way principal components analysis: theoretical bounds and simulation, *Journal of Chemometrics*, 1993, **7**, 477–494.

Henrion R, Henrion G, Onuoha GC, Multi-way principal components analysis of a complex data array resulting from physicochemical characterization of natural waters, *Chemometrics and Intelligent Laboratory Systems*, 1992, **16**, 87–94.

Henrion R, Henrion G, *Multivariate Datenanalyse. Methodik und Anwendung in der chemie und verwandten Gebieten*, Springer, Berlin, 1995.

Ho CN, Christian GD, Davidson ER, Application of the method of rank annihilation to quantitative analyses of multicomponent fluorescence data from the video fluorometer, *Analytical Chemistry*, 1978, **50**, 1108–1113.

Ho CN, Christian GD, Davidson ER, Application of the method of rank annihilation to fluorescent multicomponent mixtures of polynuclear aromatic hydrocarbons, *Analytical Chemistry*, 1980, **52**, 1071–1079.

Ho CN, Christian GD, Davidson ER, Simultaneous multicomponent rank annihilation and applications to multicomponent fluorescent data acquired by the video fluorometer, *Analytical Chemistry*, 1981, **53**, 92–98.

Hoaglin DC, Welsch RE, The hat matrix in regression and ANOVA, *The American Statistician*, 1978, **32**, 17–22.

Hoerl AE, Kennard RW, Ridge regression: biased estimation for nonorthogonal problems, *Technometrics*, 1970, **12**, 55–67.

Hopke P, Xie Y, Paatero P, Mixed multiway analysis of airborne particle composition data, *Journal of Chemometrics*, 1999, **13**, 343–352.

Hopke P, *Receptor Modeling for Air Quality*, Elsevier, Amsterdam, 1991.

Hopkins D, Derivatives in spectroscopy, *Near Infrared Analysis*, 2001, **2**, 1–13.

Horn JL, A rationale and test for the number of factors in factor analysis, *Psychometrika*, 1965, **30**, 179–185.

Hotelling H, Analysis of a complex of statistical variables into principal components, *Journal of Educational Psychology*, 1933, **24**, 417–441.

Hotelling H, Simplified calculation of principal components, *Psychometrika*, 1936, **1**, 27–35.

Hotelling H, The generalization of Student's ratio, *The Annals of Mathematical Statistics*, 1931, **2**, 360–378.

Höskuldsson A, PLS regression methods, *Journal of Chemometrics*, 1988, **2**, 211–228.

Isaksson T, Kowalski BR, Piece-wise multiplicative scatter correction applied to near-infrared diffuse transmittance data from meat products, *Applied Spectroscopy*, 1993, **47**, 702–709.

Ja'Ja' J, Optimal evaluation of pairs of bilinear forms, *SIAM Journal on Computing*, 1979, **8**, 443–461.

Jackson JE, Mudholkar GS, Control procedures for residuals associated with principal component analysis, *Technometrics*, 1979, **21**, 341–349.

Jackson JE, Quality control methods for several related variables, *Technometrics*, 1959, **1**, 359–377.

Jackson JE, *A User's Guide to Principal Components*, John Wiley & Sons, New York, 1991.

Jakobsen M, Bertelsen G, Colour stability and lipid oxidation of fresh beef. Development of a response surface model for predicting the effects of temperature, storage time, and modified atmosphere composition, *Meat Science* 2000, **54**, 49–57.

Jensen SA, Reenberg S, Munck L, Fluorescence analysis in fish and meat technology, in *Fluorescence Analysis in Foods*, Munck L (Ed.), Longman, New York, 1989, pp. 181–192.

Jiji RD, Andersson GG, Booksh KS, Application of PARAFAC for calibration with excitation-emission matrix fluorescence spectra of three classes of environmental pollutants, *Journal of Chemometrics*, 2000, **14**, 171–185.

Jolliffe IT, *Principal Component Analysis*, Springer, Berlin, 1986.

Jones C, *Visualization and Optimization*, Kluwer Academic, Boston, MA, 1996.

Jöreskog KG, Wold H, *Systems under Indirect Observation*, North-Holland, Amsterdam, 1982.

Judge GG, Griffifths WE, Carter Hill R, Lütkepohl H, Lee TC, *The Theory and Practice of Econometrics*, John Wiley & Sons, Inc., New York, 1985.

Kapteyn A, Neudecker H, Wansbeek T, An approach to n-mode components analysis, *Psychometrika*, 1986, **51**, 269–275.

Karukstis KK, Krekel DA, Weinberger DA, Bittker RA, Naito NR, Bloch SH, Resolution of the excited states of the fluorescence probe TNS using a trilinear analysis technique, *Journal of Physical Chemistry*, 1995, **99**, 449–453.

Kassidas A, MacGregor JF, Taylor PA, Synchronization of batch trajectories using dynamic time warping, *AIChE Journal*, 1998, **44**, 864–875.

Katsumoto Y, Jiang J-H, Berry R, Ozaki Y, Modern pretreatment methods in NIR spectroscopy, *Near Infrared Analysis*, 2001, **2**, 29–36.

Kiers H, Some procedures for displaying results from three-way methods, *Journal of Chemometrics*, 2000, **14**, 151–170.

Kiers HAL, A comparison of techniques for finding components with simple structure, *Multivariate Analysis: Future Directions,* 1993, **2**, 67–86.

Kiers HAL, A three-step algorithm for CANDECOMP/PARAFAC analysis of large data sets with multicollinearity, *Journal of Chemometrics*, 1998, **12**, 155–171.

Kiers HAL, Hierarchical relations among three-way methods, *Psychometrika*, 1991, **56**, 449–470.

Kiers HAL, Hierarchical relations among three-way methods, *Psychometrika*, 1991a, **56**, 449–470.

Kiers HAL, Joint orthomax rotation of the core and component matrices resulting from three-mode principal component analysis, *Journal of Classification*, 1998a, **15**, 245–263.

Kiers HAL, Krijnen WP, An efficient algorithm for PARAFAC of three-way data with large numbers of observation units, *Psychometrika*, 1991, **56**, 147–152.

Kiers HAL, Kroonenberg PM, ten Berge JMF, An efficient algorithm for TUCKALS3 on data with large numbers of observation units, *Psychometrika*, 1992, **57**, 415–422.

Kiers HAL, Recent developments in three-mode factor analysis: constrained three-mode factor analysis and core rotations, in *Studies in Classification, Data Analysis, and Knowledge Organization: Data Science, Classification, and Related Methods,* Hayashi C, Ohsumi N, Yajima K, Tanaka Y, Bock HH, Baba Y (Eds), Springer-Verlag, Hong Kong, 1998b, pp. 563–574.

Kiers HAL, Simple structure in component analysis techniques for mixtures of qualitative and quantitative variables, *Psychometrika*, 1991b, **56**, 197–212.

Kiers HAL, SIMPLIMAX: oblique rotation to an optimal target with simple structure, *Psychometrika*, 1994, **59**, 579.

Kiers HAL, Smilde AK, Constrained three-mode factor analysis as a tool for parameter estimation with second-order instrumental data, *Journal of Chemometrics,* 1998, **12**, 125–147.

Kiers HAL, Smilde AK, Some theoretical results on second-order calibration methods for data with and without rank overlap, *Journal of Chemometrics*, 1995, **9**, 179–195.

Kiers HAL, Techniques for rotating two or more loading matrices to optimal agreement and simple structure: A comparison and some technical details, *Psychometrika*, 1997a, **62**, 545–568.

Kiers HAL, ten Berge JMF, Hierarchical relations between methods for simultaneous component analysis and a technique for rotation to a simple simultaneous structure, *British Journal of Mathematical and Statistical Psychology,* 1994a, **47**, 109–126.

Kiers HAL, ten Berge JMF, The Harris–Kaiser independent cluster rotation as a method for rotation to simple component weights, *Psychometrika*, 1994b, **59**, 81–90.

Kiers HAL, Three-mode orthomax rotation, *Psychometrika,* 1997a, **62**, 579–598.

Kiers HAL, Three-mode orthomax rotation, *Psychometrika,* 1997b, **62**, 579–598.

Kiers HAL, Three-way SIMPLIMAX for oblique rotation of the three-mode factor analysis core to simple structure, *Computational Statistics and Data Analysis,* 1998b, **28**, 307–324.

Kiers HAL, Towards a standardized notation and terminology in multiway analysis, *Journal of Chemometrics*, 2000, **14**, 105–122.

Kiers HAL, TUCKALS core rotations and constrained TUCKALS modelling, *Statistica Applicata,* 1992, **4**, 659–667.

Kiers HAL, Weighted least squares fitting using ordinary least squares algorithms, *Psychometrika*, 1997, **62**, 251–266.

Kiers HAL, Weighted least squares fitting using ordinary least squares algorithms, *Psychometrika*, 1997b, **62**, 251–266.

Knudson E, Duewer D, Christian GD, Larson T, Application of factor analysis to the study of rain chemistry in the Puget sound area, in *Chemometrics: Theory and Application*, Kowalski BR (Ed.), American Chemical Society, Washington, DC, 1977, pp. 80–116.

Kosanovich KA, Dahl KS, Piovoso MJ, Improved process understanding using multiway principal component analysis, *Industrial and Engineering Chemistry Research*, 1996, **35**, 138–146.

Kourti T, Nomikos P, MacGregor JF, Analysis, monitoring and fault diagnosis of batch processes using multiblock and multiway PLS, *Journal of Process Control*, 1995, **5**, 277–284.

Kroonenberg P, De Leeuw J, Principal component analysis of three-mode data by means of alternating least squares algorithms, *Psychometrika*, 1980, **45**, 69–97.

Kroonenberg P, de Leeuw J, TUCKALS2. A principal component analysis of three-mode data. Research bulletin RB 001-77, Dept. of Data Theory, Leiden University, 1977.

Kroonenberg P, The TUCKALS line. A suite of programs for three-way data analysis, *Computational Statistics and Data Analysis*, 1994, **18**, 73–96.

Kroonenberg P, *Three-Mode Principal Component Analysis*, DSWO Press, Leiden, 1983.

Kroonenberg PM, de Leeuw J, Principal component analysis of three-mode data by means of alternating least squares algorithms, *Psychometrika*, 1980, **45**, 69–97.

Kroonenberg PM, ten Berge JMF, Brouwer P, Kiers HAL, Gram-Schmidt versus Bauer-Rutishauser in alternating least-squares algorithms for three-mode principal component analysis, *Computation Statistics Quarterly*, 1989, **2**, 81–87.

Kroonenberg PM, Three-mode principal component analysis: illustrated with an example from attachment theory, in *Research Methods for Multimode Data Analysis*, Law HG, Snyder CW, Hattie JA and McDonald RP (Eds), Praeger, New York, 1984, pp 64–103.

Kroonenberg PM, *Three-mode Principal Component Analysis. Theory and Applications*, DSWO Press, Leiden, 1983.

Kruskal J, More factors than subjects, tests and treatments: an intdeterminacy theorem for canonical decomposition and individual differences scaling, *Phychometrika*, 1976, **41**, 281–293.

Kruskal J, Three-way arrays: rank and uniqueness of trilinear decompositions with applications to arithmetic complexity and statistics, *Linear Algebra and its Applications*, 1977, **18**, 95–138.

Kruskal JB, Harshman RA, Lundy ME, How 3-MFA data can cause degenerate PARAFAC solutions, among other relationships, in *Multiway Data Analysis*, Coppi R, Bolasco S (Eds), Elsevier, Amsterdam, 1989, pp. 115–122.

Kruskal JB, Harshman RA, Lundy ME, Some relationships between Tucker's three-mode factor analysis and PARAFAC/CANDECOMP, in *Annual Meeting of the Psychometric Society*, Los Angeles, 1983.

Kruskal JB, More factors than subjects, tests and treatments: An indeterminacy theorem for canonical decomposition and individual differences scaling, *Psychometrika*, 1976, **41**, 281–293.

Kruskal JB, Multilinear methods, in *Research Methods for Multimode Data analysis*, Law HG, Snyder CW, Hattie JA, McDonald RP (Eds), Praeger, New York, 1984, pp. 36–62.

Kruskal JB, Multilinear methods, *Proceedings of Symposia in Applied Mathematics*, 1983, **28**, 75–104.

Kruskal JB, Rank, decomposition, and uniqueness for 3-way and N-way arrays, in *Multiway Data Analysis*, Coppi R, Bolasco S (Eds), Elsevier, Amsterdam, 1989, pp. 8–18.

Kruskal JB, Simple structure for three-way data: A new method intermediate between 3-mode factor analysis and PARAFAC-CANDECOMP, *paper presented at the 53rd Annual Meeting of the Psychometric Society*, Los Angeles, 1988.

Kruskal JB, Some least squares theorems for matrices and N-way arrays, 1977, Manuscript, Bell Laboratories, Murray Hill, NJ.

Lastovicka JL, The extension of component analysis to four-mode matrices, *Psychometrika*, 1981, **46**, 47–57.

Lathauwer LD, Signal processing based on multilinear algebra, Ph.D. thesis, K.U. Leuven, E.E. Dept., 1997.

Latorre G, Analysis of variance and linear models, in *Mathematics and Statistics in Chemistry*, Kowalski BR (Ed.), Reidel, Dordrecht, 1984, 377–391.

Lawton WH, Sylvestre EA, Self modeling curve resolution, *Technometrics*, 1971, **13**, 617–633.

Leardi R, Armanino C, Lanteri S, Alberotanza L, Three-mode principal component analysis of monitoring data from Venice lagoon, *Journal of Chemometrics*, 2000, **14**, 187–195.

Lee JK, Ross RT, Thampi S, Leurgans SE, Resolution of the properties of hydrogen-bonded tyrosine using a trilinear model of fluorescence, *Journal of Physical Chemistry*, 1992, **96**, 9158–9162.

Lestander T, Odén PC, Geladi P, 2- and 3-way analysis of NIR scans from seed crossings, *Near Infrared Spectroscopy. Proceedigns of the 10th International Conference*, Chichester, NIR Publications, 2002, pp. 385–388.

Leurgans SE, Ross RT, Abel RB, A decomposition for three-way arrays, *Siam Journal on Matrix Analysis and applications*, 1993, **14**, 1064–1083.

Leurgans SE, Ross RT, Multilinear models: applications in spectroscopy, *Statistical Science*, 1992, **7**, 289–319.

Levin J, Three-mode factor analysis, *Psychological Bulletin*, 1965, **64**, 442–452.

Li S, Gemperline PJ, Eliminating complex eigenvectors and eigenvalues in multiway analyses using the direct trilinear decomposition method, *Journal of Chemometrics*, 1993, **7**, 77–88.

Li S, Hamilton JC, Gemperline PJ, Generalized rank annihilation method using similarity transformations, *Analytical Chemistry*, 1992, **64**, 599–607.

Li SS, Gemperline PJ, Briley K, Kazmierczak S, Identification and quantitation of drugs of abuse in urine using the generalized rank annihilation method of curve resolution, *Journal of Chromatography B–Biomedical Applications*, 1994, **655**, 213–223.

Liang Y-Z, Kvalheim OM, Manne R, White, grey and black multicomponent systems. A classification of mixture problems and methods for their quantitative analysis, *Chemometrics and Intelligent Laboratory Systems*, 1993, **18**, 235–250.

Lickteig T, Typical tensorial rank, *Linear Algebra and its Applications*, 1985, **69**, 95–120.

Lin ZH, Booksh KS, Burgess LW, Kowalski BR, 2nd-order fiber optic heavy-metal sensor employing 2nd-order tensorial calibration, *Analytical Chemistry*, 1994, **66**, 2552–2560.

Lindgren F, Geladi P, Wold S, The Kernel algorithm for PLS, *Journal of Chemometrics*, 1993, **7**, 45–59.

Ling X, Cao YZ, Mo CY, Liu XY, Resolution of naphthalene, 1-naphthol and 2-naphthol using the parallel factor algorithm and three-way flourescence technique, *Chinese Journal of Analytical Chemistry*, 2001, **29**, 1412–1415.

Little RJA, Rubin DB, *Statistical Analysis with Missing Data*, John Wiley & Sons, Inc., New York, 1987,

Liu XQ, Sidiropoulos ND, Cramer-Rao lower bounds for low-rank decomposition of multidimensional arrays, *IEEE Transactions on Signal Processing*, 2001, **49**, 2074–2086.

Lohmöller J-B, Die trimodale faktorenanalyse von tucker, Skalierungen, Rotationen, andere modelle, *Archives of Psychology*, 1979, **131**, 137–166.

Lorber A, Faber NM, Kowalski BR, Net analyte signal calculation in multivariate calibration, *Analytical Chemistry*, 1997, **69**, 1620–1626.

Lorber A, Features of quantifying composition from two-dimensional data array by the rank annihilation factor analysis method, *Analytical Chemistry*, 1985, **57**, 2395–2397.

Lorber A, Kowalski BR, Alternatives to cross-validatory estimation of the number of factors in multivariate calibration, *Applied Spectroscopy*, 1990, **44**, 1464–1470.

Lorber A, Quantifying chemical composition from two-dimensional data arrays, *Analytica Chimica Acta*, 1984, **164**, 293–297.

Louwerse DJ, Kiers HAL, Smilde AK, Cross-validation of multiway component models, *Journal of Chemometrics*, 1999, **13**, 491–510.

Louwerse DJ, Smilde AK, Multivariate statistical process control of batch processes based on three-way models, *Chemical Engineering Science*, 2000, **55**, 1225–1235.

MacGregor JF, Jaeckle Ch, Kiparissides C, Koutoudi M, Process monitoring and diagnosis by multi-block PLS methods, *AIChE Journal*, 1994, **40**, 826–838.

MacGregor JF, On-line statistical process control, *Chemical Engineering Progress*, 1988, 21–31.

Magnus JR, Neudecker H, *Matrix Differential Calculus with Applications in Statistics and Econometrics*, John Wiley & Sons, Ltd, Chichester, 1988.

Malinowski ER, Theory of the distribution of error eigenvalues resulting from principal component analysis with applications to spectroscopic data, *Journal of Chemometrics*, 1987, **1**, 33–40.

Malinowski ER, *Factor Analysis in Chemistry*, 2nd edn, John Wiley and Sons, Inc., New York, 1991.

Mallows CL, Some comments on Mallows Cp, *Technometrics*, 1973, **15**, 661–675.

Mandel J, A new analysis of variance model for non-additive data, *Technometrics*, 1971, **13**, 1–18.

Mandel J, The partitioning of interaction in analysis of variance, *Journal of Research of the National Bureau of Standards B Mathematical Sciences*, 1969, **73B**, 309–328.

Manne R, On the resolution problem in hyphenated chromatography, *Chemometrics and Intelligent Laboratory Systems*, 1995, **27**, 89–94.

Mardia KV, Kent JT, Bibby JM, *Multivariate Analysis*, Ltd, Academic Press, New York, 1979.

Martens H, Høy M, Wise BM, Bro R, Bruce A, Brockhoff PB, Pre-whitening of data by covariance-weighted pre-processing, *Journal of Chemometrics*, 2003, **17**, 153–165.

Martens H, Martens M, *Multivariate Analysis of Quality. An Introduction*, John Wiley & Sons, Ltd, Chichester, 2001.

Martens H, Næs T, *Multivariate calibration*, John Wiley & Sons, Ltd, Chichester, 1989.

Martens HA, Dardenne P, Validation and verification of regression in small data sets, *Chemometrics and Intelligent Laboratory Systems*, 1998, **44**, 99–121.

Martins JA, Sena MM, Poppi RJ, Pessine FBT, Fluorescence piroxicam study in the presence of cyclodextrins by using the PARAFAC method, *Applied Spectroscopy*, 1999, **53**, 510–522.

Mason RL, Gunst RF, Selecting principal components in regression, *Statistics and Probability Letters*, 1985, **3**, 299–301.

McCue M, Malinowski ER, Rank annihilation factor analysis of unresolved LV peaks, *Journal of Chromatographic Science*, 1983, **21**, 229–234.

McDonald RP, A simple comprehensive model for the analysis of covariance structures: some remarks on applications, *British Journal of Mathematical and Statistical Psychology*, 1980, **33**, 161–183.

Meyers R, Montgomery D, *Response Surface Methodology. Process and Product Optimization Using Designed Experiments*, John Wiley & Sons, Inc., New York, 1995.

Miller P, Swanson RE, Heckler C, Contribution plots: a missing link in multivariate quality control, *Applied Mathematics and Computer Science*, 1998, **8**, 775–792.

Millican DW, McGown LB, Fluorescence lifetime resolution of spectra in the frequency domain using multiway analysis, *Analytical Chemistry*, 1990, **62**, 2242–2247.

Mitchell BC, Burdick DS, An empirical comparison of resolution methods for three-way arrays, *Chemometrics and Intelligent Laboratory Systems*, 1993, **20**, 149–161.

Mitchell BC, Burdick DS, Slowly converging PARAFAC sequences: Swamps and two-factor degeneracies, *Journal of Chemometrics*, 1994, **8**, 155–168.

Montgomery DC, *Design and Analysis of Experiments*, John Wiley & Sons, Inc., New York, 1976.

Moreda-Pineiro A, Marcos A, Fisher A, Hill SJ, Parallel factor analysis for the study of systematic error in inductively coupled plasma atomic emission spectrometry and mass spectrometry, *Journal of Analytical Atomic Spectrometry*, 2001, **16**, 360–369.

Murakami T, ten Berge JMF, Kiers HAL, A case of extreme simplicity of the core matrix in three-mode principal components analysis, *Psychometrika*, 1998, **63**, 255–261.

Müntz C, Solution directe de l'équation séculaire et des problèmes analogues transcedentes, *Comptes Rendus de l'Académie des Sciences, Paris*, 1913, **156**, 43–46.

Nelson PRC, Taylor PA, MacGregor JF, Missing data methods in PCA and PLS: score calculations with incomplete observations, *Chemometrics and Intelligent Laboratory Systems*, 1996, **35**, 45–65.

Neogi D, Schlags CE, Multivariate statistical analysis of an emulsion batch process, *Industrial and Engineering Chemistry Research*, 1998, **37**, 3971–3979.

Neuhaus JO, Wrigley C, The Quartimax method: an analytic approach to orthogonal simple structure, *British Journal of Mathematical and Statistical Psychology*, 1954, **7**, 81–91.

Nielson G, Schriver B, Rosenblum L, (Eds), *Visualization in Scientific Computing*, IEEE Computer Society Press, Los Alamitos, CA, 1990.

Nilsson J, De Jong S, Smilde AK, Multiway calibration in 3D QSAR, *Journal of Chemometrics*, 1997, **11**, 511–524.

Nilsson J, Homan EJ, Smilde AK, Grol CJ, Wikström H, A multiway 3D QSAR analysis of a series of (S)-N-[(1-ethyl-2-pyrrolidinyl)methyl]-6-methoxybenzamides, *Journal of Computer Aided Molecular Design*, 1998, **12**, 81–93.

Nilsson J, Multiway Calibration in 3D QSAR. Applications to dopamine receptor ligands, Ph.D. thesis, University of Groningen, 1998.

Noda I, Generalized two-dimensional correlation method applicable to infrared, Raman and other types of spectroscopy, *Applied Spectroscopy*, 1993, **47**, 1329–1336.

Nomikos P, MacGregor JF, Monitoring batch processes using multiway principal component analysis, *AIChE Journal*, 1994, **40**, 1361–1375.

Nomikos P, MacGregor JF, Multivariate SPC charts for monitoring batch processes, *Technometrics*, 1995, **37**, 41–59.

Næs T, Ellekjær MR, Cross-validation and leverage-correction revisited, *NIR News,* 1993, **4**, 8–9.

Næs T, Isaksson T, Fearn T, Davies T, *A user-friendly guide to Multivariate Calibration and Classification*, NIR Publications, Chichester, 2002.

Næs T, Leverage and influence measures for principal component regression, *Chemometrics and Intelligent Laboratory Systems,* 1989, **5**, 155–168.

Nørgaard L, Classification and prediction of quality and process parameters of beet sugar and thick juice by fluorescence spectroscopy and Chemometrics, *Zuckerindustrie*, 1995, **120**, 970–981.

Nørgaard L, Ridder C, Spectrophotometric determination of mixtures of 2-, 3-, and 4-hydroxybenzaldehydes by flow injection analysis and UV/VIS photodiodearray detector, *Talanta*, 1994, **41**, 59–66.

Nørgaard L, Ridder C, Rank annihilation factor analysis applied to flow injection analysis with photodiode-array detection, *Chemometrics and Intelligent Laboratory Systems*, 1994a, **23**, 107–114.

Nørgaard L, Ridder C, Spectrophotometric determination of mixtures of 2-, 3-, and 4- hydroxybenzaldehydes by flow injection analysis and UV/VIS photodiodearray detector, *Talanta*, 1994b, **41**, 59–66.

Osten DW, Selection of optimal regression models via cross-validation, *Journal of Chemometrics,* 1988, **2**, 39–48.

Paatero P, A weighted non-negative least squares algorithm for three-way 'PARAFAC' factor analysis, *Chemometrics and Intelligent Laboratory Systems*, 1997, **38**, 223–242.

Paatero P, Andersson CA, Further improvements of the speed of the Tucker3 three-way algorithm, *Chemometrics and Intelligent Laboratory Systems*, 1999, **47**, 17–20.

Paatero P, Tapper U, Positive matrix factorization: A non-negative factor model with optimal utilization of error estimates of data values, *Environmetrics*, 1994, **5**, 111–126.

Paatero P, The multilinear engine – a table-driven, least squares program for solving multilinear problems, including the n-way parallel factor analysis model, *Journal of Computational and Graphical Statistics*, 1999, **8**, 854–888.

Pearson K, On lines and planes of closest fit to points in space, *Philosophical Magazine*, 1901, **2**, 559–572.

Pell RJ, Seasholtz MB, Kowalski BR, The relationship of closure, mean centering and matrix rank interpretation, *Journal of Chemometrics*, 1992, **6**, 57–62.

Penrose R, On the best approximate solutions of linear matrix equations, *Proceedings of the Cambridge Philosophical Society*, 1956, **52**, 17–19.

Poe RB, Rutan SC, Effects of resolution, peak ratio and sampling frequency in diode-array fluorescence detection in liquid chromatography, *Analytica Chimica Acta*, 1993, **283**, 845–853.

Qin X, Shen LS, Wavelet transform and its application in spectral analysis, *Spectroscopy and Spectral Analysis*, 2000, **20**, 892–897.

Ramos S, Sanchez E, Kowalski B, Generalized rank annihilation method. II. Analysis of bimodal chromatographic data, *Journal of Chromatography*, 1987, **385**, 165–180.

Rao CR, Mitra SK, *Generalized Inverse of Matrices and its Applications*, John Wiley & Sons, Inc., New York, 1971.

Rao CR, *Linear Statistical Inference and its Applications*, John Wiley & Sons, Inc., New York, 1973.

Rayens WS, Mitchell BC, Two-factor degeneracies and a stabilization of PARAFAC, *Chemometrics and Intelligent Laboratory Systems*, 1997, **38**, 173–181.

Reis MM, Gurden SP, Smilde AK, Ferreira MMC, Calibration and detailed analysis of second-order flow-injection analysis data with rank overlap, *Analytica Chimica Acta*, 2000, **422**, 21–36.

Riu J, Bro R, Jack-knife for estimation of standard errors and outlier detection in PARAFAC models, *Chemometrics and Intelligent Laboratory Systems*, 2002, **65**, 35–49.

Rivals I, Personnaz L, On cross validation for model selection, *Neural Computation*, 1999, **11**, 863–870.

Ross RT, Lee C, Davis CM, Ezzeddine BM, Fayyad EA, Leurgans SE, Resolution of the fluorescence spectra of plant pigment-complexes using trilinear models, *Biochimica et Biophysica Acta*, 1991, **1056**, 317–320.

Ross RT, Leurgans SE, Component resolution using multilinear models, *Methods in Enzymology* 1995, **246**, 679–700.

Rousseeuw PJ, Leroy A, *Robust regression and outlier detection,* John Wiley & Sons, Inc., New York, 1987.

Roy R, Kailath T, ESPRIT – Estimation of signal parameters via rotational invariance techniques, *IEEE ASSP Magazine*, 1989, **37**, 984–995.

Röhmel J, Streitberg B, Herrmann WM, The COMSTAT algorithm for multimodal factor analysis: An improvement of Tucker's three-mode factor analysis method, *Neuropsychobiology*, 1983, **10**, 157–163.

Rubin DB, Titterington DM, Gilks WR, Diebolt J, Aitkin M, Smith CAB, Hinde J, Kent JT, Tyler DE, Damien P, Walker S, Chauveau D, Draper D, Dupuis JA, Fessler J, Gelman A, Green PJ, Hero AO, Lavielle M, Liu CH, Liu JS, Roberts GO, Sahu SK, Torsney B, Zaslavsky AM, The EM algorithm–an old folk-song sung to a fast new tune – discussion, *Journal of the Royal Statistical Society Series B Statistical Methodology*, 1997, **59**, 541–567.

Russell MD, Gouterman M, van Zee JA, Excitation-emission-lifetime analysis of multicomponent systems. III. Platinum, palladium and rhodium porphyrins, *Spectrochimia Acta*, 1988, **44A**, 873–882.

Sanchez E, Kowalski BR, Generalized rank annihilation factor analysis, *Analytical Chemistry*, 1986, **58**, 496–499.

Sanchez E, Kowalski BR, Tensorial calibration: I. first-order calibration, *Journal of Chemometrics*, 1988a, **2**, 247–263.

Sanchez E, Kowalski BR, Tensorial calibration: II. second-order calibration, *Journal of Chemometrics*, 1988b, **2**, 265–280.

Sanchez E, Kowalski BR, Tensorial calibration: II. second-order calibration, *Journal of Chemometrics,* 1988, **2**, 265–280.

Sanchez E, Kowalski BR, Tensorial resolution: a direct trilinear decomposition, *Journal of Chemometrics*, 1990, **4**, 29–45.

Sanchez E, Ramos S, Kowalski B, Generalized rank annihilation method. I. Application to liquid chromatography-diode array ultraviolet detection data, *Journal of Chromatography*, 1987, **385**, 151–164.

Sanchez E, Scott-Ramos R, Kowalski BR, Generalized rank annihilation methods 1. Application to liquid chromatography-diode array ultraviolet detection data, *Journal of Chromatography*, 1987, **385**, 151–164.

Sands R, Young FW, Component models for three-way data: An alternating least squares algorithm with optimal scaling features, *Psychometrika*, 1980, **45**, 39–67.

Saurina J, Hernandez-Cassou S, Quantitative determinations in conventional flow injection analysis based on different chemometric calibration statregies: a review, *Analytica Chimica Acta*, 2001, **438**, 335–352.

Saurina J, Hernandez-Cassou S, Tauler R, Multivariate curve resolution and trilinear decomposition methods in the analysis of stopped-flow kinetic data for binary amino acid mixtures, *Analytical Chemistry*, 1997, **69**, 2329–2336.

Savitzky A, Golay MJE, Smoothing and differentiating of data by simplified least-squares procedures, *Analytical Chemistry*, 1964, **36**, 1627–1639.

Scarponi G, Moret I, Capodaglio G, Romanazzi M, Cross-validation, influential observations and selection of variables in chemometric studies of wines by principal component analysis, *Journal of Chemometrics*, 1990, **4**, 217–240.

Schoenmakers PJ, *Optimization of Chromatographic Selectivity*, Elsevier, Amsterdam, 1986.

Schott JR, *Matrix Analysis for Statistics*, John Wiley & Sons, Inc., New York, 1997.

Schönemann PH, An algebraic solution for a class of subjective metrics models, *Psychometrika*, 1972, **37**, 441–451.

Searle SR, Casella G, McCulloch C, *Variance Components*, John Wiley & Sons, Inc., New York, 1992,

Searle SR, *Matrix Algebra Useful for Statistics*, John Wiley & Sons, Inc., New York, 1982.

Seasholtz MB, Kowalski BR, The parsimony principle applied to multivariate calibration, *Analytica Chimica Acta*, 1993, **277**, 165–177.

Shao J, Linear model selection by cross-validation, *Journal of American Statistical Association*, 1993, **88**, 486–494.

Sharaf M, Illman D, Kowalski B, *Chemometrics*, John Wiley & Sons, Inc., New York, 1986.

Shewart WA, *Economic Control of Quality of Manufactured Products*, Van Nostrand, New York, 1931.

Sidiropoulos ND, Bro R, On the uniqueness of multilinear decomposition of N-way arrays, *Journal of Chemometrics*, 2000, **14**, 229–239.

Sinha AE, Johnson KJ, Prazen BJ, Lucas SV, Fraga CG, Synovec RE, Comprehensive two-dimensional gas chromatography of volatile and semi-volatile components using a diaphragm valve-based instrument, *Journal of Chromatography A*, 2003, **983**, 195–204.

Smilde AK, Comments on multilinear PLS, *Journal of Chemometrics*, 1997, **11**, 367–377.

Smilde AK, Comments on three-way analyses used for batch process data, *Journal of Chemometrics*, 2001, **15**, 19–27.

Smilde AK, Doornbos DA, Simple validatory tools for judging the predictive performance of PARAFAC and three-way PLS, *Journal of Chemometrics*, 1992, **6**, 11–28.

Smilde AK, Doornbos DA, Three-way methods for the calibration of chromatographic systems: comparing PARAFAC and three-way PLS, *Journal of Chemometrics*, 1991, **5**, 345–360.

Smilde AK, Kiers HAL, Multiway covariates regression models, *Journal of Chemometrics*, 1999, **13**, 31–48.

Smilde AK, Multivariate calibration of reversed-phase chromatographic systems. Ph.D. thesis, University of Groningen, 1990,

Smilde AK, Tates AA, Boelens HFM, Oerlemans P, Ruitenberg G, Systematic investigation of process and product variations in a spindraw-winding process, *Chemical Engineering Science*, 2001, **56**, 4993–5002.

Smilde AK, Tauler R, Henshaw JM, Burgess LW, Kowalski BR, Multicomponent determination of chlorinated hydrocarbons using a reaction-based chemical sensor. 3. Medium-rank second-order calibration with restricted Tucker models, *Analytical Chemistry*, 1994a, **66**, 3345–3351.

Smilde AK, Tauler R, Saurina J, Bro R, Calibration methods for complex second-order data, *Analytica Chimica Acta*, 1999, **398**, 237–251.

Smilde AK, van der Graaf PH, Doornbos DA, Steerneman T, Sleurink A, Multivariate calibration of reversed-phase chromatographic systems. Some designs based on three-way data analysis, *Analytica Chimica Acta*, 1990, **235**, 41–51.

Smilde AK, Wang Y, Kowalski BR, Theory of medium-rank second-order calibration with restricted tucker models, *Journal of Chemometrics*, 1994, **8**, 21–36.

Smilde AK, Wang Y, Kowalski BR, Theory of medium-rank second-order calibration with restricted Tucker models, *Journal of Chemometrics,* 1994b, **8**, 21–36.

Spanjer MC, Substituent interaction effects and mathematical-statistical description of rentention in liquid chromatography. Ph.D. thesis, University of Utrecht, 1984,

Stephanopoulos G, *Chemical Process Control: an Introduction to Theory and Practice*, Prentice-Hall, Englewood Cliffs, NJ, 1984.

Stewart G, On the early history of the singular value decomposition, *SIAM Review*, 1993, **35**, 551–566.

Stoica P, Söderström T, Partial least squares: a first-order analysis, *Scandinavian Journal of Statistics*, 1998, **25**, 17–23.

Stone M, Cross-validatory choice and assessment of statistical predictions, *Journal of the Royal Statistical Society, Series B Statistical Methodology,* 1974, **36**, 111–148.

Ståhle L, Aspects of the analysis of three-way data, *Chemometrics and Intelligent Laboratory Systems*, 1989, **7**, 95–100.

Tan YX, Jiang JH, Wu HL, Cui H, Yu RQ, Resolution of kinetic system of simultaneous degradations of chlorophyll a and b by PARAFAC, *Analytica Chimica Acta*, 2000, **412**, 195–202.

Tates AA, Louwerse DJ, Smilde AK, Koot GLM, Berndt H, Monitoring a PVC batch process with multivariate statistical process control charts, *Industrial and Engineering Chemistry Research*, 1999, **38**, 4769–4776.

Tauler R, Multivariate curve resolution applied to second-order data, *Chemometrics and Intelligent Laboratory Systems*, 1995, **30**, 133–146.

Tauler R, Smilde AK, Henshaw JM, Burgess LW, Kowalski BR, Multicomponent determination of chlorinated hydrocarbons using a reaction-based chemical sensor. 2. Chemical speciation using multivariate curve resolution, *Analytical Chemistry,* 1994, **66**, 3337–3344.

Taylor AA, Gases in fresh meat packaging, *Meat World*, 1972, **5**, 3–6.

ten Berge JMF, Convergence of PARAFAC preprocessing procedures and the Deming–Stephan method of iterative proportional fitting, in *Multiway Data Analysis*, Coppi R, Bolasco S (Eds), Elsevier, Amsterdam, 1989, pp. 53–63.

ten Berge JMF, de Leeuw J, Kroonenberg PM, Some additional results on principal components analysis of three-mode data by means of alternating least squares algorithms, *Psychometrika*, 1987, **52**, 183–191.

Ten Berge JMF, Kiers HAL, Are all variaties of PCA the same? A reply to Cadima & Jolliffe, *British Journal of Mathematical and Statistical Psychology*, 1997, **50**, 368.

ten Berge JMF, Kiers HAL, Convergence properties of an iterative procedure of ipsatizing and standardizing a data matrix, with application to PARAFAC/CANDECOMP preprocessing, *Psychometrika*, 1989, **54**, 231–235.

ten Berge JMF, Kiers HAL, Explicit candecomp/PARAFAC solutions for a contrived $2 \times 2 \times 2$ array of rank three, *Psychometrika*, 1988, **53**, 579–584.

Ten Berge JMF, Kiers HAL, Optimality criteria for principal component analysis and generalizations, *British Journal of Mathematical and Statistical Psychology*, 1996, **49**, 335–345.

Ten Berge JMF, Kiers HAL, Simplicity of core arrays in three-way principal component analysis and the typical rank of $P \times Q \times 2$ arrays, *Linear Algebra and its Applications*, 1999, **294**, 169–179.

Ten Berge JMF, Kruskal's polynomial for $2 \times 2 \times 2$ arrays and a generalization to $2 \times n \times n$ arrays, *Psychometrika*, 1991, **56**, 631–636.

Ten Berge JMF, Orthogonal Procrustes rotation for two or more matrices, *Psychometrika*, 1977, **42**, 267–276.

ten Berge JMF, Sidiropoulos ND, Some new results on uniqueness in CANDECOMP/PARAFAC, *Psychometrika,* 2002, **67**, 399–409.

ten Berge JMF, Smilde AK, Non-triviality and identification of a constrained Tucker3 analysis, *Journal of Chemometrics*, 2002, **16**, 609–612.

Ten Berge JMF, The typical rank of tall three-way arrays, *Psychometrika*, 2000, **65**, 525–532.

Ten Berge JMF, *Least Squares Optimization in Multivariate Analysis*, DSWO Press, Leiden, 1993.

ter Braak CJF, de Jong S, The objective function of partial least squares regression, *Journal of Chemometrics*, 1998, **12**, 41–54.

Thurstone LL, *The Vectors of Mind*, University of Chicago Press, Chicago, 1935.

Timmerman ME, Kiers HAL, Three-mode principal components analysis: Choosing the numbers of components and sensitivity to local optima, *British Journal of Mathematical and Statistical Psychology*, 2000, **53**, 1–16.

Tracy ND, Young JC, Mason RL, Multivariate control charts for individual observations, *Journal of Quality Technology*, 1992, **24**, 88–95.

Trygg J, Kettaneh-Wold N, Wallbacks L, 2D wavelet analysis and compression of on-line industrial process data, *Journal of Chemometrics*, 2001, **15**, 299–319.

Tu XM, Burdick DS, Resolution of trilinear mixtures: Application in spectroscopy, *Statistica Sinica*, 1992, **2**, 577–593.

Tucker L, Implications of factor analysis of three-way matrices for measurement change, in *Problems in Measuring Change*, Harris C (Ed.), The University of Wisconsin Press, Madison, 1967, WI, pp. 122–137.

Tucker L, Relations between multidimensional scaling and three-mode factor analysis, *Psychometrika*, 1972, **37**, 3–27.

Tucker L, Some mathematical notes on three-mode factor analysis, *Psychometrika*, 1966, **31**, 279–311.

Tucker L, The extension of factor analysis to three-dimensional matrices, in Frederiksen N, Gulliksen H (eds), *Contributions to Mathematical Psychology*, Holt, Rinehart Winston, New York, 1964, pp. 110–182.

Tucker LR, A method for synthesis of factor analysis studies, *Personnel Research Section, Report 984, Dept. of the Army*, 1951.

Tucker LR, Some mathematical notes on three-mode fator analysis, *Psychometrika*, 1966, **31**, 279–311.

Tucker LR, The extension of factor analysis to three-dimensional matrices, in *Contributions to Mathematical Psychology*, Frederiksen N, Gulliksen H (Eds), Holt, Rinehart & Winston, New York, 1964, pp 110–182.

Tufte E, *The Visual Display of Quantitative Data*, Graphics Press, Cheshire, CT, 1983.

van Eeuwijk FA, Between and beyond additivity and non-additivity; the statistical modelling of genotype by environment interaction in plant breeding. Ph.D. thesis, University of Wageningen, 1996,

van Sprang ENM, Ramaker HJ, Westerhuis JA, Gurden SP, Smilde AK, Critical evaluation of approaches for on-line batch process monitoring, *Chemical Engineering Science*, 2002, **57**, 3979–3991.

Vandeginste BGM, Leyten F, Gerritsen MJP, Noor JW, Kateman G, Frank J, Evaluation of curve resolution and iterative target transformation factor analysis in quantitative analysis by liquid chromatography, *Journal of Chemometrics*, 1987, **1**, 57–71.

Wakeling IN, Morris JJ, A test of significance for partial least squares regression, *Journal of Chemometrics*, 1993, **7**, 291–304.

Wangen LE, Kowalski BR, A multiblock partial least squares algorithm for investigating complex chemical systems, *Journal of Chemometrics*, 1988, **3**, 3–20.

Wehrens R, Putter H, Buydens LMC, The bootstrap: a tutorial, *Chemometrics and Intelligent Laboratory Systems*, 2000, **54**, 35–52.

Wehrens R, van der Linden WE, Bootstrapping principal component regression models, *Journal of Chemometrics*, 1997, **11**, 157–171.

Weinberg JR, *A Short History of Medieval Philosophy*, Princeton University Press, Princeton, NJ, 1964, pp. 235–266.

Weisberg S, *Applied Linear Regression*, John Wiley and Sons, Inc., New York, 1985,

Wentzell PD, Andrews DT, Hamilton DC, Faber NM, Kowalski BR, Maximum likelihood principal component analysis, *Journal of Chemometrics*, 1997, **11**, 339–366.

Wentzell PD, Lohnes MT, Maximum likelihood principal component analysis with correlated measurement errors: theoretical and practical considerations, *Chemometrics and Intelligent Laboratory Systems*, 1999, **45**, 65–85.

Wentzell PD, Nair SS, Guy RD, Three-way analysis of fluorescence spectra of polycyclic aromatic hydrocarbons with quenching by nitromethane, *Analytical Chemistry*, 2001, **73**, 1408–1415.

West T, Weinstein J, Bray J, Shell disease and metal content of blue crabs, Callinectes sapidus, from the Albemarle-Pamlico estuarine system, North Carolina, *Archives of Environmental Contamination and Toxicology*, 1992, **23**, 355–362.

Westerhuis JA, Gurden SP, Smilde AK, Generalized contribution plots in multivariate statistical process monitoring, *Chemometrics and Intelligent Laboratory Systems*, 2000a, **51**, 95–114.

Westerhuis JA, Gurden SP, Smilde AK, Spectroscopic monitoring of batch reactions for on-line fault detection and diagnosis, *Analytical Chemistry*, 2000b, **72**, 5322–5330.

Westerhuis JA, Kourti T, MacGregor JF, Analysis of multiblock and hierarchical PCA and PLS models, *Journal of Chemometrics*, 1998, **12**, 301–321.

Westerhuis JA, Kourti T, MacGregor JF, Comparing alternative approaches for multivariate statistical analysis of batch process data, *Journal of Chemometrics*, 1999, **13**, 397–413.

Wilkinson JH, *The Algebraic Eigenvalues Problem*, Clarendon Press, Oxford, 1965.

Wilson BE, Kowalski BR, Quantitative analysis in the presence of spectral interferents using second-order nonbilinear data, *Analytical Chemistry*, 1989, **61**, 2277–2284.

Wilson BE, Lindberg W, Kowalski BR, Multicomponent Quantitative Analysis Using Second-Order Nonbilinear Data: Theory and Simulations, *Journal of the American Chemical Society* 1989, **111**, 3797–3804.

Wilson BE, Sanchez E, Kowalski BR, An improved algorithm for the generalized rank annihilation method, *Journal of Chemometrics*, 1989, **3**, 493–498.

Windig W, Antalek B, Direct exponential curve resolution algorithm (DECRA): A novel application of the generalized rank annihilation method for a single spectral mixture data set with exponentially decaying contribution profiles, *Chemometrics and Intelligent Laboratory Systems*, 1997, **37**, 241–254.

Windig W, Guilment J, Interactive self-modeling mixture analysis, *Analytical Chemistry*, 1991, **63**, 1425–1432.

Wise B, Gallagher N, PLS_Toolbox 2.0, Eigenvector Research, Manson, WA, 1998.

Wise BM, Gallagher NB, Butler SW, White DD, Barna GG, A comparison of principal component analysis, multiway principal component analysis, trilinear decomposition and parallel factor analysis for fault detection in a semiconductor etch process, *Journal of Chemometrics*, 1999, **13**, 379–396.

Wise BM, Gallagher NB, Martin EB, Application of PARAFAC2 to fault detection and diagnosis in semiconductor etch, *Journal of Chemometrics*, 2001, **15**, 285–298.

Wold H, Nonlinear estimation by iterative least squares procedures, in *Research Papers in Statistics, Festschrift for J. Neyman*, David F (Ed.), John Wiley & Sons, Inc., New York, 1966, pp. 411–444.

Wold H, Path models latent variables: The NIPALS approach, in *Quantitative Sociology. International Perspectives on Mathematical and Statistical Modeling*, Blalock HM, Aganbegian A, Borodkin FM, Boudon R and Capecchi V (Eds), Academic Press, New York, 1975, pp. 307–357.

Wold H, Soft modelling by latent variables; the non-linear iterative partial least squares approach, in *Perspectives in Probability and Statistics*, Gani J (Ed.), Academic Press, London, 1975.

Wold S, Antti H, Lindgren F, Öhman J, Orthogonal signal correction of near-infrared spectra, *Chemometrics and Intelligent Laboratory Systems*, 1998, **44**, 175–185.

Wold S, Cross-validatory estimation of the number of components in factor and principal components models, *Technometrics,* 1978, **20**, 397–405.

Wold S, Geladi P, Esbensen KH, Öhman J, Multi-way principal components and PLS-analysis, *Journal of Chemometrics*, 1987, **1**, 41–56.

Wold S, Kettaneh N, Friden H, Holmberg A, Modelling and diagnostics of batch processes and analogous kinetic experiments, *Chemometrics and Intelligent Laboratory Systems*, 1998, **44**, 331–340.

Wold S, Kettaneh-Wold N, Tjessem K, Hierarchical multiblock PLS and PC models, for easier model interpretation, and as an alternative to variable selection, *Journal of Chemometrics*, 1996, **10**, 463–482.

Wold S, Ruhe A, Wold H, Dunn WJ, The collinearity problem in linear regression. The partial least squares (PLS) approach to generalized inverses, *SIAM Journal of Scientific and Statistical Computing*, 1984, **5**, 735–743.

Wolfbeis OS, Leiner MJP, Mapping of the total fluorescence of human blood serum as a new method for its characterization, *Analytica Chimica Acta*, 1985, **167**, 203–215.

Wülfert F, Kok WT, Smilde AK, Influence of temperature on vibrational spectra and consequences for the predictive ability of multivariate models, *Analytical Chemistry*, 1998, **70**, 1761–1767.

Xie HP, Jiang JH, Chu X, Cui H, Wu HL, Shen GL, Yu RQ, Competitive interaction of the antitumor drug daunorubicin and the fluorescence probe ethidium bromide with DNA as studied by resolving trilinear fluorescence data: the use of PARAFAC and its modification, *Analytical and Bioanalytical Chemistry*, 2002, **373**, 159–162.

Xie YL, Baeza-Baeza JJ, Ramis-Ramos G, Second-order tensorial calibration for kinetic spectrophotometric determination, *Chemometrics and Intelligent Laboratory Systems*, 1996, **32**, 215–232.

Xie YL, Hopke P, Paatero P, Barrie L, Li SM, Identification of source and seasonal variations of Arctic aerosol by Positive Matrix Factorization, *Journal of the Atmospheric Sciences*, 1999a, **56**, 249–260.

Xie YL, Hopke P, Paatero P, Barrie L, Li SM, Locations and preferred pathways of possible sources of Arctic aerosol, *Atmospheric Environment*, 1999b, **33**, 2229–2239.

Xie YL, Hopke PK, Paatero P, Barrie LA, Li SM, Identification of source nature and seasonal variations of Arctic aerosol by the multilinear engine, *Atmospheric Environment*, 1999, **33**, 2549–2562.

Xie YL, Hopke PK, Paatero P, Barrie LA, Li SM, Identification of source nature and seasonal variations of Arctic aerosol by the multilinear engine, *Atmospheric Environment*, 1999c, **33**, 2549–2562.

Zeng Y, Hopke PK, A new receptor model: a direct trilinear decomposition followed by a matrix reconstruction, *Journal of Chemometrics*, 1992a, **6**, 65–83.

Zeng Y, Hopke PK, Methodological study applying three-mode factor analysis to three-way chemical data sets, *Chemometrics and Intelligent Laboratory Systems*, 1990, **7**, 237–250.

Zeng Y, Hopke PK, The application of three-mode factor analysis (TMFA) to receptor modeling of scenes particle data, *Atmospheric Environment*, 1992b, **26A**, 1701–1711.

Zhang J, Martin EB, Morris AJ, Process monitoring using non-linear statistical techniques, *Chemical Engineering Journal,* 1997, **67**, 181–189.

Zhao Y, Wells JH, McMillin KW, Applications of dynamic modified atmosphere packaging systems for fresh red meats: review, *Journal of Muscle Foods*, 1994, **5**, 199–328.

INDEX

Note: Figures and Tables are indicated by *italic page numbers*

Multi-way Analysis in Chemistry and Related Fields.   A. Smilde, R. Bro and P. Geladi
© 2004 John Wiley & Sons, Ltd   ISBN: 0-471-98691-7

With kind thanks to Paul Nash for compilation of this index.

Printed in the USA/Agawam, MA
December 19, 2013
583271.038